A Guide to Simulation

Paul Bratley
Bennett L. Fox
Linus E. Schrage

A Guide to Simulation

Second Edition

With 32 Illustrations

Springer-Verlag
New York Berlin Heidelberg
London Paris Tokyo

Paul Bratley
Bennett L. Fox
Département d'informatique et de
  recherche opérationnelle
Université de Montréal
Montréal, P.Q. H3C 3J7
Canada

Linus E. Schrage
Graduate School of Business
University of Chicago
Chicago, IL 60637
U.S.A.

AMS Subject Classifications: 62K99, 62N99, 90B00

Library of Congress Cataloging in Publication Data
Bratley, Paul.
  A guide to simulation.
  Bibliography: p.
  Includes indexes.
  1. Digital computer simulation. I. Fox.
Bennett L., 1938–　. II. Schrage, Linus E.
III. Title.
QA76.9.C65B73　1987　　001.4'34　　86-26287

Typeset by Composition House Ltd., Salisbury, England.
Printed and bound by R. R. Donnelley & Sons, Harrisonburg, Virginia.
Printed in the United States of America.

9 8 7 6 5 4 3 2 1

ISBN 0-387-96467-3 Springer-Verlag New York Berlin Heidelberg
ISBN 3-540-96467-3 Springer-Verlag Berlin Heidelberg New York

To

Ronald and Margaret Bratley
Daniel and Kate Fox
Alma and William Schrage

Preface to the Second Edition

Changes and additions are sprinkled throughout. Among the significant new features are:

- Markov-chain simulation (Sections 1.3, 2.6, 3.6, 4.3, 5.4.5, and 5.5);
- gradient estimation (Sections 1.6, 2.5, and 4.9);
- better handling of asynchronous observations (Sections 3.3 and 3.6);
- radically updated treatment of indirect estimation (Section 3.3);
- new section on standardized time series (Section 3.8);
- better way to generate random integers (Section 6.7.1) and fractions (Appendix L, program UNIFL);
- thirty-seven new problems plus improvements of old problems.

Helpful comments by Peter Glynn, Barry Nelson, Lee Schruben, and Pierre Trudeau stimulated several changes. Our new random integer routine extends ideas of Aarni Perko. Our new random fraction routine implements Pierre L'Ecuyer's recommended composite generator and provides seeds to produce disjoint streams. We thank Springer-Verlag and its late editor, Walter Kaufmann-Bühler, for inviting us to update the book for its second edition. Working with them has been a pleasure. Denise St-Michel again contributed invaluable text-editing assistance.

Preface to the First Edition

Simulation means driving a model of a system with suitable inputs and observing the corresponding outputs. It is widely applied in engineering, in business, and in the physical and social sciences. Simulation methodology draws on computer science, statistics, and operations research and is now sufficiently developed and coherent to be called a discipline in its own right. A course in simulation is an essential part of any operations research or computer science program. A large fraction of applied work in these fields involves simulation; the techniques of simulation, as tools, are as fundamental as those of linear programming or compiler construction, for example. Simulation sometimes appears deceptively easy, but perusal of this book will reveal unexpected depths. Many simulation studies are statistically defective and many simulation programs are inefficient. We hope that our book will help to remedy this situation. It is intended to teach how to simulate *effectively*.

A simulation project has three crucial components, each of which must always be tackled:

(1) data gathering, model building, and validation;
(2) statistical design and estimation;
(3) programming and implementation.

Generation of random numbers (Chapters 5 and 6) pervades simulation, but unlike the three components above, random number generators need not be constructed from scratch for each project. Usually random number packages are available. That is one reason why the chapters on random numbers, which contain mainly reference material, follow the chapters dealing with experimental design and output analysis.

A simulation technique is not really appreciated until a feeling for how it might be implemented in practice is acquired. We therefore give worked examples to illustrate the theory. Mastering modeling comes only by doing it; but to give inexperienced readers the proper perspective the book begins with a detailed discussion of modeling principles. The specific models given occasionally in the book are there only to illustrate technical points. They are simplistic and are not to be taken seriously as reasonable representations of real-world problems. On the other hand, we point out a common error in simulation models: excessive detail.

A simulation course should attempt to cover only the substantial amount of material that deals directly with simulation. Time does not allow it to serve also as a vehicle for teaching probability, statistics, computer programming, or data structures. For understanding most of this book one course in each of these topics should suffice. This knowledge should be complemented by a certain mathematical maturity, such as is normally acquired in a rigorous undergraduate course in advanced calculus or linear algebra. If the instructor briefly reviews a few points in class, good students can get by with fewer formal prerequisites. The rare allusions to more advanced material can be skimmed without loss of continuity. Chapter 7, on simulation programming, requires almost no mathematical background. Only Section 1.7, a few problems in Section 1.9, and Chapter 7 require computer science background beyond elementary programming to be fully appreciated.

The ordering of the material in the book reflects what we consider to be the most natural logical flow. The progression is not always from easy to hard: Chapters 2 and 3 are probably the most difficult in the book and Section 1.9 has some of the most challenging problems. Other orderings are certainly feasible, as the Precedence Diagram given after the Contents indicates. It shows only the most prominent relationships among the chapters which are all to some extent interrelated. Course structure is flexible.

For example, an instructor can structure a computer science oriented course based on this book at the last-year undergraduate level by focusing mainly on Chapters 1, 5, 6, and 7. Sections 2.1.1, 3.1, 3.2, and 4.1–4.8 can also be covered without difficulty in an undergraduate course that otherwise deals with computer science aspects of simulation. To concentrate on simulation programming languages cover Sections 1.1–1.5 and 1.8, and Chapter 7. This should be complemented by a detailed study, with programmed examples, of at least one of the special-purpose languages we discuss. Chapter 7 is not intended as a substitute for the reference manuals of the languages compared there. Rather, we aim to teach critical appreciation of general principles. To fully appreciate our critique, the student should at least have access to supplementary documentation; ideally, he should have the facilities to try several languages.

Although such a course may perhaps be thought sufficient for the computer science student or the programmer, it is not adequate for a designer with overall responsibility for a simulation project. In this wider context we believe that the student should acquire proper understanding of the theory before he rushes to get simulation programs running on a computer. The ordering of the chapters follows this principle. A course based on this book emphasizing statistical aspects of simulation can normally be expected to be at the first-year graduate level. If the students already have some computer science background, as we would expect, much of Chapters 1, 7, and 8 can be left as reading assignments.

Most of the book can be covered in a one-semester course. The instructor can flesh out a two-semester course by presenting case studies and having the students formulate and validate models, write simulation programs, and make statistical analyses. If there is time left over, a few papers from the recent research literature can be discussed, or a particular simulation language can be studied in detail. The entire book is also suitable for independent study and as a reference for practicing professionals with diverse interests and backgrounds.

The problems are often an integral part of a development. Some test comprehension. Others contain results of independent interest, occasionally used later. They vary from almost trivial to quite hard. They are placed where we think it will do the reader the most good. All should be read before proceeding: try most.

Simulation is unique in the opportunities it offers to reduce the variances of performance estimators by astute experimental design. The book explores these opportunities thoroughly. We highlight the roles of monotonicity and synchronization in inducing correlation of appropriate sign for variance reduction. One consequence is that we favor inversion as a way of generating nonuniform random numbers. This lets us choose nonuniform random number generators to discuss quite selectively among the mushrooming number of papers on the subject. Many of these papers are cited but not treated in detail.

Occasionally we take viewpoints not universally shared and emphasize certain aspects that others slight. The central role that we give to synchronization and our consequent concentration on inversion is one example of this. Probably controversial in some quarters is our position that finite-horizon models are usually appropriate, especially in a simulation context. For such models, multiple replications form the basis for fairly straightforward estimation and statistical inference. Infinite-horizon models are artifices that may occasionally be convenient fictions. Sometimes useful, they are too often used. For them, no method of constructing confidence intervals for unknown parameters that is in all respects both completely rigorous and practical has been devised. This probably explains why so many approaches have been discussed in the simulation literature. We cover the main ones: batch means and regenerative, spectral, and autoregressive methods. When pungency, not blandness, is called for our tone is sometimes frankly tendentious.

The book treats all aspects of discrete-event digital simulation in detail. We discuss so-called continuous and hybrid simulation briefly in Sections 1.4.2. and 1.4.3. The Contents, a summary of notation, and a precedence diagram follow this preface. References cited are listed in a separate section near the end of the book, ordered by author and year of publication. To avoid cluttering the text, longer programmed examples are banished to an appendix.

We acknowledge the help received from many colleagues and friends. Among those whose suggestions, comments, and criticism of earlier versions of this book were particularly valuable, we thank Robert Cooper, Eric Denardo, Luc Devroye, U. Dieter, Peter Glynn, Donald Iglehart, Donald Knuth, Patrice Marcotte, Barry Nelson, Bruce Schmeiser, and Ludo Van der Heyden. Edward Russell supplied the example program of Figure X.7.5. Kevin Cunningham wrote the programs to draw many of the figures. We are particularly grateful to those who do not share our opinions but freely gave their help anyway.

Denise St-Michel nurtured our manuscript through endless revisions on our local text-editing system. For her care and cheer, we are grateful. We also thank Yves Courcelles, whose software made life much easier. Noah Webster settled many an argument.

Our work was generously supported by the Natural Sciences and Engineering Research Council of Canada.

Contents

Notation

The following notations and abbreviations are used without further explanation throughout the book.

Section 1.2.3: Chapter 1, Section 2, Subsection 3.
Problem 1.2.3: Problem number 3 in Chapter 1, Section 2
(likewise for example, figure, lemma, proposition, table, and theorem).

$\log$	natural logarithm
$\log_2$	base 2 logarithm
$\lceil x \rceil$	ceiling of x = smallest integer $\geq x$
$\lfloor x \rfloor$	floor of x = largest integer $\leq x$
$g(n) = O(f(n))$	means that there is a finite constant M such that $\|g(n)\| \leq M\|f(n)\|$ for all sufficiently large n
$g(n) = o(f(n))$	means that $g(n)/f(n) \to 0$ (as $n \to 0$ or as $n \to \infty$, depending on the context)
$\emptyset$	the empty set
$A/B/c$ queue	a queue with arrival distribution A, service time distribution B, and c servers (replacing A or B by M indicates that the corresponding distribution is exponential)
FIFO	the first-in, first-out queue discipline (sometimes called FCFS: first-come, first-served)
LIFO	the last-in, first-out queue discipline (sometimes called LCFS).
$\Rightarrow$	convergence in distribution of random variables (and weak convergence of probability measures)

iid	independent and identically distributed
cdf	cumulative distribution function
pdf	probability density function
pmf	probability mass function

With a few exceptions we use capital letters to denote random variables, and corresponding lower-case letters to denote their values. Unless the context makes it clear otherwise, U denotes a random variable uniformly distributed on (0,1).

Precedence Diagram

1.2 = Chapter 1, Section 2
3 → 4 = 3 is a prerequisite for 4
5 ↔ 6 = 5 and 6 can be read profitably in parallel

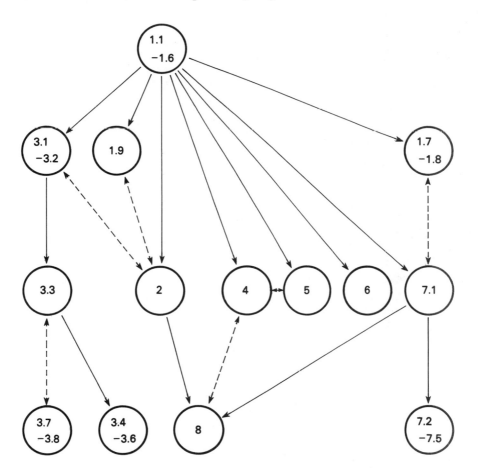

CHAPTER 1
Introduction

Every user of this book should read the first six sections of this chapter: models (§1.1); verification and validation (§1.2); states, events, and clocks (§1.3); types of simulation, with some examples (§1.4); an introduction to random numbers (§1.5); and a perspective of experimental design and estimation (§1.6). These sections set the stage for the rest of the book. Subsections 1.4.2 (continuous simulation) and 1.4.3 (hybrid simulation) are not cited in subsequent chapters and can be skipped at first reading.

Section 1.7, on clock mechanisms, and Section 1.8, on programming, could be skimmed at first reading by those interested primarily in the statistical aspects of simulation. Neither is prerequisite for other chapters, though we recommend they be read in conjunction with Chapter 7.

Section 1.7 and the optional Section 1.9 are fairly technical, which is typical of the remaining chapters. The other sections are more general: we keep mathematical notation there to a minimum to avoid cluttering our exposition. Section 1.8 gives some pointers for the management of simulation programs; further suggestions appear in Chapters 7 and 8.

1.1. Systems, Models, and Simulation

A model is a description of some system intended to predict what happens if certain actions are taken. Virtually any useful model simplifies and idealizes. Often the boundaries of the system and of the model are rather arbitrarily defined. Most forces that impinge on the system must be neglected on *a priori* grounds to keep the model tractable, even when there is no rigorous proof that such neglect is justified. Inevitably, the model is better defined than the real system. For a model to be useful, it is essential that, given a reasonably limited set of descriptors, all its relevant behavior and properties can be determined in a practical way: analytically, numerically, or by driving the model with certain (typically random) inputs and observing the corresponding outputs. This latter process is called *simulation*.

A typical *analog* simulation is wind-tunnel tests on a scaled-down airplane wing. Some writers prefer to call physical models that involve only a change of scale *iconic* models. By contrast, the differential equations describing the forces on an airplane wing can be solved on an analog computer that represents these forces by electrical potentials and then adds, multiplies, and integrates them. These equations can also be solved on a digital computer. Solving differential equations is sometimes called *continuous*, or, since the independent variable is often time, *continuous-time* simulation. Presumably, engineers use wind tunnels because they are skeptical whether the differential equations adequately model the real system, or whether the differential equation solver available is reliable.

If, as far as the logical relationships and equations of the model are concerned, everything happens at a finite number of time points, not necessarily evenly spaced, and indeed possibly random, we get a *discrete-event* simulation illustrated by

EXAMPLE 1.1.1. Suppose that a single server (such as a clerk, a machine, or a computer) services randomly arriving customers (people, parts, or jobs). The order of service is first in, first out. The time between successive arrivals has the stationary distribution F, and the service time distribution is G. At time 0, the system is empty.

We may simulate this system as follows. First, generate a random number A_1 from F, and then a random number B_1 from G. (For the moment, the reader must accept on faith that such random numbers can be generated. Random number generation is discussed in Section 1.5 and in Chapter 5.) Customer 1 enters the system at time A_1 and leaves at time $A_1 + B_1$. Next, generate two more random numbers A_2 and B_2. Customer 2 arrives at $A_1 + A_2$. If $A_1 + A_2 \geq A_1 + B_1$, he starts service right away and finishes at $A_1 + A_2 + B_2$; otherwise, he starts service at $A_1 + B_1$ and finishes at $A_1 + B_1 + B_2$. And so on. The output may be the number of customers served up to a given time, their average waiting time, etc.

This simplistic example is hardly to be taken seriously as a simulation. However, it gives a little of the flavor of how simulation works. Section 1.3 gives a more precise definition of a discrete-event simulation. Even though the distinction is not universally accepted, we shall use the term continuous-time simulation to refer to anything other than discrete-event simulation.

Since discrete-event simulations are commonly carried out using a digital computer, they are sometimes called *digital* simulations. This term may be misleading. Digital computers can also be used to solve differential equations, so that some models, such as our hypothetical airplane wing, can be manipulated either on an analog or a digital machine. If we use a combination of an analog and a digital computer to manipulate a model, the result is a *hybrid* simulation. Section 1.4 illustrates all these terms more fully.

Why are we interested in models? Some of the reasons are:

1. *Tractability.* It is too time-consuming or expensive to play with the real system to answer "what if" questions; indeed the real system may not even exist.
2. *Training purposes.* Think of flight simulators for pilots or business games for managers.
3. *The black box problem.* We are trying to guess what is going on inside some poorly understood real system. Examples here might be human reasoning, weather, and perhaps national and world economies.

One benefit of a modeling effort sometimes obtained unintentionally is that it forces an improved understanding of what is really happening in the system being modeled. Some boldly suggest that this is the principal benefit of such project management models as PERT and CPM. However, as Koopman (1977) points out, modeling a poorly understood system does not necessarily increase understanding. If a model does not reflect some real understanding and insight about the system, it is probably useless. Be prepared to accept that some important systems cannot be modeled well and some important problems cannot be solved at all.

Although simulation can be a powerful tool, it is neither cheap nor easy to apply correctly and effectively. A simulation modeling effort should not be embarked upon lightly. Wagner (1975) in his treatment of operations research methods relegates simulation to the back of the book and refers to it as a method of last resort, to be used "when all else fails." In practice, there is a strong tendency to resort to simulation from the outset. The basic concept is easy to comprehend, and hence easy to justify to management or to customers. An inexperienced or unwary analyst will usually seriously underestimate the cost of data collection and computing time required to produce results with any claim to accuracy.

Simulation is not an optimizing procedure like linear programming. It allows us to make statements like, "Your costs will be C if you take action X," but it does not provide answers like, "Cost is minimized if you take action Y." Repeated and possibly expensive simulation runs are required to identify actions or policies which are good. To identify good policies, the objective function and the constraints must be made explicit; often that is one of the hardest and most rewarding aspects of modeling.

We can use a simulation model in conjunction with a less realistic but cheaper-to-use analytic model. An analytic model gives us a mathematical formula into which we substitute the characteristics of the system in question. It can then be quickly evaluated to give a performance number for the system.

The formula is obtained by some sort of analysis: probability theory, queueing theory, or differential equation theory, for example. The mathematical sophistication required to *derive* the formula is usually substantially higher than that needed to develop a simulation model; however, once derived, a formula is much easier to *use*.

In this situation, the simulation model may be more credible: perhaps its behavior has been compared to that of the real system or perhaps it requires fewer simplifying assumptions and hence intuitively captures more of a hypothetical real system. However, the analytic model may give more insight into which policies are likely to be good. We can also use simulation to validate a more approximate analytic model. The famous square root formula of inventory control [e.g., see Wagner (1975), p. 19] falls out of a simple analytic model. For many inventory situations we would expect it to be too approximate. If, however, the recommendations of the square root formula and of a more expensive simulation are found to be consistent, then we will be more willing to rely on the "quick and dirty" square root formula.

We may consider a whole spectrum of models increasing both in realism and complexity, though we do not mean to imply that increasing the complexity necessarily increases the realism. At one end of this spectrum, the model is so simple that its behavior can be deduced on the back of an envelope; at the other, there is (presumably) little difference between the model and what is being modeled. Often we may imagine that the simpler models are imbedded in more complex ones and that the behavior of the latter can be used to check robustness to departures from the simplifying assumptions of the former. For a few concrete examples of this approach, see Ignall *et al.* (1978). Often simulation is the only way to attack these more complex models. Sometimes complex models can be broken down into analytic and simulation modules that provide input to each other. A coupled hierarchy of models with horizon, time scale, and level of detail depending on the echelon in the hierarchy may be appropriate. For example, at one echelon we may have an horizon of 1 month and time measured in days; and at a lower echelon, an horizon of 1 day and time measured in minutes.

As Disney (1980) remarks, "all applications must cut, paste, extend, rearrange or even develop (from scratch) models to make them do." Modeling a real system is largely *ad hoc*. One must always ask oneself whether the model adequately reflects the features of the real system that are important to the application in mind.

Extrapolation is the other side of the coin from idealization. We have gone from a real system to a model: what about going in the other direction? Given the behavior of the model, what can we say about the real system? Can we make reliable inferences?

This question is impossible to answer precisely. If feasible, compare the outputs of the model and of the real system driven by corresponding inputs. To avoid a self-fulfilling prophecy, use input data different from the data used to develop and calibrate the model. The situation is complicated by the fact

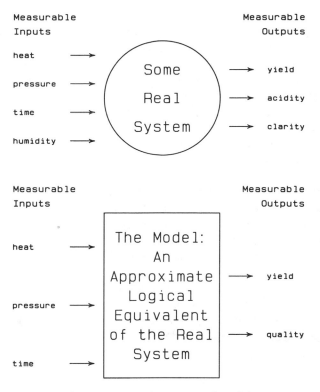

Figure 1.1.1. Real systems and models.

that, due to idealization in the model, the vectors of respective inputs and outputs may have different dimensions. For example, consider the two diagrams in Figure 1.1.1.

One system is a perfect model of another if the former produces exactly the same output measurements as the latter under the same inputs. Thus the model in Figure 1.1.1 is not perfect: it has fewer and different inputs and outputs. This is consistent with economic reality. Perfect models are too expensive, even when, in principle, possible. We must draw the boundaries of the system somewhere. Models approximate. A perfect model of an inventory system would consider the fact that demand on the industry is probably a function of the weather, which is probably a function of sunspot activity: very few economically useful models incorporate details on sunspot cycles.

Though *a priori* beliefs have no proper place in statistical inference *within* a model, the proper formal setting for analysis of any apparent discrepancies *between* a model and a real system is Bayesian statistical inference. This type of question is clouded in controversy: we have no easy answers. It is clear, however, that extrapolation ought to be a conscious process, carried out in a circumspect and guarded manner. It is a process equivalent to model *validation* (see §1.2). We seek to ensure that simplified models are not merely simple-minded. On the other hand, provided a model is valid, the simpler it is, the better: a valid model that is simple indicates real understanding and insight. Aristotle wrote [see McKeon (1941), p. 936]: "... it is the mark of an educated man to look for precision in each class of things just so far as the nature of the subject admits...." Parsimony in assumptions and parameters is good. We endorse Koopman's (1979) version of Ockham's razor: "Complications in models are not to be multiplied beyond the necessity of practical application and insight." According to Edwards [(1967), p. 307], one version of Ockham's original principle is: "What can be done with fewer [assumptions] is done in vain with more."

What we seek to learn through exercising the model, and how we hope to use this acquired knowledge, should strongly influence the development of the model itself. The first step in a simulation study is to identify its objectives: the questions to be answered, the policies or designs to be evaluated, etc. Closely involving the ultimate user ("decision-maker") in model development greatly increases the chances of successful implementation. Other keys to success are interpretation of the results in a form that the user can understand (if you must use jargon, use his, not yours) and thorough documentation of the model development and validation, data sources and statistical techniques, and the simulation program. Most published studies of models purportedly relevant to the real world are deficient in all these aspects.

A common mistake of the inexperienced is to try to build a highly-detailed model right from the start. Aside from the trite observation that such models are expensive to program, debug, and run, their complexity is self-defeating. Not only is it hard to obtain adequate data and to estimate required parameters, but also the output is difficult to explain and interpret.

A better way is to learn from experience with relatively simple models. Modularize them so that particular modules may be specified in more detail as required. As a bonus, modularized models lead to structured programming: models, and the programs implementing them, are developed by stepwise refinement of the details. Give considerable thought to overall program design before coding begins.

Once a model has been programmed and is running correctly, we face the problem of interpreting its output. Taking account of statistical fluctuations in simulation output requires care. To make any statistically meaningful statement, many observations are needed—perhaps hundreds of thousands. For the moment, we simply remark that these fluctuations are one reason why simulation is no panacea and why, when it seems reasonable, we try to make do with simpler models that can be handled analytically.

Do not underestimate the magnitude of these fluctuations, particularly when using a model to examine extreme cases (the limiting capacity of a system, for example). The results from a single run of a good simulation model often have a specious plausibility which can entrap the novice and the unwary into a quite unwarranted confidence in their accuracy. Simulations carried out without adequate statistical support are not merely useless but dangerous.

Not only is the output from a simulation subject to random fluctuations, it can also, all statistical aspects left aside, be plain wrong. Of course, one does not ordinarily expect that the discrepancies between a real system and the model, or between parameters and their estimates, should be nil. However, there is little point in carrying out a formal simulation or optimization study without reasonable confidence that such discrepancies are not gross. If this is not the case, abort any formal study and simply try to get by with seasoned judgment.

Even if we acknowledge that there may be modest "errors" inherent in the model or the data, this does not excuse a willingness to accept further, more serious errors in the "solution." If a formal study is worth doing, it is worth doing well. In the simulation context, gross errors can well arise if the design and analysis of the experiments do not give high (statistical) confidence in the estimates produced.

PROBLEM 1.1.1. It is fashionable to derive worst-case bounds on the quality of solutions delivered by heuristic algorithms for combinatorial optimization problems. Here quality means the ratio or difference between the delivered and the optimal solution values. Is a bound on the ratio not within a few percent of unity of practical interest? What about bounds on the difference?

PROBLEM 1.1.2. Is an estimator with standard deviation more than three times its mean of significant practical value?

On the other hand, there are always approximations in the model and inaccuracies in its input-parameter estimates. Reducing the error below a tiny tolerance often requires a huge number of runs. At best this wastes work and at worst it misleads about the resulting precision relative to the real system and about the competitiveness of simulation relative to deterministic approaches.

Morse (1977) laments that many models studied in scholarly journals are detached from observational underpinning. He goes on to say, rightly, that "the practice of devising a model and then looking for an operation that fits it is putting the cart before the horse." Fox (1968) points out that gratuitous assumptions may simply give a model a spurious air of objectivity and illustrates this with models of outlier contamination and reliability growth. It is perhaps not obvious that implicit assumptions are made when constraints are imposed. Quite often constraints *can* be violated—at a price. In these cases it may be appropriate to exchange certain explicit constraints for a modified objective function. A particularly dishonest practice is to massage the model, e.g., by imposing artificial constraints, until the simulation output supports preconceived or existing policies or doctrines; that is dissimulation.

PROBLEM 1.1.3. Find a number of models in articles in professional journals (e.g., *Management Science, Operations Research, JASA*). Examine these models critically and decide whether or not the assumptions made seem intuitively likely to correspond to reality. What about robustness with respect to departures from these assumptions? What parameters must be estimated? If the assumptions seem unnatural, what other reasons might the author have had for making them? Skepticism about assumptions is healthy; validate!

A prospective model-builder should consider using Koopman (1979) as a paradigm.

1.2. Verification, Approximation, and Validation

Whether the model and the program implementing it accurately represent the real system can be checked in two stages.

Verification. Checking that the simulation program operates in the way that the model implementer thinks it does; that is, is the program free of bugs and consistent with the model? Such checks are rarely exhaustive. As De Millo *et al.* (1979) remark, "... verification is nothing but a model of believability."

Validation. Checking that the simulation model, correctly implemented, is a sufficiently close approximation to reality for the intended application. As already indicated, no recipe exists for doing this. Due to approximations made in the model, we know in advance that the model and the real system do not have identical output distributions; thus, statistical tests of model validity have limited use. The real question is the *practical* significance of any disparities.

Although the concepts of verification and validation are distinct, in practice they may overlap to a considerable extent. When a simulation program produces nonsense, it is not always clear whether this is due to errors in the conceptual model, programming errors, or even the use of faulty data. Model construction, verification, and validation often are in a dynamic, feedback loop. A failed attempt at validation usually suggests modifications to the model, at least to those not blindly in love with their creation. Except perhaps where there is no question in the user's mind that the model structure is correct and that the required input distributions (see Chapter 4) and parameters have been estimated with great precision (a rare circumstance), he must be convinced that the results are robust to mild misspecifications in all these aspects. If he is not, the model must be successively modified or the estimates successively refined until he is.

Sargent (1980) points out that validation should include validation of the data used to develop the conceptual model, and the data used for testing. It is usually time-consuming and costly to obtain accurate and appropriate data, and initial attempts to validate a model may fail if insufficient effort has been applied to this problem. A model that *merely* accommodates known facts is just an *ad hoc* fit. A worthwhile model must generate predictions that are then corroborated by observations of, or experiments with, the real system. If this process leads to really new insight, then the model takes on the status of a theory.

1.2.1. Verifying a Program

All the standard tools for debugging any computer program apply to debugging a simulation. A useful technique is to include throughout the program at the outset statements which can all be turned on with a single "switch." These statements show in detail (numerically or, via a plotter, graphically) how the system state changes over time. Sometimes this is called a *trace*. With this capability present, perform the following kinds of verification:

(1) *Manual verification of logic*. Run the model for a short period by machine and by hand and then compare the results.
(2) *Modular testing*. Individually test each subroutine to verify that it produces sensible output for all possible inputs.
(3) *Check against known solutions*. Adjust the model so that it represents a system with a known solution and compare this with the model results; e.g., in queueing systems, solutions are frequently known if all the probability distributions are exponential.
(4) *Sensitivity testing*. Vary just one parameter while keeping all the others fixed, and check that the behavior of the model is sensible.
(5) *Stress testing*. Set parameters of the model to strange values (e.g., replace probability distributions with the Cauchy distribution, set the load on the system in excess of 100%, etc.) and check if the model "blows up" in an understandable fashion. Some of the bugs which are both most embarassing and difficult to locate are ones which do not manifest themselves immediately or directly but may show up under stress.

1.2.2. Approximation and Validation

The validation problem arises because various approximations to reality are made in creating the model. We always restrict the boundary of the model, ignoring everything outside that is not an explicit input, and neglect factors believed to be unimportant. Other types of approximation typically made are:

(1) *Functional*. Highly nonlinear functions, themselves often approximate, are approximated by simpler ones, e.g., piecewise linear. The important consideration is that the simpler function should be close to the "correct" one in the region where the system is likely to operate. If the program is forced to operate in a region where the fit is poor, it should print a warning message.

(2) *Distributional.* Real-world probability distributions which are known only approximately are frequently approximated by simple distributions, such as the normal or exponential, which have nice features. This is discussed in Chapter 4. The most extreme distributional approximation is to replace a random variable by a constant. This tends to be justified only if the outputs are linear functions of the random variables, though there are some nonlinear "certainty equivalence" situations [Theil (1961) and Problem 1.9.18] where this substitution is legitimate.

(3) *Independence.* The model is frequently simplified if various components (random variables) are assumed to be statistically independent.

(4) *Aggregation.* The most pervasive type of approximation is aggregation. By this we mean that several of something are treated as one. Some typical types of aggregation are:

Temporal aggregation. All discrete time models do temporal aggregation. An interval of time such as a day is treated as a single period. All events occurring during the day are assumed to have occurred simultaneously.
Cross-sectional aggregation. Several divisions, firms, product lines, etc., are treated as one.
Resource aggregation. Several resources are treated as one. For example, a computer system with two cpu's in parallel is treated as having one "twice-as-fast" cpu; a series of machines through which a product must pass are treated as a single machine with capacity equal to that of the lowest-capacity machine in the series.
Age aggregation. In tracking individuals who move through a system it is convenient to treat those within a given age range as a single group. The age ranges are determined by commonality of characteristics such as death rate and working capacity. For people these characteristics might suggest the ranges (0–2), (3–5), (6–18), (19–22), (23–65), (66–120).

The combination of age aggregation and temporal aggregation can pose problems in a discrete-time model if the unit of time is not chosen to be about the same as the span of a typical age group. For the above age grouping, suppose that our model transfers a fixed proportion of the members from one group to the next at each time interval. Then if the unit of time is 1 year, our model would have the unrealistic feature that a sudden increase in the birth rate would ripple up in four periods to the 19–22 age group. In modern societies, it is not usually the case that a 4-year-old child is already entering the work force.

Mulvey (1980) presents a useful procedure for aggregating taxpayers in a simulation model used to evaluate the impact of proposed changes in U.S. federal tax policy. The procedure aggregates taxpayers into one of about 75,000 classes.

(5) *Stationarity.* It simplifies matters to assume that parameters and other features of the system do not vary over time. This may be reasonable if it can be legitimately argued that any changes over the relatively limited

period of interest are negligible. For certain physical processes, such as some astronomical phenomena, this may be an acceptable first approximation. However, in the political, economic, and social domains everyday experience indicates that many phenomena are grossly nonstationary and, consequently, that the notion of steady state is often vacuous. Most forecasting methods, based on regression, moving averages, or exponential smoothing, assume stationarity either of the original process or of a specific transformation of it, which suggests that these heuristics should be taken with a large grain of salt. In the dynamic programming context, Hinderer (1978) notes that real problems have finite horizons, "unlimited stationarity of the data is quite unrealistic," and "the influence of the terminal reward cannot be neglected." Kleijnen (1978) emphatically states that "in practical simulation studies there is usually no interest in steady-state behavior; start-up and end effects do form part of the relevant output." Kleijnen agrees with Keynes: "In the long run we are all dead."

Despite this, models based on a steady-state approximation can occasionally be useful. For instance, when invocation of the steady state permits an otherwise unobtainable, intuitively understandable, analytic characterization of certain phenomena or an order of magnitude reduction in computer time needed to produce numerical or graphical descriptions, the steady state may sometimes be regarded as a convenient fiction. A vital proviso is that any attempts to extrapolate steady-state results to real systems—in which stationarity is not assumed *a priori*—be regarded with great skepticism unless this extrapolation has been carefully validated. Such validation may be via numerical calculations [e.g., see Neuts (1977a)] or, more often, via simulation. It has also been argued that invoking the steady state reduces the number of parameters which the model-builder has to estimate. Such parameters may be needed to specify boundary conditions or nonstationarities. The implicit assumption that the parameters thus neglected are in fact unimportant should not be accepted uncritically.

A steady-state hypothesis typically simplifies the treatment of analytic models but complicates the analysis of simulation models. In Chapter 3 we cover both transient (finite-horizon) and steady-state approaches and continue the discussion of the appropriateness of steady-state simulations.

1.3. States, Events, and Clocks

One key to building good models is the adequate definition of states. Built into the definition must be enough information about the history so that, given the current state, the past is statistically irrelevant for predicting all future behavior *pertinent to the application at hand*. This is consistent with the definition of state in the Markov-chain setting, though the italicized qualification is often not emphasized in books on probability. We move from state to state according to one-step probabilities that depend only on the current state. In many cases the process of interest is not a Markov chain, but it often contains an *imbedded* Markov chain that governs transitions among a distinguished subset S of states. As far as the imbedded chain is concerned, when the system state is not in S the system is in limbo; furthermore, all transitions in the chain take unit time, no matter what the real clock says. Often it suffices to specify only the states in S, resurrecting the real clock to keep track of the transition times. This characterizes discrete-event simulation. It also leads to Markov renewal theory [e.g., see Çinlar (1975) and Disney (1980)], to which we refer occasionally. During transitions, other things may happen that the simulation program has to keep track of, summarize, and analyze; for example, customers may arrive or leave and costs may be incurred. Table 1.3.1 lists, not necessarily exhaustively, state components for various systems. The times at which the system enters a state in S are called *event epochs*. The corresponding state changes are called *events*.

Table 1.3.1 Components of System State for Various Systems.

System	State
Industrial Jobshop	Number of jobs at each work center, with their characteristics (e.g., remaining work content).
Inventory System	Stock on hand and on order for each product at each inventory center.
Airport	Number of planes stacked for landing and awaiting takeoff on each runway.
Elevator System	Number of people waiting at each floor; number on each elevator; position of each elevator; indicators for which buttons have been pressed and as yet not answered.
Fire Department	Location and status of each piece of equipment; location of current fire alarms and fires.
Ambulance System	Location and status of each piece of equipment; location of current requests for help.
Computer	Number of jobs waiting for the central processor, with their characteristics; resources claimed by jobs currently in process: number of jobs waiting for input/output processor, with their characteristics.
Message Switching System	Position, destination, and length of each message in the system.
Distribution System	Inventory levels at various centers in the system.
U.S. Economy	Stocks of various types of money; capacities and inventory levels in various industries; price levels.

In Example 1.1.1 the events are arrivals, departures, and starts of service; at each arrival epoch, we generate the next (in time) arrival epoch.

Several of the approximations made in going from a real system to a model have the effect of reducing the number of states. This tends to make the model more tractable. Schruben (1983a) appropriately emphasizes that state variables can be "event conditioning or performance monitoring or both." The former cause the scheduling or canceling of future events and can be related pictorially by what Schruben calls "event incidence graphs." Performance monitoring variables correspond to "fictitious" events. They are used to collect statistics for output analysis. Example 1.4.1 illustrates this.

Most simulation clocks are driven by a dynamic event list so that time skips from one event epoch to the next. Nothing happens between event epochs. There are two main ways to view these clocks.

(1) *Event orientation.* For each possible event, there is a subroutine that determines what other events get triggered and their respective provisional activation dates. The clock mechanism then puts these activation dates on the event list with pointers to the corresponding events, and then it activates the next event due.

(2) *Process orientation.* All the code that concerns one process (e.g., customer) is grouped together in one subroutine. An event sublist is kept for each process, and a master list contains the next event due in each sublist. The head of the master list points to the next event due. The clock mechanism activates it. Suppose that this event concerns process Q and that the immediately preceding event concerns process P. If $P = Q$, process P continues. If $P \neq Q$, the clock mechanism suspends P and then passes control to Q.

We illustrate these views in Section 1.4 and discuss how to structure clock mechanisms in Section 1.7.

PROBLEM 1.3.1. For Example 1.1.1 show how to arrange that the number of pending event notices is at most two, no matter what the queue length. Under what circumstances can the number of pending event notices be one? zero?

PROBLEM 1.3.2. Give several different models where the number of pending event notices is generally greater than one hundred.

If we explicitly model the system as a Markov chain, in discrete or continuous time, then the need for an event list disappears. Instead, we require a compact generator that produces quickly any row of the transition matrix as needed. Designing a matrix (row) generator for Markov chains is a special case of designing a matrix (column) generator for linear programming. It may be possible to use the latter virtually off the shelf. Recoding a multidimensional state space as a (one-dimensional) vector, using standard devices, may be desirable.

1.4. Simulation—Types and Examples

Throughout this book, we are chiefly concerned with discrete-event digital simulation. Many of our examples will be hung on the following peg:

EXAMPLE 1.4.1. A bank employs three tellers. Each works at exactly the same rate and can handle exactly the same customers. Most days, all three tellers report for work. However, 15% of the time only two tellers are present, and 5% of the time only one teller turns up at work.

The bank opens at 10 A.M., and closes at 3 P.M. From 9:45 A.M. until 11 A.M., and from 2 P.M. until 3 P.M., one customer arrives every 2 minutes on average, and from 11 A.M. until 2 P.M. one customer arrives every minute. More precisely, the intervals between successive arrivals are independent and exponentially distributed with the appropriate mean. Those customers who arrive before 10 A.M. wait outside the door until the bank opens; at 3 P.M. the door is closed, but any customers already in the bank will be served.

Customers form a single queue for the tellers. If, when a customer arrives, there are n people ahead of him in the queue (not counting the people receiving service), then he turns around and walks out ("balks," in queueing terminology) with the following probability:

$$P[\text{balk}] = \begin{cases} 0, & n \leq 5, \\ (n-5)/5, & 6 \leq n \leq 9, \\ 1, & n \geq 10. \end{cases}$$

The customer at the head of the queue goes to the first teller who is free. Customer service times are distributed according to an Erlang distribution with parameter 2 and mean service time 2 minutes. We want to estimate the expected number of customers served on an arbitrary day. We might be interested in the effect of changing the number of tellers, changing their speed, etc.

Our model is, of course, chosen to highlight certain aspects and techniques of discrete-event simulation. Even at this early stage, however, the reader should begin to be aware of two things. First, our model is quite typical of models used to solve real-life problems, particularly in its drastic simplification of complex phenomena. Second, even the most rudimentary validation of such a model—and indeed, even to transform vague *a priori* ideas into such a model in the first place—demands an arduous and costly process of data collection.

The model is programmed in detail in Chapter 7, and various techniques for accelerating its convergence to a sufficiently trustworthy answer are illustrated in Chapter 8.

We consider the system only at the event epochs corresponding to the following events:

 (i) the arrival of a customer;
 (ii) the end of a period of service;
 (iii) opening the door at 10 A.M.;
 (iv) closing the door at 3 P.M.

To characterize the system at these instants, the following are sufficient:

 (a) an integer giving the number of tellers who are free;
 (b) an integer giving the number of people in the queue.

The number of possible states of the system is therefore quite small: certainly less than 4×11. This characterization of the system is not the simplest possible. We could keep just one integer giving the number of people in the bank: with this, and knowing how many tellers turned up for work, the state of the system is sufficiently specified for our purposes. For some applications, we may have to include in the state description the times when the customers in service started service because the service distribution has memory. Here we do not include them because the balking probability does not depend on them and because the corresponding end-of-service events are already posted on the pending event list. Between event epochs the system might as well not exist: neither of the state variables we have chosen will change, so such intermediate periods are irrelevant. We keep track of time but, somewhat arbitrarily, do not consider time to be a state variable (see Problem 1.4.2). For these reasons, we do not consider the instants when the arrival rate changes as event epochs.

For each possible event, we shall write a subroutine to keep track of what happens as we move from one state to the next. For instance, the following things may occur when a customer arrives:

 (i) he may find the queue too long and leave immediately (in which case the system state does not change, although we should probably want to count how many times this happens);
 (ii) he may find a teller free (in which case we decrement the number of free tellers, changing the system state, and draw a random number to decide how long it will be until this service period ends);
 (iii) if no teller is free, he may join the queue (in which case the queue counter is incremented, changing the state of the system).

Then (and only then) we generate the next arrival time.

As mentioned briefly in Section 1.3, a simulation organized in this fashion is *event-oriented*. In a high-level language, there is usually one subroutine corresponding to each possible kind of event. An alternative organization is to make the simulation *process-oriented*. In our example, this would suggest, perhaps, that we group together in one subroutine all the code that concerns one customer: the state change on his arrival, the state change when he begins to receive service, and the state change when he leaves. Such a process-oriented simulation requires a more sophisticated implementation, because it is now necessary to suspend one subroutine temporarily while another takes a turn. In programming terms, we need a co-routine structure. In practice, an event-oriented simulation can easily be written in any general-purpose high-level programming language, because all such languages allow for subroutines. A process-oriented simulation usually requires a special-purpose language including a co-routine mechanism: such languages are typically intended specifically for simulation, or else for writing operating systems, although other applications for co-routines exist.

Neither the choice of states to be represented nor the choice of events to be simulated is entirely trivial. Both may depend on the results we wish to obtain, the type of statistical analysis we wish to perform, and on our desire to build a simple program. For instance, we have represented the state of the tellers by a single integer, because for our purposes it is irrelevant whether a teller is busy or whether he is absent: we only need to know that he is unavailable. If we wished to study, say, the utilization of each teller, it might be necessary to use a vector of three elements, one for each teller, saying whether he is free, busy, or absent.

Similarly, we have included an event corresponding to closing the door of the bank at 3 P.M. Because we are not gathering waiting time statistics and we know that everyone in the queue at 3 P.M. will eventually be served, we might as well stop the simulation at this point. It would be nearly as easy, however, to omit this event, to refuse new arrival events which occur after 3 P.M., and to let the model run until the bank is empty. The event corresponding to opening the door is, in contrast, essential. Important changes happen at 10 A.M.: the first customers occupy the available tellers, service begins, and so on.

If we want to gather statistics at regular intervals—to see how many people are in the bank at 10 A.M., 11 A.M., noon, . . . , say—then it may be convenient to include "fictitious" events scheduled to occur at these times (rather as though we had an event "the clock strikes"). Although these fictitious events cause no change in the state of the model, they afford a convenient opportunity for collecting the required statistics. The alternative is to test, at every real event, whether it is the first to occur since the clock struck: if so, we must record the state of the system before we change it. This usually leads to a much less transparent program.

PROBLEM 1.4.1. Suppose we wish to adapt Example 1.4.1 to measure waiting-time statistics. What state variables should we add?

The distinction between "state variables" and "statistics being gathered" can be rather arbitrary. In our example, we could consider that the number of customers served so far is a state variable of the model, because it is certainly changed by the end-of-service event. However, this variable in no way affects the *behavior* of the model. It is better to consider it simply as an output: the number of states should not be multiplied unnecessarily.

PROBLEM 1.4.1. Suppose we wish to adapt Example 1.4.1 to measure waiting-time of the model, though clearly the model's behavior will be different at different (simulated) moments. Discuss the special status of this variable.

1.4.1. Synchronous and Asynchronous Discrete-Event Simulation

The above example is an *asynchronous* simulation: events, such as arrivals, can occur at any time. It is occasionally convenient to push the idea of fictitious events occurring at fixed, regular intervals (like the hourly gathering of statistics in the example above) to its limit: now, instead of simulating events whenever they occur, we look at the model *only* at regular intervals $0, \delta t, 2\delta t, \ldots$. Events which occur "in the cracks" are taken to happen at the next regular time. This approach, called *synchronous* simulation, has a number of drawbacks, but one major advantage.

The first drawback is that such models are hard to program. Despite the apparent simplicity of the clock mechanism, the fact that nonsimultaneous events may be treated as simultaneous if they fall in the same interval usually leads to complicated problems of priority and sequencing. Next, to obtain a sufficiently accurate estimate it may be necessary to take δt very small. The number of times that the event list must be checked goes up as δt goes down. Unless we take special steps to pick out and skip moments when nothing will happen, thus complicating the program, its speed goes down as estimation accuracy improves. An asynchronous simulation does not have this failing.

On the other hand, in this kind of simulation it is comparatively easy to implement so-called *wait-until* statements: in general, we want a certain event to occur at the time a particular condition is satisfied. The condition may include a reference to simulated time (for example, wait until one of the tellers has been idle for ten consecutive minutes; wait until the first customer in the queue has been held up for 15 minutes, etc.). This may be difficult or impossible to implement correctly in an asynchronous model because we may not become aware that the condition is true until the next event comes along: indeed by then the condition may be false again. With the synchronous approach, we simply scan all the elements of the model at every time interval to see if any of them is waiting on a condition which has now become true.

Although we mention synchronous simulation again briefly in the section on clock mechanisms (§1.7), its disadvantages usually outweigh its benefits. We accordingly pay it rather little attention.

1.4.2. Continuous Simulation

Continuous simulation is the name given to the use of a digital computer to solve models with continuously changing states. Such models are generally composed of a number of differential equations, which are solved numerically to see how the model behaves. The name simulation is applied to such studies because the independent variable in the equations to be solved is usually time: we want to calculate the state of the system at some time in the future, starting from a known initial configuration. Even more explicitly, this may be called continuous-time simulation. Some may feel that it is stretching a point to speak of (continuous) simulation when we mean solving a system of deterministic differential equations, and that the term should be reserved for solving stochastic differential equations. However, it is perfectly possible to use a so-called continuous simulation language to solve any system of deterministic differential equations. Solving stochastic differential equations is far harder.

Planning a continuous simulation is rather different from planning a discrete-event simulation. In general, the differential equations in the model can include random variables, to represent noise in a channel, for instance, or other random disturbances. These random variables are generated using exactly the same techniques as in discrete-event simulation. For continuous simulations with random variables, statistical theory to guide the simulation design and the interpretation of its output scarcely exists. In the case of continuous simulation the analyst has a second concern, namely to choose an appropriate algorithm for carrying out the necessary numerical integrations, and to avoid instability and numerical errors in his results. The trade-off he has to make is between simple algorithms and long step sizes, which can given highly inaccurate results, and, on the other hand, sophisticated algorithms and short step lengths, which can lead to expensive computations. Computation speed may also be important if the model is part of a real-time system. In other words, continuous simulation generally poses problems not only of statistics but also of numerical analysis.

Several good introductions to continuous simulation can be found elsewhere: see, for instance, Chu (1969), which gives, besides a number of true simulations, many examples of the application of a continuous simulation package to purely mathematical problems.

No continuous simulation language has achieved the popularity of Simscript (§7.2) or GPSS (§7.3) in the area of discrete-event simulation. The examples below illustrate two available languages, MIMIC and SL/1. Because most languages have a strong family resemblance, these may be taken as typical of what presently exists. Alternatively, one can use with roughly comparable effort a general purpose language such as Fortran when appropriate subroutine packages are available. The IMSL (1980) library provides routines for differential equations. A small number of languages, such as GASP IV [Pritsker (1974) and SLAM [Pritsker (1986)] provide facilities for both discrete-event and continuous simulation.

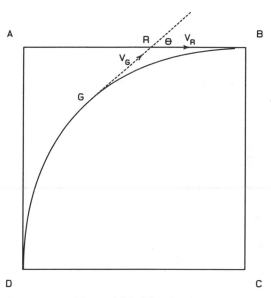

Figure 1.4.1. The chase.

EXAMPLE 1.4.2. Figure 1.4.1 shows the system we wish to simulate. A rabbit (animal, not car) R and a greyhound (dog, not bus) G are in the square field $ABCD$, where $AB = 100$ ft. At time $t = 0$, the rabbit is at A and the dog at D. The rabbit's hole is at B. Subsequently, the rabbit runs directly for its hole at a speed V_R, and the greyhound runs directly towards the rabbit at a speed V_G. We wish to simulate the system for different values of V_G.

Taking the origin at D, and with the obvious notation, the equations of the model are

$$x_R = V_R t,$$
$$\dot{x}_G = V_G \cos \theta, \qquad \dot{y}_G = V_G \sin \theta,$$
$$\tan \theta = (100 - y_G)/(x_R - x_G),$$

with the condition $x_G = y_G = 0$ at $t = 0$.

PROBLEM 1.4.2. Show analytically that the dog catches the rabbit if

$$V_G/V_R > (1 + \sqrt{5})/2.$$

(Hint: Take polar coordinates r and θ of the rabbit relative to the dog. Resolve along and at right angles to the line GR, eliminate θ from the resulting equations, and integrate.)
    Verify this result using a continuous simulation system.

Figures 1.4.2(a) and 1.4.2(b) give the corresponding programs in MIMIC [CDC (1972)] and in Fortran, respectively. Little explanation is needed. In the MIMIC program, CON introduces a constant, which is read once and never changed. PAR introduces a parameter: for each value read, a complete

```
                              CON(VR)

                              PAR(VG)

                  XR      =  VR*T

                  THETA   =  ATN(100.0-YG,XR-XG)

                  XG      =  INT(VG*COS(THETA),0.0)

                  YG      =  INT(VG*SIN(THETA),0.0)

                              HDR(T,XR,XG,YG)

                              OUT(T,XR,XG,YG)

                              PLO(XG,YG)

                              FIN(XR,100.0)

                              FIN(XG,XR)

                              END
```

Figure 1.4.2(a). A program in MIMIC for the chase.

```
C    MAIN PROGRAM FOR THE CHASE.
C
      REAL Y(2), YPRIME(2)
      REAL C(24), W(2,9)
C
C    TELL THE COMPILER THAT FCN IS A FUNCTION, NOT A VARIABLE.
C
      EXTERNAL FCN
C
C    ERROR TOLERANCE
C
      DATA TOL /.00001/
      DATA NW /2/
C
C    TIME INCREMENT IN SECONDS FOR RESULTS.
C
      DATA STEP /.02/
C
C    OUTPUT UNIT NUMBER
C
      DATA LOUT /5/
C
C    START AT TIME = 0.001
C
      X = 0.001
      XEND = 0.001
C
C    GREYHOUND STARTS AT POSITION
C
      Y(1) = 0.
      Y(2) = 0.
C
C    CHOOSE DEFAULT ALGORITHM PARAMETERS.
C
      IND = 1
```

```
C
C  SIMULATE UNTIL XEND.
C
   100 XEND = XEND + STEP
C
       CALL DVERK(2,FCN,X,Y,XEND,TOL,IND,C,NW,W,IER)
       WRITE (LOUT,200) Y(1), Y(2)
   200 FORMAT(X,F9.3, ', ', F9.3)
C
C  CONTINUE THE CHASE IF GREYHOUND HAS TRAVELED
C  LESS THAN 99.5 FEET IN Y DIRECTION
C
       IF (Y(2) .LT. 99.5) GO TO 100
C
       STOP
       END

       SUBROUTINE FCN(N,T,Y,YPRIME)
C
       REAL Y(N), YPRIME(N)
C
C  INTERPRETATION OF VARIABLES
C    T = CURRENT TIME
C    XR = X COORDINATE OF RABBIT
C    Y(1) = X COORDINATE OF GREYHOUND
C    Y(2) = Y COORDINATE OF GREYHOUND
C    YPRIME(1) = RATE OF CHANGE OF Y(1)
C    YPRIME(2) = RATE OF CHANGE OF Y(2)
C    THETA = DIRECTION OF GREYHOUND
C    VR = SPEED OF RABBIT
C    VG = SPEED OF GREYHOUND
C
       DATA VR /20./, VG /30./
C
C  CURRENT POSITION OF RABBIT
C
       XR = VR * T
C
C  CURRENT DIRECTION OF GREYHOUND
C
       THETA = ATAN ((100. - Y(2))/(XR - Y(1)))
C
C  RATE OF CHANGE OF GREYHOUND'S COORDINATES
C
       YPRIME(1) = VG * COS(THETA)
       YPRIME(2) = VG * SIN(THETA)
C
       RETURN
       END
```

Figure 1.4.2(b). A Fortran program for the chase. (Note: Figures 1.4.1 and 1.4.3 were drawn to machine accuracy using the **IMSL** (1980) routine DVERK, called above, for numerical integration.)

simulation is executed, then another value is read, and so on. The equations of the model are given in any convenient order: MIMIC will rearrange them as required to carry out the computation.

ATN (arctangent) and INT (integration) are supplied by MIMIC. At times $t = 0, 0.1, 0.2, \ldots$, the values of t, x_R, x_G, and y_G will be printed by the function OUT. HDR prints appropriate headings and PLO plots a graph

Figure 1.4.3. The flight.

showing the greyhound's track. A simulation ends when the rabbit reaches its hole ($x_R \geq 100$) or the dog bowls the rabbit over ($x_G \geq x_R$). If more values of the parameter V_G remain to be read, a new simulation will then begin.

MIMIC is easy to use, but the programmer has little control over what is going on. In particular, he cannot change the integration algorithm which will be used. He can change the print interval for OUT, put limits on the step size used by the integration algorithm, and choose between a relative and an absolute error criterion, but that is all.

EXAMPLE 1.4.3. Figure 1.4.3 shows the system we wish to simulate. A golfer drives an American ball, which flies off at 80 mph in a direction 30° above the horizontal. The ball moves under the influence of gravity, and of air resistance equal to kv^2, where v is the speed of the ball, and

$$k = \tfrac{1}{2}\rho C_d S$$

(ρ is the density of the atmosphere, S the cross-sectional area of the ball, and C_d a drag coefficient depending on the exact geometry of the ball). If the ball swerves neither left nor right, and the course is exactly horizontal, where will the ball come to earth?

Resolving horizontally and vertically, so $v^2 = \dot{x}^2 + \dot{y}^2$, $\dot{x} = v \cos \theta$, $\dot{y} = v \sin \theta$, the equations of the system are

$$m\ddot{x} = -kv^2 \cos \theta$$
$$= -kv\dot{x},$$
$$m\ddot{y} = -kv^2 \sin \theta - mg$$
$$= -kv\dot{y} - mg.$$

At $t = 0$ we have $x = y = 0$, $\dot{x} = v_0 \cos 30°$, $\dot{y} = v_0 \sin 30°$. The necessary constants are:

$$\rho = 0.07647 \text{ lb/ft}^3,$$
$$g = 32.174 \text{ ft/s}^2,$$
$$s = 0.01539 \text{ ft}^2,$$
$$m = 0.10125 \text{ lb},$$

and we shall guess quite arbitrarily that $C_d = 0.2$.

Figure 1.4.4 gives the corresponding program in SL/1 [Xerox (1972)]. It is divided into three regions: INITIAL, DYNAMIC, and TERMINAL. As the names imply, the INITIAL region is executed once at the beginning of a simulation, and the TERMINAL region is executed once at the end. The DYNAMIC region is executed repeatedly until the programmer explicitly transfers control to the TERMINAL region.

Within the dynamic region, the differential equations of the model are found in the DERIVATIVE section. When control enters this section, the integration routines of SL/1 take over. If the current values of the state variables of the model represent the state at time t_0, the integration routines advance their values so they now represent the state at $t_0 + \delta$, where δ, called the communication interval, is defined by the programmer. When this is done, the integration routines return control to the instructions, if any, following the derivative section. When the end of the dynamic region is reached, control returns to the beginning of the region for another pass. On this new pass, the derivative section advances the model from $t_0 + \delta$ to $t_0 + 2\delta$, and so on.

In our example, the communication interval CI is defined to be 0.1 time units (seconds, in our interpretation). The values of t, x, and y are printed each

```
PROGRAM GOLF
        INITIAL
            CONSTANT RHO = 0.07647, CD = 0.2, G = 32.174
            CONSTANT S = 0.01539, M = 0.10125
            PARAMS VO
            PAGE EJECT; TITLE 20, THE FLIGHT
            PRINT FMTA, VO
            FMTA: FORMAT (/, 1X, 'INITIAL SPEED', F6.1, 'MPH')
            K = 0.5*RHO*CD*S ; VO = VO*22/15
            XDOTO = VO*SQRT(3.0)/2 ; YDOTO = VO/2
            SKIP 1 ; HDR T, X, Y
        END
        DYNAMIC
            OUTECI T, X, Y ; IF (Y.LT.0.0) GO TO FIN
            PX = X ; PY = Y
            DERIVATIVE
                CINTERVAL CI = 0.1
                V = SQRT(XDOT**2 + YDOT**2)
                X2DOT = -K*XDOT*V/M
                Y2DOT = -K*YDOT*V/M - G
                XDOT = INTEG(X2DOT, XDOTO)
                YDOT = INTEG(Y2DOT, YDOTO)
                X = INTEG(XDOT, 0.0) ; Y = INTEG(YDOT, 0.0)
            END
        END
        TERMINAL
            FIN: HIT = PX + (X-PX)*PY/(PY-Y)
            PRINT FMTB, HIT/3
            FMTB: FORMAT (/, 1X, 'DISTANCE', F6.1, 'YARDS')
        END
    END
```

Figure 1.4.4. A program in SL/1 for the flight.

communication interval, i.e., at each pass through the dynamic region. The variables px and py save the old values of x and y before the integration routines update the latter. The dynamic part of the simulation ends when y becomes negative. At this point we enter the terminal region and estimate where the ball hit the ground using linear interpolation between its previous position (px, py), $py \geq 0$, and its newly calculated position (x, y), $y < 0$.

Except within the derivative section, the instructions in an SL/1 program are executed in their order of appearance. Like MIMIC, however, SL/1 accepts the differential equations of the model (the derivative section) in any order, and rearranges them as required. In the above example, the integration routine is chosen by default to be a fixed-step Runge–Kutta algorithm of order 3, with one step per communication interval. The programmer can change the number of steps, or choose any one of six available integration procedures, or provide his own. By grouping his equations in different derivative sections, he can specify different algorithms for different equations.

SL/1 offers a number of other features which we shall not discuss. One, however, is worth mentioning. On a computer with the appropriate hardware, the rate of execution of the derivative section, i.e., the communication interval, can be controlled by a real-time clock. Two predefined arrays called ADC and DAC are connected to analog–digital and digital–analog converters. Then, for instance, whenever the program refers to the variable ADC(3), it is supplied automatically with the value found at that instant on A-to-D channel 3. Similarly, if the program assigns a value to the variable DAC(7), the value is transmitted to channel 7 of a D-to-A converter. These facilities simplify the construction of hybrid simulations, where part of the simulation is carried out by digital equipment, and part by analog.

There exist many other continuous simulation languages. As we have already remarked, their family resemblance is strong. We therefore leave the topic of continuous simulation here. We shall not return to it.

1.4.3. Hybrid Simulation

To complete our survey of simulation types, we briefly discuss hybrid simulation. A modern hybrid computer is a digitally based system utilizing a low-cost, high-speed CPU to control and augment parallel analog processors [Landauer (1976)]. Proponents of such systems make several claims for them. First, the computing speed of a hybrid system can, for typical engineering design studies, surpass considerably the speed attainable with a pure digital system. Second, they claim that the cost of such studies is lower, partly because of the increased speed available, and partly because a hybrid machine costs less to make than its digital counterpart. Finally, a hybrid computer is inherently a real-time device. This makes it relatively easy to interface to the existing components of a system under development to simulate the behavior of missing elements.

For more details see Landauer (1976), where several examples of hybrid simulation are mentioned: simulation of the rocket engine for the space shuttle, of a nuclear generating station, of a heat exchanger, and so on.

Simulation on a hybrid machine should not be confused with simulation carried out entirely on a digital computer using a language which allows both discrete-event and continuous modules to be incorporated into a model (such as GASP IV and SLAM, mentioned in §1.4.2).

1.5. Introduction to Random Numbers

Remarkably, random numbers from virtually any distribution can be obtained by transforming $(0, 1)$-uniform random numbers. For the latter we reserve the symbol U, possibly subscripted or superscripted. This notation sometimes indicates a string of such numbers. The context makes the meaning evident.

Proposition 1.5.1. *Let* $P[X = a] = q$ *and* $P[X = b] = 1 - q$. *Generate* U. *If* $U < q$, *output* a; *otherwise, output* b. *Then the output and* X *have the same distribution.*

PROBLEM 1.5.1. Write down the (trivial) proof. When is the output a nondecreasing function of U?

Proposition 1.5.2 (Inversion). *Let* X *have distribution* F. *Suppose that* F *is continuous and strictly increasing. Generate* U. *Then* $F^{-1}(U)$ *has distribution* F.

PROOF. Because U is uniform on $(0, 1)$, $P[U \leq x] = x$, $0 \leq x \leq 1$. Hence $P[F^{-1}(U) \leq t] = P[U \leq F(t)] = F(t)$. $\qquad\square$

PROBLEM 1.5.2. Illustrate this method graphically. Show that the output is an increasing function of U. What is the distribution of $F^{-1}(1 - U)$? Suppose now that F has jumps or flat spots. Show how to modify the method, making the above two propositions special cases.

PROBLEM 1.5.3. What is the distribution of $-\lambda \log U$?

Chapter 5 shows a number of other ways to transform uniform random numbers. These transformations should be considered in the light of remarks in the introduction to Chapter 2.

Generating *truly* random numbers is generally both impractical and in fact undesirable. Chapter 6 describes generators of $(0, 1)$-uniform *pseudorandom* numbers. As detailed there, these generators have certain number-theoretic properties which show *a priori* that their outputs behave very much like strings of truly random numbers, in precisely defined ways. In addition, their outputs have been subjected to a battery of statistical tests of "randomness." For most purposes, the strings generated can be regarded as genuinely random. This is discussed briefly in Chapter 6 and at length by Knuth (1981) and Niederreiter (1978).

Pseudorandom number generators have the great advantage that the random number strings can be reproduced exactly, without storing them. Starting from the same *seed*, the first number in the string, suffices. This can be used to improve the design of certain simulation experiments, as indicated in Chapter 2. It is also of great importance when debugging simulation programs, because runs can be repeated in case of aberrant behavior. With some simulations, it may be convenient to generate all the uniform and nonuniform random numbers in a separate input phase, followed by the actual simulation: see Section 1.8. In Example 1.1.1 we would probably transform uniform random numbers to nonuniform random numbers A_i and B_j as required, but it may be helpful to think of the A_i's and B_j's as being generated in a preliminary phase and then, in a second phase, being further transformed to the final output.

1.6. Perspective on Experimental Design and Estimation

Much of the vast literature on experimental design and estimation is relevant to simulation. We shall not try to recapitulate in a few pages the extensive work done over a number of decades by many eminent statisticians. It is only a slight oversimplification to distill the questions they ask as follows:

(1) How do we get good estimates of some measure of performance?
(2) How do we get good estimates of the goodness of these estimates?

In the simulation setting, we have absolute control over all factors—because no *real* randomness enters (cf. §1.5). This unique situation leads to special problems and opportunities. General statistics books are not concerned with handling such situations and so they ignore or at best pass rapidly over many of the topics we cover. Our coverage complements theirs. The rest of this section builds a philosophical framework for the mainly theoretical Chapters 2 and 3. Illustrations are given in Chapter 8. These are helpful for fixing ideas. More importantly, they show that the techniques of the preceding chapters can be applied in practice to produce significant gains in statistical reliability and savings in computer time over an obvious but inefficient approach.

In Question 2 above, *goodness* was left undefined. Perhaps oversimplifying, we assume that a scalar measure adequately summarizes performance.

Definition. The *goodness* of an estimator is the reciprocal of its theoretical mean square error.

This definition is imperfect, seriously so if the distribution of the estimator is highly skewed.

PROBLEM 1.6.1. Why? We shun vector performance measures because then we would sometimes be unable to say which of two systems we prefer. Give an example.

Anyone who believes that his estimator is approximately normally distributed would certainly not balk at the above definition and could fearlessly construct confidence intervals. However, even if there is an asymptotic normality theory in the background, we believe that users who invoke normality are often on shaky terrain. Our reasons will become clearer below, as we outline two ways to evaluate goodness.

A. Take an *idealized* situation where the parameter to be estimated is known or can be readily calculated. A popular testbed is a simple textbook queue (stationary $M/M/1$) with the parameter taken to be mean waiting time or mean queue length. Simulate the queue, evaluate the (generally biased) estimator and an estimator of its *asymptotic* variance, invoke normality, construct a confidence interval, record its length, and note whether or not the interval covers the true parameter value. Repeating this procedure over many runs leads to a measure of performance. The final, usually implicit step is to *extrapolate* this performance in an idealized situation to hypothetical performance in generally unspecified real cases. In the simulation literature, methods for analyzing infinite-horizon simulations are judged in this way. We cite examples in Chapter 3 with cautionary notes.
B. Simulate the *real* situation of interest. Estimate the goodness of the estimator by its *sample* variance or something akin to it (see Chapters 2 and 3).

Both these methods are open to criticism:

A. There is little or no statistical basis for such extrapolation. Schruben (1978), probably reflecting widespread opinion, attempts to rebut this position by asserting that "this [extrapolation] seems rather descriptive [of] applied statistics itself." In statistics, one does extrapolate, but from a sample to the population from which it was drawn. The extrapolation above is radically different: it goes from a sample from one population (the idealized model) to a hypothetical sample from a distinct population (the real system). To some, such bold extrapolation may seem plausible, reasonable, and legitimate. This is subjective. However, defending it as in line with standard statistical practice seems stretched. The procedure may have some value in screening out really bad estimators [e.g., see Law and Kelton (1984)] but it has trouble distinguishing reliably among those that pass the screen.
B. The observations from which the sample variance is calculated are often correlated. In such cases, the usual variance estimators are *biased*

estimators of the true variance. Even when the estimators used are un-biased estimators of mean square error, it is not always the case that the estimator with the smallest theoretical mean square error will give the smallest sample variance in any particular situation. Figure 8.8.2 shows the variability of the sample variance in a situation where it is an unbiased estimate of the true variance. In general, however, when no better criterion is available we advocate comparison of candidate estimators using their observed sample variances.

When certain variance reduction techniques, such as stratification (§2.4) or conditional Monte Carlo (§2.6) are applicable, the *true* variance of the resulting estimator is necessarily reduced. With other techniques, such as common random numbers (§2.1) and antithetic variates (§2.2), there is a guaranteed variance reduction only when the output is a monotone function of the driving uniform random number stream. When one knows that the true variance of one estimator is less than that of another, there is no need to compare their respective sample variances; in fact, such a comparison may give a false indication.

Although B is certainly not perfect, we view it as a lesser evil than A. Suppose that we have several candidate estimators with no *a priori* ranking. A two-stage procedure in which the estimator used in stage 2 is the "best" of the estimators tried in stage 1 seems reasonable. At stage 1, the screening stage, we pick the "best" on the basis of the ranking of sample variances, provided we can show that these are consistent estimators of the true variance (usually this is obvious one way or the other).

And so, reader, choose your poison.

Implicit in Question 1 above is determining *what* to estimate. Mean "response" (e.g., queue length, waiting time, critical path length (Problem 1.9.8)) is often not enough. It hardly captures and summarizes response, especially when the response distribution is highly skewed. This is certainly so if the profit or cost, perhaps not explicitly specified, is strongly nonlinear in the response. A graph of its empirical distribution is more informative than its empirical mean. In this book we always work with means, but that word should be interpreted freely: for instance, if X is an output (response) from a simulation, put

$$Z_\alpha = \begin{cases} 1, & \text{if } X \leq \alpha, \\ 0, & \text{otherwise.} \end{cases}$$

PROBLEM 1.6.2. What is $E[Z_\alpha]$?

The mean of Z_α is a scalar; however, by choosing a number of different constants α, we may, in effect, estimate the whole distribution of X or, by interpolation, quantiles.

The exact form of the objective function with respect to its parameters is unknown *a priori*. Nevertheless, for sensitivity analysis or optimization, we must estimate its gradient. To do this, several authors haved proposed "perturbation analysis" of sample paths (simulation runs). Counterexamples show that such schemes do not work universally. Their (rigorous) application appears limited, as Heidelberger (1986) details. Glynn's (1986a) approach, via likelihood ratios, to gradient estimation has much wider applicability. Problem 2.5.8 gives an elementary version. The objective function is typically not convex. So even perfect gradient estimation does not by itself solve the problem of global optimization.

A heuristic approach postulates a (usually convex) form for the objective function, successively updates estimates of its parameters, and explores the region near the estimated optimum. This begs the question, for the postulated form may be grossly inaccurate, even in the neighborhood of the optimum; goodness-of-fit tests have low power against many alternatives. If the postulated form has more than, say, three parameters, the number of runs needed to estimate them accurately may be prohibitive. This approach, called *response surface methodology* in the statistical literature, is surveyed by Kleijnen (1974), (1975).

Schruben (1986a) uses response surfaces less ambitiously. Estimating the coefficients of a response surface amounts to gradient estimation. Instead, Schruben assesses the relative importance of input parameters by identifying which coefficients appear significantly nonzero, saying nothing about their signs. Since the response surface is usually based implicitly on a truncated Taylor series that may not be globally valid, relative importance of input parameters can be judged only locally. Schruben varies input parameters throughout each run, each at a carefully selected frequency. This lets him relate input and output spectra. Peaks signal important input parameters, at least for steady-state simulations. With his technique, we can screen out locally unimportant variables. Only the gradient of the remaining variables has to be estimated by some other technique; see Problem 2.5.8. In exceptionally favorable cases, Schruben's technique may indicate that enough random variables can be replaced by their expectations that a model tractable by mathematical programming emerges. Assuming such a model *a priori* is often risky.

The mathematical programming literature (e.g., see Avriel [(1976), Chapter 8] or Bazaraa and Shetty [(1979), §8.4]) contains several heuristic multidimensional search algorithms that use only function evaluations. An appealing approach is to evaluate the objective function over a *low-discrepancy* sequence of points, defined in Section 6.2. Niederreiter and McCurley (1979), Niederreiter and Peart (1986), and Sobol' (1982) study such schemes. Once a neighborhood of an apparent optimum is found, gradient information expedites further probing. Adapting these techniques to situations where a point is sought that approximately maximizes or minimizes the expectation of a random variable, as in simulation, is an open problem.

When considering situations where some alternatives are discrete, such as network topology, consult the classical literature on experimental design. A good reference is Box *et al.* (1978).

Chapters 2 and 3 discuss general principles for experimental design and estimation, but obviously no systematic treatment of *ad hoc* techniques is possible. Section 1.9 illustrates a few of the latter. Common sense helps.

Remove logical dependencies among inputs. Remember that a simulation for a given horizon supplies information for all shorter horizons.

Sequential sampling procedures bound *a priori* the length of the confidence intervals produced. Often this more than compensates for their complexity. Some of the subtle statistical issues generated by sequential experimentation are taken up in Problem 1.9.15 and in Chapter 3.

1.7. Clock Mechanisms

Roughly, there are two types of discrete-event simulation: *static* and *dynamic*.

Consider a critical path determination in a network with random arc lengths (see, for example, Problem 1.9.8). All arc lengths can be generated simultaneously, at least conceptually: the passage of time is irrelevant to program logic. This characterizes static simulation. Once all the arc lengths have been generated, the critical path is found by a sequence of computations that proceed, say, from left to right; although this does have a clock-like flavor, it is superficial.

Example 1.4.1 is quite different. Just after the activation of an event (see §1.3) we need to know which event must be activated (simulated) next and when, in simulated time, it must happen. We proceed directly from the activation of one event to the next. As far as the simulation logic is concerned, *nothing* happens between successive event activations. That is the reason for the term discrete-event simulation.

In most simulations, the activation of one event triggers the provisional scheduling of future related events. For instance, to an arrival corresponds a start of service and a departure. The schedule is held in the form of an *event list*. As the name implies, this is the list of events scheduled to occur, together with their respective times of occurrence. More precisely, the event list is a list of pointers to so-called *event notices*. Before these future events are activated, their activation times or even their nature may be modified as a result of what goes on after they are originally entered into the event list. In particular, the next event and its scheduled time can change dynamically. Managing this event list efficiently is central to the logic of a simulation program. We frequently need to scan this list to find the next event due or to insert newly scheduled events. Sometimes this list is long, in the hundreds or thousands. The required *clock mechanism* must be specially constructed if the simulation program is written in a general-purpose programming language. In special simulation languages a clock mechanism is already there, at least as a default option, and may appear as a black box to the user. This simplifies matters. On the other hand, hiding an inflexible and possibly inefficient clock mechanism in a black box can lead to unnecessarily long running times. Our brief discussion below helps to evaluate black boxes, if one can get inside, or to write a specially tailored clock mechanism from scratch. The need for such a mechanism characterizes dynamic simulation.

In a small number of cases it may be more efficient neither to hold an event list nor to schedule future events, but instead, when one event has been simulated, to scan the whole model to decide what can happen next, and when. This *activity-scanning approach* is generally useful only in models where the number of things that can happen is rather small, but the conditions under which they can happen are somewhat complicated.

The remainder of this section is based mainly on Fox's (1978b) survey article.

In *asynchronous* simulations, where events can occur at arbitrary times, *heaps* are often the appropriate data structure for event lists. This idea is frequently rediscovered. It has long been part of the folklore. Ties are broken using a secondary key (e.g., to maintain FIFO). McCormack and Sargent (1981) find that heaps perform well compared to other clock mechanisms, contrary to a number of published claims. When the event list is small or events tend to be added to the event list approximately in the order in which they are scheduled to occur, it is easier simply to maintain an *ordered* list. A convenient way to do this is to use a circular buffer, with the "tail" joined to the "head." Such lists are updated by interchanging elements to restore order. If instead of a circular buffer sequential storage is used ("linear" arrays), shift elements clockwise or counter-clockwise as convenient to open a space for insertion. Linked lists avoid shifts, but maintaining pointers requires overhead and storage. To find where to insert an event, scanning from tail to head usually works better than from head to tail.

In a uniform-width *bucket* system, each bucket i has width w and contains keys (i.e., event times) in the half-open interval $[iw, (i + 1)w)$. A bucket's contents can be sorted as it is emptied. If the range of possible key values is large, keep the keys bigger than a (possibly dynamic) given value in an "overflow" bucket. Vaucher and Duval (1975), in their study of event-list algorithms, find that "indexed" lists are the "most promising" structures considered; these lists are, in our terminology, buckets.

PROBLEM 1.7.1. Given a key, how would you determine the appropriate bucket? On a binary computer, how would you implement this if w is a power of 2?

PROBLEM 1.7.2. How would you structure a bucket's contents so as to empty it easily on a FIFO or LIFO basis? (Assume that the sorting algorithm maintains the order of tied elements.) How could you partition its contents into narrower sub-buckets?

With *synchronous* simulations, where time is advanced in fixed increments, buckets are the right data structure for event lists—generally, one bucket for each possible event-occurrence time.

With either synchronous or asynchronous simulations, keeping a heap of nonempty buckets may be worthwhile; this requires a "dictionary" that gives the location (if any) in the heap of each bucket. Another possibility is a multiechelon bucket system, similar to that developed by Denardo and Fox (1979) for finding shortest paths. This system is a dynamic k-ary tree where at any given moment most nodes are inactive. Because buckets are recycled after being emptied and only active nodes use buckets, the tree is sparse. A system of buckets with nonuniform widths might also be considered. These widths

(i) can be based on *a priori* estimation of key distribution; or
(ii) starting from the current time, the kth bucket can have width 2^{k-1}.

PROBLEM 1.7.3 [Fox (1978b), pp. 704–705]. In case (i) above, given a key, how would you determine the appropriate bucket? (Hint: Binary search is better than linear search but a combination with a higher echelon of equal-width buckets accessed by address calculation is even better.) In case (ii), how would you manage the dynamic bucket system?

PROBLEM 1.7.4. A two-bucket clock mechanism uses a dynamic partition between buckets equal to an empirical median of the pending event due dates. Each bucket's contents is kept sorted. Show how to implement this system efficiently. Show that among all two-bucket systems this division according to a median is minimax: it minimizes the maximum number of comparisons to find where to insert a new event notice. Quantify the following: relative to a simple linear list, this two-bucket system requires from about half as many comparisons to about the same number plus a small amount of overhead. Under what circumstances would you recommend the latter? [Davey and Vaucher (1980) and Davey (1982) study the average behavior of variants of a two-bucket scheme that use a median as a partition.]

The literature contains several highly idealized probabilistic models of event list dynamics aimed at optimizing clock mechanisms; e.g., see Vaucher (1976). Possibly a good approach is adaptive *ad hoc* adjustment performed by the clock mechanism itself as the simulation runs proceed; Davey and Vaucher (1980) and Davey (1982) detail a method along these lines. It is supported by analysis of a probabilistic model and by empirical results.

If pending events can be modified or deleted, we need a *dictionary* with pointers from events to their respective locations on the event list. Such a dictionary is also needed if, for example, we want to activate event B ten minutes after event A is scheduled to occur.

1.8. Hints for Simulation Programming

At the most basic level, simulation programming is no different from any other kind of programming, and all the standard techniques of software engineering for ensuring good design can and should be used. Keep in mind Hoare's (1981) admonition:

> "... there are two ways of constructing a software design: One way is to make it so simple that there are *obviously* no deficiencies and the other way is to make it so complicated that there are no *obvious* deficiencies."

Hoare was decrying overly complex programming languages, compilers, and operating systems: his criticism applies equally well to simulation programs. To simplify program logic, modularize: each event or process routine, each random number generator, the clock mechanism, and each data collection routine usually falls naturally into a separate module. If the conceptual model to be implemented is lucid and complete, then writing the code will be easy. On the other hand, a deficient model forces the programmer to improvise dubious techniques to paper over the cracks; then he has to wonder whether strange output is due to a faulty program or a faulty model.

Choose the programming language with care. Chapter 7 discusses in detail the relative merits of special-purpose simulation languages and non-specialized high-level languages. There are arguments both ways. It is generally easier to write correct programs in a well-adapted special language provided the user is intimately familiar with all its quirks; otherwise, using a special language may lead to subtle errors and misconceptions. Maintenance of a program written in a special-purpose language may also prove troublesome. Like any other program, a simulation program should be self-documenting, with meaningful names for the variables and frequent comments.

It is usually easier to test random number generators in isolation before they are incorporated into a model. As far as possible, test the actual non-uniform random number generators to be used, not merely the underlying uniform random number generators. Graphical methods are highly recommended: a simple plot of the desired and the actual distribution is often enough to catch gross programming errors. (If a programmer can't program a generator properly, his version of a complicated goodness-of-fit test is likely to be wrong, too.) If several streams of random numbers are being generated from the same uniform generator, choose starting seeds to avoid using overlapping series of values. Beginners often feel that using the computer's clock to provide a truly random starting seed is better than choosing seeds with some care. This is usually untrue, rarely necessary, and it makes debugging an order of magnitude harder.

Programs are usually easier to understand and to change if data collection and data analysis are separated. Except in the simplest cases, the simulation program proper should simply dump relevant measurements and observations onto a file to be analyzed later by a separate statistics program. One can thus be certain that changes to the model do not introduce errors into the data analysis; furthermore, it is possible to experiment with different statistical techniques without necessarily repeating the simulation run. Finally, the raw data can be examined if need be when it is unclear whether errors are in the model or in the analysis program. A similar technique may also be useful to separate the generation of random variables from the simulation proper. If all the required values are generated beforehand and saved on a file, the main program may be much simplified, particularly when

comparative runs of slightly different models are to be "synchronized":
see Chapter 2. Two special instances of this technique are worth noting:

(i) to trace the evolution of a model, one simple tactic is to introduce
artificial events at the required interval, whose only effect is to dump the
relevant values onto a file for later analysis;

(ii) in long simulations, restart information (the current values of random
number seeds, system state variables, etc.) can be saved at intervals: this
is often simpler than using the operating system's restart facilities.

Code all simulation programs to allow an execution-time trace of the
events being simulated, changes in the event-list, measurements being
collected, and so on. Whatever other means of program verification are used,
there is no substitute for a detailed, event-by-event study of at least parts of a
simulation run. Do not be afraid to check the "obvious." Does the clock
ever go backwards? Does the simulation end at the right time? Can you
account for all the events still pending in the event-list when the simulation
stops? Do all the customers who come in eventually go out again? And so
on. This check, though likely tedious, should be repeated after any major
changes in the program. Patient and painstaking examination of such a
trace can help not only to verify the program, but also, surprisingly often, to
validate (or invalidate) the underlying model.

Finally, the best advice to those about to embark on a very large simulation is often the same as Punch's famous advice to those about to marry:
Don't! If problems are encountered with programs that take too much time
or with memory requirements that cannot be met, the first stage should
always be to re-think the conceptual model. Remember that complexity and
accuracy are not synonymous: there is *no* merit in using enormously detailed
sub-models if the global result desired is insensitive to such details. Ask
yourself if the data available justifies the computations being based on it:
if a few minutes' observation are being extrapolated to the simulation of
several hours of a hypothetical system, how justified is this? Consider using
better statistical techniques: straightforward replication of an uncomplicated
model is easy to understand and explain, but in many cases even the simplest
variance reduction techniques offer great gains of computer time. There may
be examples of very large, very long, very complex simulations which produce
better results than could be obtained in any other way, but they are rare.
We are skeptical about the value of any simulation study that appears to be
primarily a giant programming exercise.

1.9. Miscellaneous Problems

We give below some problems that are not conveniently placed elsewhere in
the book. They can be tackled without further preparation. Though they by
no means span the range of problems that simulation can attack, they are

interesting in their own right and set the stage for Chapter 2. They show that there is more to simulation than one might initially suspect.

PROBLEM 1.9.1 (Monte Carlo integration). Suppose that we want to estimate

$$\theta = \int_0^1 f(x)\, dx.$$

Further, assume that the integral cannot be evaluated analytically and that numerical evaluation would be difficult. So we try simulation. (In this essentially deterministic context, simulation is often called Monte Carlo, after that well-known mecca where chance phenomena are intensively studied. Monte Carlo methods are generally not recommended to evaluate one-dimensional integrals; in high-dimensional cases, however, they may offer the only practical approach. If you prefer, the above integral can be interpreted as being over the unit hypercube rather than over the ordinary unit interval.) Put $V_i = f(U_i)$ and $\hat\theta = (V_1 + V_2 + \cdots + V_n)/n$. Find $E[\hat\theta]$ and $\mathrm{Var}[\hat\theta]$. You should conclude that $\hat\theta$ is unbiased and its expected square error is $O(1/n)$. Except for an implicit constant factor, the latter is independent of the dimension. Recently, trapezoidal-like methods that yield expected square errors $O(1/n^4)$ and $O(1/n^2)$ in one and two dimensions, respectively, have been proposed by Yakowitz et al. (1978). In more than three dimensions, they show that the new methods are not competitive with the above crude method. Niederreiter (1978) gives a comprehensive survey of integration methods, including sophisticated schemes that reduce the square error to an order of magnitude below $1/n$ in any dimension. Monte Carlo and other approaches to evaluating integrals often need to check whether a given point is in a given complex region. Lee and Requicha (1982a, b) discuss ways to do this efficiently.

PROBLEM 1.9.2 (continuation). Suppose that $0 \le f(x) \le 1$ for $0 \le x \le 1$. Put

$$g(x, y) = \begin{cases} 1, & \text{if } y \le f(x), \\ 0, & \text{otherwise,} \end{cases}$$

$$\tilde\theta = \frac{1}{n} \sum_{i=1}^n g(U_i, U_{n+i}).$$

[Hammersley and Handscomb (1964) call $\tilde\theta$ a "hit-or-miss" estimator.] Interpret $\hat\theta$ and $\tilde\theta$ geometrically. Find $E[\tilde\theta]$ and $\mathrm{Var}[\tilde\theta]$. Show that $\mathrm{Var}[\hat\theta] \le \mathrm{Var}[\tilde\theta]$ and that $f(1 - f) \ne 0$ on a set of positive probability implies $\mathrm{Var}[\hat\theta] < \mathrm{Var}[\tilde\theta]$.

PROBLEM 1.9.3 (continuation). Show how to estimate π, the area of the unit circle, using $\hat\theta$ and $\tilde\theta$. Define *efficiency* of an estimator as the reciprocal of the product of the expected mean square error and the computation time. [Glynn and Whitt (1986c) refine and justify this notion of efficiency.] Comment on the relative efficiency of $\hat\theta$ and $\tilde\theta$. Bear in mind that square roots are needed for $\hat\theta$ but not for $\tilde\theta$.

PROBLEM 1.9.4 (Shuffling and sampling). Essentially the following shuffling algorithm is presented in Knuth [(1981), p. 139], for example:

0. Input symbols $S_1, S_2, \ldots, S_N$.
1. For i running from N down to 2, do the following:
   (a) Generate a random number U_i.
   (b) Set $K \leftarrow \lceil iU_i \rceil$, so K is a random integer between 1 and i.

(c) Exchange S_K and S_i. This randomly selects one of the symbols in positions $1, 2, \ldots, i$, all currently unassigned, to put in position i. After this exchange symbols have been randomly assigned to positions $i, i + 1, \ldots, N$.
2. Stop.

Show that this produces all permutations of $S_1, S_2, \ldots, S_N$ with equal probability. (Hint: The problem reduces to verifying that the comment at step 1(c) is correct.) To randomly select a sample of size n from the N symbols, change the loop on i to run down to $N - n + 1$. Show that, if the N symbols are in random access memory, this method requires $O(n)$ work and no extra storage.

PROBLEM 1.9.5 [Estimating the cost of backtracking: Knuth (1975)]. Suppose we want to estimate the cost (for example, the computer time) required for a given backtrack algorithm that implicitly enumerates all solutions to a particular problem. Let T be a (hypothetical) complete enumeration tree, rooted at node 0, and let $c(i)$ be the cost to inspect node i. Let S be the set of those nodes i in T with the property that inspection of node i would reveal that it may be the root of a subtree of T containing a solution. Typically, an inspection checks a necessary, but seldom sufficient, condition for a given subtree to contain a solution. Let $d(i)$ be the number of sons of i which are contained in S. To estimate the cost of backtracking, we randomly choose a branch of T to go down, stopping when we know that no solution can be found below and recording an appropriately weighted sum (see below) of inspection costs incurred during this exploration. This process is then repeated as often as desired for statistical reliability. We now detail the method.

1. Set $C \leftarrow c(0)$, $m \leftarrow 0$, $D \leftarrow 1$.
2. Repeat the following until $d(m) = 0$:
   (a) Choose a node n at random from the set of sons of m that are elements of S. Any such son is chosen with probability $1/d(m)$.
   (b) Set $D \leftarrow Dd(m)$.
   (c) Set $C \leftarrow C + Dc(n)$, $m \leftarrow n$.
3. Output C and then stop.

Show that $E[C]$ is the actual cost of the backtrack algorithm. (Hint: Let node i have ancestors $0, a_1(i), \ldots, a_k(i)$. Then $[d(0)d(a_1(i)), \ldots, d(a_k(i))]c(i)$ is a term in the sum leading to C with probability equal to the reciprocal of the bracketed term, if the latter is positive. But $c(i)$ is a term in the *actual* cost if and only if the bracketed term is positive.) Can this estimation procedure be adapted to branch-and-bound? [For recent results on backtracking, see Purdom (1978) and Brown and Purdom (1981).]

PROBLEM 1.9.6. (Network reliability). Suppose that each arc of a network has probability p of working. For fixed p, give a straightforward simulation procedure to estimate the probability $h(p)$ that the network is connected, i.e., that every node of the network is connected to every other node by a path of operational arcs. In other words, estimate the probability that a spanning tree exists. Problem 2.2.8 indicates a better estimator.

PROBLEM 1.9.7 [continuation, Kershenbaum and Van Slyke (1972)]. Now suppose that we want to estimate the entire function h, without having to repeat the procedure of the preceding problem parametrically in p. A clever approach exploits the fact that a minimal spanning tree (a spanning tree with minimal total arc length) can be constructed with a greedy algorithm:

> Start by putting the shortest arc into the tree. At the general step, add the shortest available arc to the tree provided that it does not form a loop; otherwise discard it.

Show that the following adaptation works:

> On a given simulation run, suppose that arc i has length U_i, distributed uniformly on $(0, 1)$. Suppose that the greedy algorithm above adds an arc of length Q last, forming a minimal spanning tree. On this run, estimate $h(s)$ by 0 if $Q > s$ and by 1 if $Q \leq s$.

(Hint: If $Q \leq s$, all arcs in the minimal spanning tree just constructed work. If $Q > s$, the last arc added fails; conclude that no spanning tree of working arcs exists. For the latter, show that for any other spanning tree T there are only two cases to consider: (i) T contains the last arc added above; and (ii) T does not contain the last arc added above but that arc joined to T would form a loop of arcs containing at least one other arc of length at least Q.) For more information on minimal spanning trees, see, for example, Ford and Fulkerson [(1962), pp. 174–176]. Typically, greedy algorithms can be implemented efficiently with a heap. Show that this holds here.

PROBLEM 1.9.8 (Critical path length). Suppose we are given an acyclic network with random arc lengths, not necessarily independent. A longest path between given start and finish nodes is called a *critical path*. Any such path and its length are random. Show that the expected value of the length L_A of a critical path A is greater than the length L_E of any longest path B in the corresponding network with all arc lengths replaced by their respective expected values. The constant L_E generally differs from L_B, the (random) length of B in the original network. Hint: The short argument is $E[L_A] \geq E[L_B] = L_E$. A longer, but (to us) more revealing, argument follows. Let $L_1, L_2, \ldots, L_n$ be (possibly dependent) random variables with joint distribution $H(t_1, t_2, \ldots, t_n)$. Then

$$E[\max(L_1, L_2, \ldots, L_n)] = \int \max(t_1, t_2, \ldots, t_n) \, dH(t_1, t_2, \ldots, t_n)$$

$$\geq \int t_j \, dH(t_1, t_2, \ldots, t_n) \quad \text{[where } j \text{ is arbitrary]}$$

$$= E[L_j].$$

Hence

$$E[\max(L_1, L_2, \ldots, L_n)] \geq \max(E[L_1], E[L_2], \ldots, E[L_n]).$$

Show by example that $E[L_A] - E[L_B]$ can be arbitrarily large and that $|\text{Corr}(L_A, L_B)|$ can be arbitrarily small. Indicate how to estimate $E[L_A]$ by simulation, even when the arc lengths may be dependent. How would you estimate $P[L_A > t]$, where t is a fixed number? [See also Devroye (1979), Fulkerson (1962) and Meilijson and Nádas (1979).] Anticipating Chapter 2, show (trivially) that L_A is a monotone function of the arc lengths. Problem 2.2.8 exploits this observation.

PROBLEM 1.9.9 [continuation, Van Slyke (1963)]. We want to estimate the probability θ that a certain arc is critical, i.e., on a longest path between given start and finish nodes in an acyclic network. One way to do this is to generate all arc lengths independently across simulation runs but possibly dependently within a run. Our estimate is then the proportion of runs where the arc in question is on a longest path. Suppose that we are really interested only in arcs with a reasonable probability of being critical. Then another possibility is to divide the runs into two parts. In the second part, all arcs that were critical less than a certain small fraction of the time in the first part get fixed at their respective expected values. Clearly, this introduces correlation between runs. Will this procedure increase or decrease the mean square error $E[(\hat{\theta} - \theta)^2]$ of our estimator? Will $\hat{\theta}$ be biased? This procedure could allow more runs in a fixed amount of computer time because fewer random arc lengths have to be generated. On the whole, is the procedure to be recommended or not? Justify your answer.

PROBLEM 1.9.10. (Solving linear equations by Monte Carlo). Consider a finite-state Markov chain with one-step transition matrix P, one-step reward vector r, discount factor α, $0 < \alpha < 1$, and infinite-horizon expected return function v. The ith component of r or v corresponds to starting in state i. Clearly, $v = r + \alpha P v$. Putting $Q = \alpha P$, we get $v = (I - Q)^{-1} r$. Cycling repeatedly through all possible starting states, simulating the process for a fixed (large) number of transitions, and accumulating the discounted rewards, we can estimate v. If the process runs n transitions, given an explicit formula for the natural estimator of a fixed component of v and show that the absolute value of its bias is at most $(\max_k |r_k|)\alpha^{n+1}/(1 - \alpha)$. (If the corresponding starting state is recurrent, it can be shown that the expected square error goes to 0 as n goes to ∞.)

Given any Q, not necessarily arising as above, with nonnegative elements and row sums strictly less than one, we can renormalize the rows to get a Markov matrix P and a "discount factor" α. For this reduction, is it necessary that all the original row sums be equal?

(Answer: No, though perhaps this is not obvious. One can transform to an equal row-sum problem as follows:

Set

$$\alpha = \max_i \left[\left(\sum_j Q_{ij} - Q_{ii} \right) \middle/ (1 - Q_{ii}) \right],$$

$$\tilde{Q} = DQ + E \geq 0,$$

$$\tilde{r} = Dr,$$

where D and E are diagonal matrices with

$$D_{ii} = (1 - \alpha) \middle/ \left(1 - \sum_j Q_{ij} \right),$$

$$E_{ii} = 1 - D_{ii}.$$

It is easily checked that $0 \leq \alpha < 1$. Since also $\sum_j Q_{ij} < 1$ for all i, it follows that D^{-1} exists. Now

$$\tilde{r} + \tilde{Q}z = Dr + (DQ + E)z$$
$$= D(r + Qz) + Ez,$$

and hence

$$z = \tilde{r} + \tilde{Q}z \Leftrightarrow (I - E)z = D(r + Qz)$$
$$\Leftrightarrow Dz = D(r + Qz)$$
$$\Leftrightarrow z = r + Qz.$$

That $z = v$ follows from the invertibility of $I - Q$. Federgruen and Schweitzer (1978) consider this transformation in a more general setting.)

Show explicitly how to estimate $(I - Q)^{-1}r$ by Monte Carlo. Solving systems of linear equations by Monte Carlo is an old idea. Halton (1970) considers when and how the problem of solving $Ax = b$ can be reduced to the calculation of $(I - Q)^{-1}r$, as above. Generating a transition amounts to multinomial sampling. For each component, this can be done in constant time via the *alias* method described in Chapter 5, if enough memory is available. So, if Q has size m, the overall work for one run through every starting state is $O(mn)$. In some cases, this could be competitive with conventional methods.

PROBLEM 1.9.11 (Generating random numbers). Let the conditional distribution of X given Y be $G_y(x)$. Let the unconditional distributions of X and Y be F and H, respectively. For each fixed y, suppose that $G_y(\cdot)$ is continuous and strictly increasing. Generate Y from H and generate a uniform random number U. Show that $G_Y^{-1}(U)$ has distribution F. Hint: Mimicking the proof of Proposition 1.5.2, verify that

$$P[G_Y^{-1}(U) \le t] = \int P[U \le G_Y(t)| Y = y] \, dH(y)$$

$$= \int G_y(t) \, dH(y) = F(t).$$

Sometimes this indirect method of generating random numbers from F is faster than any direct method.

If $H(t) = t^n$, $0 < t < 1$, $G_Y(t) = (t/Y)^{n-1}$, $0 < t < Y$, and U and V are independent uniform random numbers, show that $U^{1/n}V^{1/(n-1)}$ is a random number from F, the distribution of the second largest of n uniform random numbers.

PROBLEM 1.9.12 (continuation). Give a one-pass algorithm that produces a sorted list of uniform random numbers in linear time. Note that it can be implemented on line: the kth element in the sorted list can be generated without knowing its successors. This result is implicit in Schucany (1972) and explicit in Lurie and Hartley (1972) and in Bentley and Saxe (1980). Show how to go from a sorted list of uniform random numbers to a sorted list of random numbers with an arbitrary common distribution. (Hint: Use inversion.) See also Gerontidis and Smith (1982).

PROBLEM 1.9.13 (Random knapsacks). Consider the knapsack problem

$$\max \sum c_i x_i,$$

$$\text{s.t.} \sum a_i x_i \le b,$$

$$x_i = 0 \text{ or } 1.$$

Suppose that a_i and c_i are randomly generated. We look at three cases:

(i) a_i and c_i are uncorrelated;
(ii) a_i and c_i are positively correlated;
(iii) a_i and c_i are negatively correlated.

In every case, if $i \ne j$, the pairs (a_i, c_i) and (a_j, c_j) are independent. Case (ii) is most typical of real-world problems. Why? Which case is the hardest to handle (say, by branch-and-bound)? Why? What practical conclusions can be drawn by testing the relative efficiency of various algorithms solely on problems representative of case (i)? Relate your answer to the discussion in Section 1.6 about extrapolation beyond the class of problems tested.

PROBLEM 1.9.14 (Random sampling without random access). Suppose we want to select exactly n items from a population of known size N, e.g., n records from a file of length N, so that the items selected are a random sample. Problem 1.9.4 gave an efficient method when the file is stored in random access memory. By contrast, here we assume that the file is on tape. Show that the algorithm R1 below works.

1. Set $t \leftarrow 0, m \leftarrow 0$. This initializes the counters t and m which keep track of the current record number and the number of records already selected, respectively.
2. Repeat the following until $m = n$:
   (a) Generate a uniform random number U.
   (b) If $(N - t)U \geq n - m$, skip the next record; otherwise select the next record, and set $m \leftarrow m + 1$.
   (c) Set $t \leftarrow t + 1$.
3. Stop.

Now suppose that N is unknown but that the file ends with an end-of-file marker. Remarkably, a one-pass algorithm to randomly select n items still exists. Associate random number U_j with record j. Show that records $[1], [2], \ldots, [n]$ form a random sample, where $[i]$ is the index of the ith smallest of the N random numbers generated. Show that the following algorithm R21 correctly implements that idea.

1. Set up a record buffer S capable of holding n records, and a key buffer K capable of holding n random numbers. Each record in S has a corresponding key in K.
2. Read the first n records of the file into S. (If $N < n$, the required sample does not exist, and the end-of-file marker will be encountered during this initialization.) For these n records generate n random keys $U_1, U_2, \ldots, U_n$.
3. Repeat the following:
   (a) Read the next record. If it is the end-of-file marker, stop.
   (b) Otherwise, generate a random number U.
   (c) Find the largest key in K, say M.
   (d) If $U < M$, replace M by U and the corresponding record in S by the record just read.

Show how to implement steps 3(c) and 3(d) efficiently by keeping K as a heap. Give an alternative implementation R22 where step 3(d) is modified to remove nothing from K and S, which now become "reservoirs"; at the finish, a second pass (this time through the reservoirs only) selects the items corresponding to the n smallest U_i. Which implementation do you prefer? Why? {References: Fan et al. (1962), Kennedy and Gentle [(1980), §6.6.2], Knuth [(1981), §3.4.2].}

The above algorithms are slow: the first runs in $O(N)$ time and the second in at least $O(N)$ time. Devroye (1981b) pointed out to us an efficient variant of algorithm R21. The key observations are:

(i) the number Y_M of U's generated until one of them is less than M has the geometric distribution with parameter M;
(ii) the last U generated, i.e., the one which is less than M, is in fact uniform on $(0, M)$.

Section 5.4.5 shows how to generate Y_M in $O(1)$ time; for now, just assume it can be done. To obtain the modified algorithm, R2D1 say, replace step 3 of R21 by

3'. Repeat the following:
   (a) Find the largest key in K, say M.
   (b) Generate Y_M.
   (c) Skip $Y_M - 1$ records and read the Y_Mth. If the end-of-file marker is encountered during this operation, stop.
   (d) Otherwise generate U uniform on $(0, M)$, replace M by UM [using observation (ii) above], and replace the corresponding record in S by the record just read.

Show that this modified algorithm works. On many machines records in a sequential file can be skipped by the device controller, the central processor time necessary to initiate and terminate the skip being negligible. If we consider just the load on the central processor, Devroye shows that R2D1 runs in $O[(n \log n)(1 + \log(N/n))]$ average time. Prove that his answer is correct. If only computing time is considered, R2D1 dominates R1, R21, and R22; if the processor's free time cannot be used productively while the device controller is busy skipping records, this may be of little importance in practice.

Devroye [(1986), Chapter 12] surveys other interesting algorithms to pick n records from a file of length N. What is R2D2?

PROBLEM 1.9.15 (Confidence intervals). Suppose that for a fixed number n of iid replications one could construct a confidence interval I_n of random length L_n such that the probability that I_n covers the true parameter is at least α, where α is specified in advance. Now suppose instead that we continue replicating until $L_n \leq d$, where d is also fixed in advance. Is n now random? (Answer: Yes.) Does n depend stochastically on the observations? (Answer: Yes.) Is the probability that I covers the true parameter now necessarily at least α?

(Answer: Bearing in mind the answers to the preceding two questions, No. However, if asymptotic normality is legitimately being invoked, it would not be far wrong to say "Yes." For small d, hence stochastically large n, Chow and Robbins (1965) give the appropriate limit theorem. Callaert and Janssen (1981) show that the rate of convergence of the coverage probability to α is $O(\sqrt{d})$ provided that all moments of the observations exist. Feller [(1971), pp. 264–265], gives a central limit theorem for partial sums with the number of terms random and independent of the summands; the latter condition makes it inapplicable in the present context.)

PROBLEM 1.9.16 (Walking in n-space). A convex region R in n-space is defined by

$$R = \left\{ (x_1, \ldots, x_n) : \sum_{j=1}^{n} a_{ij} x_j \leq b_i, i = 1, \ldots, m \right\}.$$

Assume that R has finite volume and that x^0 is in the interior of R. Generate a sequence of points $\{x^k\}$ as follows:

(i) Given $x^k = (x_1^k, \ldots, x_n^k)$, generate a random direction d^k. One way to generate d^k is to use the algorithm in Knuth [(1981), p. 130] for getting a random point on the n-dimensional unit sphere centered at the origin: $d_i^k = z_i/r$ where

$$r = (z_1^2 + \cdots + z_n^2)^{1/2}$$

and $z_1, \ldots, z_n$ are independent normal variates with mean 0 and variance 1. Can you think of a better way?

(ii) Points on the line passing through x^k with direction d^k have the form $x^k + \lambda d^k$. Find the points $x^k + \lambda^+ d^k$ and $x^k - \lambda^- d^k$ where the line intersects the boundary of R. [Set

$$s_i = \left(b_i - \sum_{j=1}^{n} a_{ij} x_j^k\right) \Big/ \sum_{j=1}^{n} a_{ij} d_j^k$$

and show that $\lambda^+ = \min_i \{s_i : s_i > 0\}$, $\lambda^- = \max_i \{s_i : s_i < 0\}$.]

(iii) Generate U uniform on $[0, 1]$ and set

$$x^{k+1} = x^k + U\lambda^+ d^k + (1 - U)\lambda^- d^k.$$

[Show that x^{k+1} is uniformly distributed on the line segment $(x^k - \lambda^- d^k, x^k + \lambda^+ d^k)$. Show that in step (i) we could have put $d_i^k = z_i$ without computing r and without changing x^{k+1}.]

Justify the following statements heuristically and then, if possible, rigorously.

(a) With probability one, the sequence $\{x^k\}$ is dense in R.
(b) x^k is asymptotically uniform on R. (Hint: The stochastic kernel governing the transition from x^{k-1} to x^k is symmetric.)
(c) If x^0 is uniform on R, then so is x^k for each k.

Let f be a real-valued, continuous function on R. How could you use $\{x^k\}$ to find an approximate maximizer of f? See also Problem 3.6.13. For more information on random search methods using this and other techniques, see Boender *et al.* (1982), Boneh (1983), Smith (1984), and Solis and Wets (1981).

PROBLEM 1.9.17 (Channel contention). At time n a certain number of stations are ready to transmit a packet over a channel. Only one packet can be transmitted. If several stations try to transmit simultaneously, none of the transmissions succeeds and all the stations involved must wait until time $n + 1$ when they may choose to try again. If only one station tries to transmit its packet, it succeeds, taking unit time.

Suggest reasonable strategies for choosing times, perhaps *randomly*, at which to attempt packet transmissions for a station that tries to maximize its rate of successful transmissions. The other stations' strategies and the number of other stations using the channel are unknown and possibly random. No collusion between stations is possible. Show how simulation could be used to evaluate strategies. Describe a useful strategy which can be analyzed without simulation. Do so.

Tannenbaum (1981) discusses this problem in detail.

PROBLEM 1.9.18 (Certainty equivalence). Using policy π to run a system, suppose that the profit is

$$w(\pi) = a(\pi) + \sum_{i=1}^{n} b_i(\pi)[f_i(\pi) + e_i]$$

$$+ \sum_{i=1}^{n} \sum_{j=1}^{n} c_{ij}[f_i(\pi) + e_i][f_j(\pi) + e_j],$$

where a, b_i, and f_i are deterministic functions, c_{ij} and n are constants, and e_i is a random variable independent of policy for $i, j = 1, 2, \ldots, n$.

(i) Suppose that we are interested in $E[w(\pi) - w(\sigma)]$, the expected relative performance of policies π and σ. Show that we can replace each e_i by its expectation.

(ii) Now suppose that we are interested in $E[w(\pi)]$, the expected absolute performance of π. Under what circumstances, if any, is it legitimate to replace each e_i by its expectation?

PROBLEM 1.9.19 (Fractional parts). If $U_1, U_2, \ldots, U_n$ are iid uniform on (0, 1), show that the fractional part of $U_1 + U_2 + \cdots + U_n$ is uniform on (0, 1). [Wichmann and Hill (1982) use the result to combine uniform random number generators. L'Ecuyer (1986), with the same goal, generalizes this result to a sum of discrete variates distributed uniformly on possibly *different* sets of consecutive integers.]

PROBLEM 1.9.20 (Simulated annealing). Suppose we have a heuristic for a discrete-optimization problem. Starting from a feasible solution, it iteratively improves through a sequence of neighboring solutions until it reaches a *local* optimum. One way to search for a *global* optimum is to restart this procedure at many randomly chosen feasible solutions. Another way to climb out of local valleys is to switch occasionally to inferior neighbors, with probabilities that decrease over time. A large literature is developing on the latter scheme, called *simulated annealing*. Pick your favorite combination of problem and heuristic; then empirically compare random restarts and simulated annealing.

PROBLEM 1.9.21 (Outliers). An *outlier* is an observation from a population different from that ostensibly sampled. For example, in a chemistry experiment a stray substance may contaminate an observation. In simulation experiments such contamination is impossible; all observations come from one population. Why? Can one legitimately discard *a posteriori* anomalous observations in a simulation experiment? Would *a priori* truncation of input distributions (legitimately) reduce anomalies? (See the introduction to Chapter 4.)

Variance Reduction

Placing variance reduction techniques before output analysis, discussed in Chapter 3, may appear to be putting the cart before the horse. Reversing the order of Chapters 2 and 3, however, would increase the number of forward references.

To read Chapter 2, all the reader needs to know about output analysis is the following: we seek to estimate some measure of performance, almost always using a sample mean, and to estimate in turn the goodness of that estimate, almost always using the variance of the same sample. Variance reduction techniques try to reduce the true variance of the sample mean. Usually the sample variance reflects this attempt. Marshall (1956) remarks that: "In most cases, a little thought pays big dividends in variance reduction and the use of straightforward sampling is presumptive evidence of a lack of thought or ingenuity in design of the sampling." This chapter provides a framework for such thought but leaves plenty of room for ingenuity.

Simulation offers unusual opportunities for deliberately and advantageously inducing correlation, positive or negative, among observations. When comparing policies, using random number streams common to all of them offers a fairer comparison, intuitively, than would statistically independent streams: one source of variability has been removed by testing all policies under the same conditions. Section 2.1 supports this intuition mathematically by giving a monotonicity condition under which introduction of such positive correlation is good. Throughout this chapter, "good" is used in a precisely defined way consistent with our discussion of the criterion of mean square error in Section 1.6. So, when our estimators are unbiased, it suffices to direct our efforts at variance reduction. Common random numbers (§2.1), antithetic variates (§2.2), and control variates (§2.3) all reflect a single idea: introduce correlation to reduce variance. After reading these three sections, pause to review these techniques from this perspective. In retrospect, the strong links between them will be apparent.

The other variance reduction techniques that we discuss do not depend on inducing correlation. Stratification (§2.4) and importance sampling (§2.5) use weighted sampling based on *a priori* qualitative or quantitative information in an attempt to reduce variance. An unweighting correction eliminates bias. Stratification embodies the precept that recasting an estimation problem so that random variables are replaced by their expected values, whenever known, often is a good idea. *When valid*, this replacement reduces variance without injecting bias. This idea is exploited to the hilt in a powerful technique called conditional Monte Carlo (§2.6); sometimes the terms virtual measures, certainty equivalence, or even conditional expectations are associated with this device. When applicable, it demonstrably reduces variance. It may require more analysis or calculation than many unsophisticated simulators are accustomed to doing. Andrews *et al.* [(1972), §§14, 15, and 40] use conditional Monte Carlo to advantage, without naming it, in their study of robust estimates of location. Our treatment of conditional Monte Carlo aims at simulation in the wide sense rather than just at integration. Judging from the scarcity of allusions to conditional Monte Carlo in the literature, this technique is underutilized. Many simulation applications discussed in the literature could have used it profitably. Sometimes continuous variates can be replaced easily by their conditional expectations given the sequence of states visited, as in (possibly uniformized) continuous-time Markov chains.

With some variance reduction techniques, one cannot estimate the resulting variance without bias. If the bias of a straightforward estimator has a certain general form, it can be reduced via jackknifing (§2.7). This technique has wide applicability.

Some readers may find the style of this chapter dry. Parallel reading of Chapter 8, where the techniques are illustrated, may add spice although the programming considerations have not yet been touched upon. Just because some combinations of variance reduction techniques are not examined in the text does not mean that they are not worthwhile in practice. Of the variance reduction techniques discussed, stratification and (strict) conditional Monte Carlo provably reduce the true variance. Under a monotonicity condition (see §2.1) so do either common random numbers or antithetic variates. This condition can be made to hold more often than some misleading folkore indicates. Even when it does not demonstrably hold, using common random numbers, antithetic variates, or control variates often is a good heuristic; in such cases, we recommend screening runs to check their efficacy. When it seems to be grossly violated in an unavoidable way, we would not even try using them. Nothing general can be proved about the efficacy of control variates and importance sampling. However, we believe that control variates often are attractive but that importance sampling can be useful only in rather special cases. Siegmund (1976) documents one such case: estimating acceptance probabilities in sequential testing. See also Asmussen (1985). In writing this chapter, Kleijnen (1974, 1975) has been helpful. See also Wilson's (1984) excellent survey.

All the variance reduction techniques discussed in this chapter can expedite "classic" Monte Carlo integration. The term "classic" indicates an explicit, fixed-dimensional integrand, whereas in general simulations a computer program gives the integrand implicitly. The reader will find it instructive to invent simple "classic" examples and apply variance reduction techniques discussed here to them. On the other hand, we do not discuss certain specialized techniques that seem to apply only to classic cases. Deák (1980), for example, combines antithetic variates and orthonormalization to estimate multivariate normal integrals efficiently. Orthonormalization seems feasible only when the integrand is explicit. Hammersley and Handscomb (1964) discuss variance reduction, mainly with "classic" applications in mind. They also discuss special techniques for particle physics simulations. Under this heading they discuss *splitting* and *Russian roulette*. Even in more mundane settings, we can *split* a run into subruns when (and if) a simulation run reaches an "interesting" state. We terminate all but one of these subruns when (and if) they become "uninteresting." This collapse corresponds to *stratified* Russian roulette. In forthcoming joint papers, Fox and Glynn optimize this scheme adaptively—converging with probability one to the optimal split factor. For concreteness, think of an interesting state as one that begins a peak-load period and of an uninteresting state as one that ends it.

To conclude this introduction to Chapter 2, we now give a more detailed overview of its first three sections.

Let X and Y be the random outputs of two simulations driven by pseudo-random number streams U and V, respectively. Note that $\text{Var}[X \pm Y] = \text{Var}[X] + \text{Var}[Y] \pm 2\,\text{Cov}[X, Y]$. For fixed marginals, maximizing covariance is equivalent to maximizing correlation; likewise for minimizing. If X and Y are *increasing* functions of U and V, respectively, then

(i) $U = V \Rightarrow \text{Corr}(X, Y) \geq 0$.

(ii) $\left. \begin{array}{l} U = (U_1, U_2, \ldots, U_m) \\ V = (V_1, V_2, \ldots, V_m) \\ V_i = 1 - U_i \end{array} \right\} \Rightarrow \text{Corr}(X, Y) \leq 0$.

Also, if $m = 1$ and X and Y have distributions F and G, respectively, then:

(iii) to maximize $\text{Corr}(X, Y)$, generate (X, Y) as $(F^{-1}(U), G^{-1}(U))$;

(iv) to minimize $\text{Corr}(X, Y)$, generate (X, Y) as $(F^{-1}(U), G^{-1}(1 - U))$.

Unless $Y = aX + b$, (iii) and (iv) are essentially the only ways of getting extremal correlations. Assuming that $Y = aX + b$ often is unduly restrictive, as discussed in the introduction to Chapter 4.

(The proofs of these four properties are rather involved and could be skimmed on first reading.)

Properties (i) to (iv) and the basic relation

$$\text{Var}[X \pm kY] = \text{Var}[X] + k^2\,\text{Var}[Y] \pm 2k\,\text{Cov}[X, Y]$$

underpin the variance reduction techniques below:

(a) *Common random numbers.* Compare policies corresponding to X and Y, respectively, using $X - Y$ as an estimator of $E[X - Y]$.
(b) *Control variates.* For a fixed policy, let X be a principal output, Y a control variate with $E[Y] = 0$, and use $X - kY$ as an estimator of $E[X]$.
(c) *Antithetic variates.* For a fixed policy, let X be the output of run 1, Y the same output from run 2, and use $(X + Y)/2$ as an estimator of $E[X]$.

The correspondence between variance reduction techniques and correlation properties is

$$(a), (b) \leftrightarrow (i), (iii),$$

$$(c) \leftrightarrow (ii), (iv).$$

To maximize (or minimize) $\mathrm{Corr}(X, Y)$, the inputs U and V must be properly *synchronized*—as we show by examples. Among other things, this means designing the simulation so that X and Y are increasing functions of U and V, respectively. Roughly, this monotonicity ensures that the correlations induced have the right sign for variance reduction. This is greatly facilitated if the nonuniform random numbers needed are generated via inversion. Only this method simultaneously

- transforms uniform random numbers *monotonely*;
- requires exactly one uniform random number per nonuniform random number;
- induces extremal correlations via (iii) and (iv) between pairs of nonuniform random numbers (which are then combined with other pairs of nonuniform random numbers and further transformed to produce the respective outputs X and Y).

We often have strong intuitive grounds for believing that X and Y are monotone, even if we do not have a formal proof. Careful design of the simulation helps. Transform uniform random numbers monotonely by inversion. Then check whether the nonuniform random numbers thus generated are themselves transformed monotonely to generate the final output.

Aiming to assure monotonicity, *synchronize*:

1. Use inversion.
2. Drive distinct classes of events (e.g., arrivals) and attributes (e.g., service times, priorities, routings) by disjoint random number streams. This folklore is certainly a reasonable heuristic. In queueing networks, give each node its own disjoint random number streams for service times and one-step transitions.
3. Generate all the attributes of a transaction (e.g., a job) as it enters the system and assign it an identification number, as discussed further in Example 2.3.2 and Section 3.4. When this is not feasible, avoid using transaction-based observations (defined in Chapter 3).

Steps 1 and 2 seem always worthwhile. Step 3 seems almost always worth-while, but Problem 2.1.1 shows what can go wrong in a pathological case.

In a simulation experiment, increasing statistical precision generally requires increasing running time. For finite-horizon problems, running time is the product of the number of runs and the average time per run. Coupling inversion with variance reduction techniques reduces the number of runs needed to achieve any fixed statistical precision, if the synchronization attempt really does synchronize; otherwise, it usually does no harm. Usually, this is the dominant consideration because relatively little can be done to reduce average time per run. The conclusion is the same for infinite-horizon simula-tions (Chapter 3): simply replace "run" by "batch" or "cycle." Generating nonuniform random numbers rarely dominates overall running time; logical operations (including maintenance of the event-sequencing list) and statistical calculations play a significant role. We return to this point in Chapter 5.

2.1. Common Random Numbers

We begin our treatment of common random numbers with an informal discussion (§2.1.1). Next we formalize these intuitive ideas (§2.1.2). Last, we give two lemmas (§2.1.3) of independent interest; they are used in Section 2.1.2 and in Sections 2.4 and 2.6. The theorems in Section 2.1.2 have counterparts in Sections 2.2 and 2.3. Casual users of this book should read Section 2.1.1 carefully and skim Sections 2.1.2 and 2.1.3; mathematically inclined readers should do the reverse.

2.1.1. Informal Approach

To obtain good performance estimates, we must distinguish between absolute and relative performance. Estimating the Gross National Product (GNP) four years hence under a given policy is an example of absolute performance estimation; estimating the difference in GNP using two different policies is an example of relative performance estimation. Relative performance can usually be estimated more accurately than absolute performance. Fortunately, this is usually what interests us. For example, to determine whether one computer system gives faster response than another is a matter of relative performance. Likewise, in checking sensitivity to changes in model structure or parameters, it is relative performance that counts.

Assume initially that we are comparing only two policies Π_X and Π_Y for operating some system and we have made N simulations of each policy. In Sections 2.1.2 and 3.2.2 we compare several policies. Let X_i be the performance in the ith simulation under policy Π_X, with Y_i defined similarly. We want to estimate $E[X_i - Y_i]$. With a given amount of computer time and a properly designed simulation we can usually do this more accurately than we can estimate $E[X_i]$ alone. Use common random numbers, as explained below. Let

$$D_i = X_i - Y_i,$$

$$\overline{X} = \sum_{i=1}^{N} X_i/N,$$

$$s_X^2 = \sum_{i=1}^{N} (X_i - \overline{X})^2/(N - 1),$$

and define $\overline{Y}$, s_Y^2, $\overline{D}$, and s_D^2, similarly. A reasonable estimate of the expected difference in performance is $\overline{D}$.

If the D_i are normally distributed, then the form of the confidence interval is

$$\overline{D} \pm k\sqrt{s_D^2/N},$$

where k is the appropriate value from the t-distribution with $N - 1$ degrees of freedom. In this case s_D^2/N is an estimate of the variance of $\overline{D}$. If $\overline{D}$ is approximately normally distributed, as can often be expected in view of the central limit theorem, then the above confidence interval is approximately correct.

In general,

$$\text{Var}[D_i] = \text{Var}[X_i - Y_i]$$

$$= \text{Var}[X_i] + \text{Var}[Y_i] - 2\,\text{Cov}[X_i, Y_i].$$

Thus making the $\text{Cov}[X_i, Y_i]$ term positive reduces the variance of D_i and hence of $\overline{D}$. There is a common-sense way of achieving this: make the two simulation runs generating X_i and Y_i as similar as possible. In particular, use the same set of random events in each pair of simulations. Although this is difficult to achieve in real-world experiments, it is possible in simulation: reproduce the same "random" numbers from one run to its mate. Intuitively, if a particularly devastating set of simulated events occurs in this pair of runs, both policies appear bad. Thus, differences in performance can be attributed to policies and not to the chance occurrence of a devastating event during the simulation of one policy but not of the other.

To illustrate this, we carried out twelve pairs of simulations of policies Π_X and Π_Y, using the same random numbers for each simulation in a pair. We assume that the square of the empirical standard deviation is an unbiased estimate of the true variance. The performance statistics were:

	Mean performance	Standard deviation
X	9.70	1.80
Y	10.70	1.21
$X - Y$	-1.00	1.27

If X_i and Y_i are (falsely) assumed to be independent, then the natural estimate of the standard deviation of $\overline{X} - \overline{Y}$ is

$$\sqrt{(1.80^2 + 1.21^2)/12} = 0.626.$$

At a 0.95 confidence level, still (falsely) assuming 24 *independent* normally-distributed observations, the appropriate ordinate from a t-distribution with 22 degrees of freedom is 2.074; so we get a confidence interval

$$-1 \pm 2.074 \times 0.626 = -1 \pm 1.298,$$

which includes zero. Based on this (incorrect) interval, we cannot conclude that the difference is significant.

The X_i and Y_i are plotted in Figure 2.1.1. It is apparent that the performance of Π_Y is (statistically) significantly higher than Π_X, since Y_i is higher than X_i on every simulation run except the first and eighth.

If we use the method based on differences, then the estimate of the standard deviation of $\overline{X} - \overline{Y}$ is $\sqrt{1.27^2/12} = 0.367$. A 0.95 confidence interval is

$$-1 \pm 2.201 \times 0.367 = -1 \pm 0.808.$$

This (correct) confidence interval does not include zero: we conclude that the difference is significant.

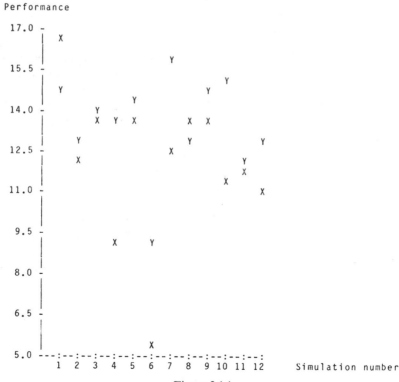

Figure 2.1.1

Probably the single most important statistical consideration in designing a simulation experiment is to use the same random events to drive each policy. For a queueing simulation, for example, all policies should see the same sequence of arrivals unless it is obvious that the choice of policy affects the sequence of arrivals; likewise for service times.

Attaining full stochastic comparability is not always straightforward. In this chapter we give some guidelines. Simply using the same sequence of "raw" random numbers may be insufficient. In a queueing simulation, for example, one policy may serve customers in a different order from another. Thus, if random numbers are generated and used when required, a number used to generate an interarrival time under one policy may be used to generate a service time under another. This may have the opposite of the desired effect.

If the choice of policy affects the distribution of some random variable, then comparable (monotone, one-to-one) methods should be used for generating variables for each policy. Therefore, we recommend inversion (see Proposition 1.5.2 and Chapter 5). This maintains comparability even though one policy, say, may require log-normally distributed variables while another uses exponentially distributed variables instead.

PROBLEM 2.1.1. In a queueing simulation, we may either: (a) generate service times as customers arrive, so that each customer is "tagged" with his service time; or (b) generate each service time when service begins. We want to compare two policies which result in the customers' being served in different orders. The arrival distribution will not change. How would you choose between methods (a) and (b) in general? (In some cases the choice may be obvious. If queue discipline depends on service time, only method (a) is applicable. When comparing the two disciplines "shoot the odd-numbered customers" and "shoot the even-numbered customers" method (b) gives essentially perfect synchronization, method (a) none at all. However, most cases are not so clear-cut.)

2.1.2. Formal Development

Suppose we wish to compare several policies for operating a system. Let U^i be the string of uniform random numbers used with policy i and, switching notation for convenience, let $V^i = T^i(U^i)$ be the corresponding output. We make a technical assumption, harmless in practice, that T^i is integrable; this always holds in real problems.

The example below serves several purposes. First, it illustrates T^i. Second, it shows how to synchronize. Third, it shows that synchronization can assure monotonicity. Theorem 2.1.1 below explains why we want this.

EXAMPLE 2.1.1 (continuation of Example 1.1.1). Let $N_B(t)$ be the number of customers served up to time t using service distribution B. For fixed t and service distributions F and G, we want to estimate $E[N_F(t) - N_G(t)]$. Clearly $N_F(t)$ and $N_G(t)$ are decreasing functions of the arrival spacings and of the service times. Identify T^B with N_B for $B = F, G$. To make T^F and T^G increasing functions of the random number stream U^B, generate the *same* arrival spacings for the two systems with transformations of the form $A^{-1}(1 - U_i)$, using random number stream U^{B1}, and the respective service times with transformations of the form $F^{-1}(1 - U_j)$ and $G^{-1}(1 - U_j)$, using random number stream U^{B2}; by symmetry, we can replace $1 - U_j$ by its stochastic equivalent U_j.

We prefer (by definition) policy i to policy j if $E[V^i - V^j] > 0$. Trivially, the empirical difference $V^i - V^j$ is an unbiased estimator of this expected difference. Our confidence in it depends on its variance

$$\text{Var}[V^i - V^j] = \text{Var}[V^i] + \text{Var}[V^j] - 2\,\text{Cov}[V^i, V^j].$$

Obviously, if U^i and U^j are generated independently, then $\text{Cov}[V^i, V^j] = 0$. However, we have the intuitive (folklore) result below:

Theorem 2.1.1. *Suppose that T^i and T^j are bounded, not constant functions of any argument (single uniform random number) on the open interval $(0, 1)$, and are either both increasing or both decreasing. Then $U^i = U^j \Rightarrow \text{Cov}[V^i, V^j] > 0$.*

Note. The reader not interested in proofs should skip to the statement of Theorem 2.1.2.

PROOF OF SPECIAL CASE NO. 1. Let $U^i = U^j = U$ be a *single* uniform random number. Suppose that V^i and V^j are uniformly distributed on the respective values $a_1 < a_2 < \cdots < a_n$ and $b_1 < b_2 < \cdots < b_n$. Because of the monotonicity of T^i and T^j, the joint output (V^i, V^j) is uniformly distributed on $(a_1, b_1), (a_2, b_2), \ldots, (a_n, b_n)$. For example, if $(k - 1)/n \leq U < k/n$ (where k is one of the numbers $1, 2, \ldots, n$), then the output is (a_k, b_k). Hence

$$\text{Cov}[V^i, V^j] = \sum_p a_p b_p/n - \left(\sum_p a_p\right)\left(\sum_q b_q\right)\bigg/n^2$$

$$= \sum_p a_p b_p/n - \sum_p \sum_q a_p b_q/n^2.$$

By Lemma 2.1.2, given at the end of this section, it follows that $\text{Cov}[V^i, V^j] > 0$. $\square$

PROOF OF SPECIAL CASE NO. 2. Let $U^i = U^j = (U_1, U_2)$. Suppose that, given U_1, the outputs V^i and V^j are uniformly distributed on the respective values $c_1 < c_2 < \cdots < c_n$ and $d_1 < d_2 < \cdots < d_n$. The c_p's and d_q's depend on U_1. Now suppose, in addition, that c_p and d_q are uniformly distributed on $c_{p1}, \ldots, c_{ps}$ and $d_{q1}, \ldots, d_{qs}$, respectively. By the covariance-decomposition lemma 2.1.1, given at the end of this section,

$$\text{Cov}[V^i, V^j] = E(\text{Cov}[V^i, V^j | U_1])$$
$$+ \text{Cov}(E[V^i | U_1], E[V^j | U_1]).$$

By the monotonicity in U_2 of T^i and T^j, the (conditional) joint output is uniformly distributed on $(c_1, d_1), (c_2, d_2), \ldots, (c_n, d_n)$. Following the argument for case no. 1, $\mathrm{Cov}[V^i, V^j | U_1] > 0$ and hence $E(\mathrm{Cov}[V^i, V^j | U_1]) > 0$. Since

$$E[V^i | U_1] = (c_1 + \cdots + c_n)/n$$

and

$$E[V^j | U_1] = (d_1 + \cdots + d_n)/n,$$

we have

$$\mathrm{Cov}(E[V^i | U_1], E[V^j | U_1]) = \sum_p \sum_q \mathrm{Cov}[c_p, d_q]/n^2.$$

By the monotonicity in U_1 of T^i and T^j for any fixed U_2, it follows that c_p and d_q are increasing functions of U_1. Reapplication of the argument for special case no. 1 shows that $\mathrm{Cov}[c_p, d_q] \geq 0$. Combining our results, we have shown that $\mathrm{Cov}[V^i, V^j] > 0$. $\qquad\square$

SKETCH OF PROOF OF SPECIAL CASE NO. 3. Let $U^i = U^j = (U_1, U_2, \ldots, U_m)$. Following the pattern of case no. 2, suppose that the appropriate conditional output distributions are discrete and uniform and use induction on the number of variables in the conditioning to show that $\mathrm{Cov}[V^i, V^j] > 0$. $\qquad\square$

SKETCH OF GENERAL PROOF. To see intuitively how this special case leads to a general proof, let V^i and V^j have distributions R and S, respectively, and think of $\{a_1, a_2, \ldots, a_n\}$ and $\{b_1, b_2, \ldots, b_n\}$ as "grids" for R and S, respectively, such that

$$a_i = \sup\{x: R(x) \leq i/(n + 1)\},$$

$$b_i = \sup\{x: S(x) \leq i/(n + 1)\}.$$

To a first approximation, we can regard V^i and V^j as being uniformly distributed over their respective grids. This amounts to approximating R and S by step functions with a jump of height $1/n$ at each (respective) grid point. Approximate the conditional distributions used in special cases no. 2 and no. 3 similarly. This can be made rigorous. The above remarks can be used as a guide to bootstrap this result for special T^i and T^j to general integrable functions. In a related setting, Whitt (1976) shows how to bootstrap. $\qquad\square$

The proofs in Lehmann (1966), Mitchell (1973), and Ross [(1985), §11.6] differ from that outlined above.

PROBLEM 2.1.2. Supply the missing details in our proof.

Reviewing the literature on bivariate distributions with given marginals, Whitt (1976) proves

Theorem 2.1.2. *For marginal cdf's F and G with finite variances, the joint distribution with maximum correlation and these marginals is the distribution of*

$$[F^{-1}(U), G^{-1}(U)];$$

to get minimum correlation, replace the argument of G^{-1} by $1 - U$.

Note. The reader not interested in technical details should skip to the end of Problem 2.1.3.

HEURISTIC JUSTIFICATION. That the marginal distributions are correct follows from Proposition 1.5.2. For fixed marginals, maximizing (minimizing) correlation is equivalent to maximizing (minimizing) covariance. We argue that the (bivariate) distribution maximizing covariance is the distribution of $[F^{-1}(U), G^{-1}(U)]$. The argument for the other part of the theorem is completely symmetric.

Approximate F and G by step functions with breakpoints $\{a_1, a_2, \ldots, a_n\}$ and $\{b_1, b_2, \ldots, b_n\}$, respectively, where each jump has height $1/n$. Define pseudoinverses

$$F^{*n}(x) = a_i \quad \text{if } (i-1)/n < x \leq i/n,$$

$$G^{*n}(x) = b_i \quad \text{if } (i-1)/n < x \leq i/n.$$

Let Y and Z be uniformly distributed on $\{a_1, a_2, \ldots, a_n\}$ and $\{b_1, b_2, \ldots, b_n\}$, respectively. The proof of special case no. 1 of Theorem 2.1.1 shows that $\text{Cov}(Y, Z)$ is maximized if $\{Y = a_i\} \Leftrightarrow \{Z = b_i\}$. The distribution of $[F^{*n}(U), G^{*n}(U)]$ does this. One can show rigorously how to approximate the distribution of $[F^{-1}(U), G^{-1}(U)]$ by distributions of the form $[F^{*n}(U), G^{*n}(U)]$. □

PROBLEM 2.1.3. Let U and V be independent uniform random numbers. What is the distribution of $-\log(UV)$? Among bivariate distributions with this distribution as marginals, does the distribution of $\{-\log(UV), -\log[(1-U)(1-V)]\}$ have minimum correlation? Let X and Y have distribution F. Suppose that F has a point of symmetry. With $Y = aX + b$, pick a and b to make $\text{Corr}(X, Y)$ minimal.

In principle, assuming that the expectations of the outputs V^i and V^j exist (and hence that V^i and V^j have bounded variation), we can decompose them to get monotonicity "in pieces," even when these inputs have more than one random number; e.g., see [Hobson (1957), pp. 343–344]. Specifically, we get

$$V^k = A^k - B^k, \qquad k = i, j,$$

where A^k and B^k are increasing functions of U^k.

EXAMPLE 2.1.2. In an inventory problem with ordering policy k. suppose that V^k is the sum of shortage and holding costs and that the shortage (resp., holding) cost is an increasing (resp., decreasing) function of the demand. Also, assume that demand is an increasing function of U^k. Now identify A^k with the shortage cost and B^k with the negative of the holding cost.

To estimate $E[V^i - V^j]$, it suffices to estimate $E[A^i - A^j]$ and $E[B^i - B^j]$. This can be done on independent series S_1 and S_2 of paired runs, using common random numbers on each pair and deliberately throwing away information about $A^i - A^j$ (resp., $B^i - B^j$) on S_2 (resp., S_1). Screening runs could be

used to check the effectiveness of this type of strategy as well as to determine
the sizes of S_1 and S_2. If $\text{Var}[A^i - A^j] = \text{Var}[B^i - B^j]$, it is a simple exercise
to show that these sizes should be the same and hence that a pooled estimate
should be formed instead of throwing anything away; thus, in this case, we are
effectively back to a straightforward strategy. So the strategy tentatively
suggested here can be worthwhile only when $\text{Var}[A^i - A^j]$ and $\text{Var}[B^i - B^j]$
differ significantly, as could well be the case in the inventory example above.
Wright and Ramsay (1979) aptly point out that, when V^k itself is not a monotone function of U^k, direct attempts at synchronization may be ineffective;
they do not mention the above strategy. Mechanizing this decomposition is
an open problem in general.

PROBLEM 2.1.4. Let X have distribution F, $Y = G^{-1}(F(X))$, $Z = G^{-1}(1 - F(X))$, and
suppose that G is a distribution function. Show that Y and Z have distribution G and the
pairs (X, Y) and (X, Z) each have extremal correlation. If X is generated in a monotone,
one-to-one way, show that the above method for producing extremal correlations is
equivalent to that indicated in Theorem 2.1.2. (Remark: If X were generated by some
method faster than inversion and if F were easier to compute than F^{-1}, then generating
(X, Y) or (X, Z) as above could be faster than by the method of Theorem 2.1.2; however,
it would be incompatible with synchronization.)

In practice, $E[V^i]$ generally will be estimated by an average over, say, R
simulation runs: $\bar{V}^i = (V^{i1} + \cdots + V^{iR})/R$, where V^{ir} is the output of run r
of policy i. We estimate $\text{Var}[V^i - V^j]$ by the empirical variance

$$Z^{ij} = \frac{1}{R - 1} \sum_{r=1}^{R} [(V^{ir} - V^{jr}) - (\bar{V}^i - \bar{V}^j)]^2.$$

PROBLEM 2.1.5. Find $E[Z^{ij}]$ and $\text{Var}[Z^{ij}]$ when U^i and U^j are independent and when
$U^i = U^j$, defining quantities as needed, and compare the respective expressions. Under
the conditions of Theorem 2.1.1, show that $E[Z^{ij}]$ is less when $U^i = U^j$. (Hint: Show that
$E[Z^{ij}] = \text{Var}[V^i - V^j]$.)

PROBLEM 2.1.6. In Example 2.1.1 let $L_B(t)$ be the number of customers in the system at
time t, in queue or in service. Show that $L_B(t)$ is not a monotone function of the arrival
spacings. In estimating $E[L_F(t) - L_G(t)]$ directly do we now necessarily get a variance
reduction using the same synchronization? Let $M(t)$ be the number of arrivals up to time
t. Consider the relation

$$L_B(t) = M(t) - N_B(t).$$

Exploit it to estimate $E[L_F(t) - L_G(t)]$ indirectly while using the same synchronization
effectively. Let $W_B(t)$ be the average system wait (queueing time plus service time) per
customer up to time t. Explain why $E[W_B(t)] \approx tE[L_B(t)]/E[M(t)]$. (Hint: See §3.5 and
references cited there.) How would you estimate $E[W_F(t) - W_G(t)]$? (Hint: $E[M(t)]$ is
known.)

PROBLEM 2.1.7. Consider a queueing network where the routing probabilities are independent of the system state. Let $N(t)$ be the number of system departures up to time t,
$L(t)$ the number of customers in the system at time t, $M(t)$ the number of exogenous

arrivals up to time t, and $W(t)$ the average system wait per customer up to time t. Argue that $N(t)$ is a decreasing function of exogenous arrival spacings and of service times. (Hint: The latter correspond to endogenous arrival spacings.) What in Problem 2.1.6 has analogs here? State them.

PROBLEM 2.1.8. In our banking example (Example 1.4.1) a slightly longer service time might "intercept" an arriving customer and (indirectly) cause him to balk when he otherwise would have joined the queue. From the bank's viewpoint, this might be good if that customer would have required an exceptionally long service time. This suggests that the number of customers served may not be a monotone function of the input random number stream. Why? Is there a way to synchronize the simulation to insure monotonicity?

2.1.3. Auxiliary Results

We close Section 2.1 with two lemmas to which we have already referred. The first of these is used in subsequent sections to show that several variance reduction techniques actually work; in fact, this lemma could be used to motivate them. These facts are well known except perhaps for Lemma 2.1.1(iii).

Lemma 2.1.1 (Variance decomposition). *Let X, Y, and Z be random variables with a joint distribution such that all the expectations, variances, and covariances indicated below exist. Then*

(i) $E[X] = E(E[X|Y])$;
(ii) $\text{Var}[X] = E(\text{Var}[X|Y]) + \text{Var}(E[X|Y])$;
(iii) $\text{Cov}[X, Y] = E(\text{Cov}[X, Y|Z]) + \text{Cov}(E[X|Z], E[Y|Z])$.

PROBLEM 2.1.9. Give a proof. (Hint: Simply use the definitions.)

Lemma 2.1.2. *If $a_1 < a_2 < \cdots < a_n$ and $b_1 < b_2 < \cdots < b_n$, the permutation π maximizing $\sum_j a_j b_{\pi(j)}$ is the identity permutation $\pi(j) = j$.*

PROBLEM 2.1.10. Give a proof. (Hint: Try a contradiction-interchange argument.)

2.2. Antithetic Variates

Suppose that we have a fixed policy and that in run r we use a string U^r of random numbers. In this section, superscripts index runs, not policies.

Let the output be $X^i = T(U^i)$. We want to estimate $E[X^i]$. With R runs, our estimator is

$$X = (X^1 + \cdots + X^R)/R$$

and

$$\begin{aligned}
\text{Var}[X] = (\text{Var}[X^1] + &\cdots + \text{Var}[X^R] \\
&+ 2\,\text{Cov}[X^1, X^2] + \cdots + 2\,\text{Cov}[X^{R-1}, X^R])/R^2.
\end{aligned}$$

For convenience we suppose that R is even and for future reference we set $Y_i = (X^i + X^{i+1})/2$, i odd, and $n = R/2$.

Plainly it is worth trying to make the covariance terms negative. The strategy adopted in this section, except for Problem 2.2.9, is to work on the pairs (X^1, X^2), (X^3, X^4), ..., (X^{R-1}, X^R). Within pairs we try to induce negative correlation but across pairs we leave the observations independent. For a more general strategy, related to stratification, see Problem 2.4.4. Another type of strategy, possibly helpful in queueing simulations, is to let the service times in run i be the arrival spacings in run j, and vice versa. In fixed-dimensional problems, a series of *randomly shifted* grids introduces correlation across *all* terms. In each coordinate all points of a regular grid get circularly shifted the same random distance. Sometimes this scheme is called *systematic sampling* or *rotation sampling*. A better alternative is to use low-discrepancy sequences, defined in Section 6.2, in which randomness plays no role. Except in one dimension, grids have high discrepancy. Sobol' (1974) indicates the appropriateness of low-discrepancy sequences for simulating Markov chains, even when the number of transitions is random.

Set $U^k = (U^k(1), U^k(2), \ldots)$. The counterpart of Theorem 2.1.1 is

Theorem 2.2.1. *Suppose that T is not constant on the interior of its domain and is a bounded, monotone function of $U^k(h)$ for all h. If $U^j(h) = 1 - U^i(h)$, $h = 1, 2, \ldots$, then $\mathrm{Cov}[x^i, x^j] < 0$.*

PROBLEM 2.2.1. Prove this theorem.

PROBLEM 2.2.2. If T is linear, what is

$$\mathrm{Corr}[X^i, X^j]? \quad (\text{Answer}: -1).$$

If U and $f(U)$ are uniform random numbers with $\mathrm{Cov}[U, f(U)] < 0$, then U and $f(U)$ are called antithetic variables. The choice $f(U) = 1 - U$ considered above seems sufficient when T is monotone, in view of Theorems 2.1.2 and 2.2.1.

PROBLEM 2.2.3. Give an example of T where $f(U) = 1 - U$ would be bad and $f(U) = U + \frac{1}{2} \pmod 1$ might be good. (Hint: Consider symmetric functions.)

PROBLEM 2.2.4. Suppose that we are comparing two policies using common random numbers across policies and antithetic random numbers on successive pairs of runs. Show that this may be counterproductive, even if using each technique separately does reduce the variance.

This problem offers a valuable lesson: two techniques, good individually, may make a bad team. With certain experimental designs, Schruben and Margolin (1978) show how to combine common random numbers and antithetic variates so that the team is demonstrably good. We give another example below.

EXAMPLE 2.2.1. Consider a problem of service system design, balancing the number of servers against the work rate of each server. Using one server with (random) service time T is generally better than using k servers with iid service times $T_1, T_2, \ldots, T_k$, respectively, where T and kT_1 are stochastically equivalent; i.e., $P[T \leq t] = P[kT_1 \leq t]$ for all t. Intuitively, when there are less than k customers in the system, some of the k slow servers are idle. However, there exist systems where k servers are better than one, even when, as we assume, they are fed by a common arrival stream. In addition, scaling up the system (increasing arrival and service rates by the same factor) generally improves system performance. For steady-state criteria, Brumelle (1971), Kleinrock [(1976), §5.1], and Stidham (1970) give precise analyses. For finite-horizon or nonstationary problems, no corresponding analytic results are available. To investigate such problems via simulation, the one-server and k-server systems can be synchronized as follows:

   (i) Let run r for the single-server system correspond to runs r' and r'' for the k-server system.
  (ii) Use the same arrival stream for runs r, r', and r''.
 (iii) On runs r' and r'', let (slow) server 1 have service times equal to k times the respective service times of the fast server on run r.
  (iv) For $i = 2, 3, \ldots, k$. generate the service times of server i on runs r' and r'' antithetically, using inversion.
   (v) Compare the output of run r with the average of the outputs of runs r' and r''.

We leave it to the reader to try to improve this design.

Our next task is to estimate $\mathrm{Var}[X]$, given in the second paragraph of this section. The Y_i's defined there are independent and so, with $\bar{Y} = (Y_1 + \cdots + Y_n)/n$,

$$S^2 = \sum_{i=1}^{n} (Y_i - \bar{Y})^2/n(n-1)$$

is an unbiased estimator of $\mathrm{Var}[X]$.

PROBLEM 2.2.5. Show from first principles that $E[S^2] = \mathrm{Var}[X]$.

PROBLEM 2.2.6. Show that the straightforward empirical variance of the unpaired observations $(X^1, X^2, \ldots, X^R)$ is biased and that the bias is positive if the hypotheses of Theorem 2.2.1 hold. (Remark: This biased estimate could well be greater than the empirical variance of R independent observations.)

PROBLEM 2.2.7. Consider the stochastic program

$$z = \max\{f(x) + E[\max\{g(x, y): h(y) \leq B, y \in Y\}]: x \in X\},$$

where B is the only random variable in sight. Assume: (i) that the inner maximization is not too hard for any fixed x and B; (ii) that X is a small discrete set; but (iii) that the indicated expectation is hard to compute for any fixed x. Give a naive Monte Carlo procedure to estimate z. Show that using either antithetic variates for fixed x or common random numbers across X works better. Open question: Under what conditions is it profitable to use both these techniques simultaneously?

PROBLEM 2.2.8. Show that using antithetic variates is sure to reduce the variance of the estimators considered in Problems 1.9.6 and 1.9.7; likewise, in Problem 1.9.8 if the arc lengths are generated by inversion. What about Problem 1.9.9?

PROBLEM 2.2.9 [Arvidsen and Johnsson (1982)]. Let $FP(x)$ denote the fractional part of x. Suppose $n > 2$. Generate U_1. For $i = 2, 3, \ldots, n - 1$, put $U_i = FP(2^{i-2}U_1 + \frac{1}{2})$. Set $U_n = 1 - FP(2^{n-2}U_1)$. Show that U_i is uniform on $(0, 1)$ for $i = 2, \ldots, n$. Prove that $U_1 + U_2 + \cdots + U_n = n/2$. (Hint: Set $U_1 = 2^{1-n}A + B$ where A is a nonnegative integer less than 2^{n-1} and B is a fraction less than 2^{1-n}. Use induction and analyze the cases depending on A and B.) Now let N be a positive integer. Show that $\mathrm{Corr}(U_1, FP(NU_1 + c)) = (1 - 6c(1 - c))/N$ and that the minimum, $-1/2N$, occurs at $c = \frac{1}{2}$. (Hint: Set $U_1 = A/N + B$ where A is a nonnegative integer less than N and B is a fraction less than $1/N$. Write $E[U_1 \cdot FP(NU_1 + c)]$ as the sum of N terms indexed by A running from 0 to $N - 1$, each summand being the sum of two integrals over B. The upper limit of the first integral and the lower limit of the second equal $(1 - c)/N$. The first integral corresponds to $NB + c < 1$; the second, to $NB + c \geq 1$.) Compute $\mathrm{Corr}(U_i, U_j)$ and display a matrix giving these correlations. State the obvious variant of antithetic variates based on these observations. (Arvidsen and Johnsson get good empirical results with it.) Open question: Is there an analog of Theorem 2.2.1 for this method?

PROBLEM 2.2.10 (Canonical convergence rate). Suppose that the work per run is a constant, independent of how correlation is induced across runs, and that the computer time allowed for all the runs together is t. You induce correlation within successive, nonoverlapping groups of $r(t)$ runs each but no stochastic dependence across these groups. Unless the variance of the estimator corresponding to each group is zero, show that when $r(t)$ is constant the average of these group estimators can't converge to its expectation faster than $O(t^{-1/2})$. What if $r(t)$ is not constant? Under what conditions can the variance of within-group estimators be zero? In answering these two questions, consider the one-dimensional and two-dimensional cases separately. (Hint: Refine randomly circularity-shifted grids when the dimension is fixed.) See also Problem 2.4.4 and the discussion of low-discrepancy sequences in Section 6.2.

2.3. Control Variates

Think of X as the primary output and of Y, the *control variate*, as a positively correlated secondary output for which $E[Y]$ is known. The positive correlation is usually achieved by generating X and Y from the same random number stream.

EXAMPLE 2.3.1 (continuation of Example 2.1.1 and Problem 2.1.6). Let the arrival process be Poisson and let G be the exponential distribution with mean one. Suppose that F also has mean one, that $F \neq G$ but that the shapes of F and G are not "too" different, and that variates from F (as well as G) can be generated easily by inversion. Set the control variate $Y = L_G(t) - E[L_G(t)]$. With effort, the last term can be calculated numerically from formula (2.163) of Kleinrock (1975) or formula (15) of Cooper [(1981), p. 122]. (To the extent that the "closure approximations" of Rothkopf and Oren (1979) or Clark (1981) for possibly nonstationary $M/M/s$ queues can be considered exact, an analogous control variate for handling possibly nonstationary $G/G/s$ queues is available.) Synchronize the simulation of these two queues ($M/F/1$ and $M/M/1$) as in Example 2.1.1. Intuitively, $L_F(t)$ and Y are strongly positively correlated; clearly, $L_F(t) - Y$ is an unbiased estimate of $E[L_F(t)]$. In Problem 2.1.7 we compared $L_F(t)$ and $L_G(t)$; here, we compare $L_F(t)$ and Y. This illustrates the close connection between common random numbers and control variates.

EXAMPLE 2.3.2. We consider open "Jackson"-like networks. (Readers unfamiliar with these should either consult Cooper [(1981), §4.2] or Kleinrock [(1975), §4.8], say, or skip this example.) The transition probabilities r_{ij} and service times T_i depend only on (i, j) and i, respectively, and not, for example, on the number of customers at node i. (Relaxing this condition would make it difficult to synchronize effectively.) We want to estimate the expected number of customers that leave the system in the time interval $[0, H]$, the "throughput." We have collected a huge amount of data on an existing system and therefore suppose that its transition probabilities, service time distributions, and expected throughput are effectively known. A new system is proposed; at first, we suppose that it differs from the existing system only in the distribution of T_5. The control variate is the existing system's (random) throughput minus its expected value.

The challenge in synchronization design is that a customer's path through the network is random. Following the third synchronization step in Section 2.1, generate all his transitions and service times as he enters the system. If r_{5j} also differs between the two systems, the second synchronization step in Section 2.1 becomes especially important: use a separate service-time generator and transition generator for each node. If this synchronization does not make it likely that the respective sets of nodes visited by a given customer in the two systems overlap strongly, then using the existing system's throughput as a control variate is probably inappropriate.

PROBLEM 2.3.1. Theorems 2.1.1 and 2.1.2 are relevant here. Why?

Suppose that X is based on a model M. Either Y may be some other, usually simpler, output (sometimes called an internal control variate) from the same model or else it may be an output from a closely-related simplified version of M, as in Example 2.3.1 above. In the former case, synchronization problems sometimes disappear because synchronization is automatic. An internal control variate tries to compensate for the difference between the theoretical and observed values of some parameter, say λ and $\tilde{\lambda}$, respectively. To compensate, we set $Y = \tilde{\lambda} - \lambda$ and then subtract kY from the straightforward estimator $X(\tilde{\lambda})$ to estimate $E[X(\lambda)]$. We generalize this idea below. Section 8.4 and Rothery (1982), for example, use internal control variates.

Centering Y at its expectation, without loss of generality assume that $E[Y] = 0$. Let $E[X] = \theta$ (to be estimated) and let k be a positive constant. Then

$$E[X - kY] = \theta,$$
$$\text{Var}[X - kY] = \text{Var}[X] + k^2 \text{Var}[Y] - 2k \text{Cov}[X, Y].$$

PROBLEM 2.3.2. When is $\text{Var}[X - kY] < \text{Var}[X]$?

PROBLEM 2.3.3. What value of k minimizes $\text{Var}[X - kY]$? (This is an academic exercise, because $\text{Cov}[X, Y]$ is virtually never known. At best, we may know its sign.)

At this point, the reader whose knowledge of regression is rusty should review a standard text, like Draper and Smith (1981).

PROBLEM 2.3.4. In the (theoretical) least-squares regression $X = \theta + cY + \text{error}$, find c by minimizing $E[(X - \theta - cY)^2]$. (The underlying regression model is nonstandard because the "independent" variable Y is, presumably, correlated with the error term. Here we assume nothing about the structure of this correlation and, perhaps cavalierly, ignore it in the calculation of c.)

PROBLEM 2.3.5. Compare the answers to Problems 2.3.3 and 2.3.4. (Conclusion: They are the same.)

PROBLEM 2.3.6. How would you estimate the optimal k *a posteriori* from outputs $(X_1, Y_1), \ldots, (X_R, Y_R)$ of R simulation runs? (Answer: Via regression.) Call your estimator $\hat{k}$. Is $E[\hat{k}] = c$? (Hint: The ratio of expectations typically does not equal the expectation of the ratio.) Is $E[X - \hat{k}Y] = \theta$? (The answer is no unless you used a technique like splitting, defined below.)

To eliminate bias, one possibility is to use, for run r, an estimate $\hat{k}_r$ from a regression in which the output from run r is deleted. This is sometimes called *splitting*, akin to *jackknifing* (§2.7).

PROBLEM 2.3.7. Outline an efficient computational procedure to implement splitting, showing how to compute $\hat{k}_r$ given $\hat{k}$. Show that $E[X_r - \hat{k}_r Y_r] = \theta$.

When the weight on a control variate is a function of the observations, as is the case if it is determined via regression with or without splitting, then finding a good estimator of the mean square error of the corresponding estimator of θ is difficult. Dussault (1980) proposes an approach based on "conditional" regression where $E[X \mid Y] = \theta + cY$. One can split more crudely, with slightly less computation, by partitioning the runs into just two groups A and B and using for group A (B) an estimate $\hat{k}_A$ ($\hat{k}_B$) from a regression in which the output from group A (B) is deleted. Based on his theoretical and empirical results, he recommends splitting into two groups and, as a lesser evil, estimating the variance of the resulting estimator by the usual empirical variance, without jackknifing. He reports that the mean square error of the empirical variance seems less than that of the other variance estimators considered, including unbiased estimators.

Set

$$\hat{\theta}_1 = \sum_{i=1}^{R}(X_r - \tilde{k}Y_r)/R,$$

$$\hat{\theta}_2 = \sum_{i=1}^{R}(X_r - \hat{k}Y_r)/R,$$

$$\hat{\theta}_3 = \sum_{i=1}^{R}(X_r - \hat{k}_r Y_r)/R,$$

$$\hat{\sigma}_j^2 = \text{Var}[\hat{\theta}_j],$$

where $\tilde{k}$, $\hat{k}$, and $\hat{k}_r$ are computed as in Problems 2.3.3, 2.3.6, and 2.3.7, respectively. Put $\hat{\sigma}_j^2$ equal to the empirical variance estimate of σ_j^2. From elementary statistics, we know that $(\hat{\theta}_1 - \theta)/\hat{\sigma}_1$ has the t-distribution with $R - 1$ degrees of freedom when the summands are normally distributed. If (X_r, Y_r) is multivariate normal, then Lavenberg and Welch (1981) indicate that $(\hat{\theta}_2 - \theta)/\hat{\sigma}_2$ has the t-distribution with $R - 2$ degrees of freedom. Under the same condition, $(\hat{\theta}_3 - \theta)/\hat{\sigma}_3$ has the same distribution asymptotically. Lavenberg and Welch point out that σ_2/σ_1 is proportional to $(R - 2)/(R - 3)$. The same holds for σ_3/σ_1 asymptotically. These facts are proved by Lavenberg et al. (1982) or in references they cite.

Taking proper account of covariance across runs is hard. One possible choice for the regression coefficient of the control variate is zero. Therefore, the sample variance of an estimator with a control variate whose coefficient is determined by least-squares regression is at most equal to the sample variance of the corresponding estimator without the control variate. This holds regardless of the ordering of the true mean square errors. Thus, any spurious correlation in the data artificially deflates the sample variance.

The generalization to several control variates is immediate, but Lavenberg et al. (1982) show why caution is needed. Generalizing the definitions of $\hat{\theta}_j$ and σ_j to situations with R runs and Q control variates and assuming multivariate normality, they show that $\sigma_2/\sigma_1 = (R - 2)/(R - Q - 2)$. Moral: Take Q small relative to R.

PROBLEM 2.3.8. Suppose that a large set of candidate control variates is available. Via stepwise multiple regression, we may select a relatively small subset to actually use. No problems arise if this is done in a separate series of runs that do not directly contribute to our final estimate of θ. If these runs are not separated, show that the sample variance (across runs) of the estimators $\hat{\theta}_1, \hat{\theta}_2, \ldots, \hat{\theta}_R$ underestimates the true variance of $\hat{\theta} = (\hat{\theta}_1 + \cdots + \hat{\theta}_R)/R$, possibly by an arbitrarily large amount. (Hint: The sample variance neglects an important source of variance, namely the *random* subset of chosen control variates.) Discuss the pros and cons of using control variate weights depending only on pilot runs that do not directly affect the estimate of θ. Show that the *sample* variance of $\hat{\theta}$ based on runs where the control-variate weights are found by least-squares regression is a decreasing function of the number of control variates. Show that this does not necessarily reflect the trend of the *true* variance.

PROBLEM 2.3.9 (cf. Hammersley and Handscomb [(1964), p. 19]). Suppose that $\hat{\theta}_1$, $\hat{\theta}_2, \ldots, \hat{\theta}_k$ are unbiased estimators of θ. Put

$$\hat{\theta} = \sum_{i=1}^{k} a_i \hat{\theta}_i, \quad \text{where} \quad \sum_{i=1}^{k} a_i = 1.$$

Show that $E[\hat{\theta}] = \theta$. What (theoretical) choice of $a_1, a_2, \ldots, a_k$ minimizes $\mathrm{Var}[\theta]$? Relate this to constrained multiple regression. (Hint: Write

$$\hat{\theta} = \tilde{\theta} + \sum_{i=1}^{k} b_i \hat{\theta}_i,$$

where

$$\tilde{\theta} = (\hat{\theta}_1 + \cdots + \hat{\theta}_k)/k \quad \text{and} \quad \sum_{i=1}^{k} b_i = 0.$$

Consider the regression $\hat{\theta} = \theta + \sum_{i=1}^{k} b_i \hat{\theta}_i + \text{error}$, where the b_i's are to be estimated subject to the constraint $\sum_{i=1}^{k} b_i = 0$. From this perspective, we can view $\tilde{\theta}$ as our primary estimator and the $\hat{\theta}_i$'s as analogs of control variates.) If the choice of a_i's has to be based on the *sample* covariance matrix of the $\hat{\theta}_i$'s, how would you estimate the mean square error of the corresponding $\hat{\theta}$? Under these circumstances, is using $\hat{\theta}$ necessarily a good idea?

PROBLEM 2.3.10 (Cf. Problem 2.2.4). Show that using both antithetic variates and control variates can be counterproductive. Can this happen with Cheng's (1981, 1982) procedure? (Despite large empirical efficiency increases with Cheng's method, Wilson (1984) shows by example that the theoretical answer is yes. This illustrates the danger in extrapolating from experimental evidence.)

2.4. Stratification

The classic examples of stratification arise in the design of sample surveys. Suppose one wants to ask a certain question to a random sample of the population of Montreal, recording the answers for statistical analysis. A straightforward procedure selects person i with probability inversely proportional to the population of Montreal. For simplicity, we do not worry about choosing the same person twice.

   A more sophisticated approach partitions the population into "strata," where the number n_i of Montrealers in stratum i can be estimated with negligible error from census statistics. For instance, one such stratum could comprise all women between 21 and 30 whose native language is French. If there are m strata, then to select a sample of size S randomly select $Sn_i/(n_1 + \cdots + n_m)$ people, modulo rounding, from stratum i.

The intuitive underlying idea is that stratification legitimately replaces a random variable (the number of times a stratum is sampled) by a constant (its expectation). If we know the size of each stratum, this replacement *must* reduce variance. For a proof that this intuition is correct, consider:

PROBLEM 2.4.1. Let X_{ij} be the response of person j in stratum i. Let $E[X_{ij}] = \mu_i$ and $\text{Var}[X_{ij}] = \sigma_i^2$. Let A and B be the overall average response using, respectively, the naive and the sophisticated method. Find $E[A]$, $E[B]$, $\text{Var}[A]$, and $\text{Var}[B]$. (Hint: Use the variance-decomposition lemma 2.1.1. In that lemma, let X be A and then B and let $Y = (Y_1, \ldots, Y_S)$ where Y_k is the stratum to which the kth person picked belongs.) Show that $\text{Var}[E(X \mid Y)]$ is nonnegative and zero, respectively, in the expressions for $\text{Var}[A]$ and $\text{Var}[B]$. If (and this is a big if) all the σ_i^2 were known in advance but not necessarily equal, suggest a better survey design. (Hint: Formulate a one-constraint optimization problem and use a Lagrange multiplier to show that, modulo rounding, the number of people chosen from stratum i should be

$$n_i \sigma_i S \bigg/ \sum_k n_k \sigma_k.$$

This agrees with intuition: the larger σ_i, the more stratum i should be sampled.) Generalize to the case where the unit cost of sampling stratum i is c_i. What happens if all the σ_i are equal? If the σ_i^2 are unknown, suggest a two-phase sampling procedure that exploits the foregoing result. (Remark: Variance estimates generally have high variance; if variance estimates are used, variance reduction cannot be guaranteed. By contrast, if we delete σ_i and σ_k from the displayed formula, we are sure to reduce variance—though possibly not to the optimal level.)

PROBLEM 2.4.2. Suppose that, in a stochastic system, initial condition i occurs with known probability P_i. On each run, the simulation ends at a given finite horizon—say one day. Suggest two ways of generating initial conditions in a simulation of this system where you can afford to make R runs. (Hint: Try using initial condition i on $\lfloor P_i R \rfloor$ runs, $i = 1, 2, \ldots$; on the remaining runs, show that selecting initial condition i with probability proportional to $P_i R - \lfloor P_i R \rfloor$ produces no overall bias.) Stratification introduces logical but *not* statistical dependence among runs. State a preference between the two methods and justify it. (Remark: If there are significant carry-over effects from one day to the next, this suggests that the horizon should not be one day but, say, one week, and that proper accounting of any remaining end effects should be made.)

PROBLEM 2.4.3. Suppose we are interested primarily in simulating "rush-hour" (peak-load) behavior. Directly estimating the distribution of "initial" conditions at the start of the rush hour may be impractical, particularly when we cannot simply assume that the system is memoryless without introducing an uncountable state space incorporating history. Is there an alternative to the seemingly wasteful indirect approach of simulating buildups of load starting from an empty or near-empty system? (Forthcoming joint papers by Fox and Glynn show constructively that the answer is yes.) Compare efficiencies of stratified and unstratified versions of Russian roulette. For example, with an unstratified version runs which hit an uninteresting set are discarded independently with probability $(m - 1)/m$; the stratified version keeps every mth such run. In both cases the runs continued after hitting an uninteresting set are weighted by m.

PROBLEM 2.4.4. Set $U_j = (U + j)/m, j = 0, 1, 2, \ldots, m - 1$, where U is a uniform random number. Thus, U_j is uniformly distributed on the interval $(j/m, (j + 1)/m)$ and generating the block $[U_0, U_1, \ldots, U_{m-1}]$ corresponds to stratified sampling. Why? Can this scheme be adapted for variance reduction using an approach akin to antithetic variates (§2.2)? (See Hammersley and Handscomb [(1964), pp. 60–66].)

PROBLEM 2.4.5 (continuation of Problem 2.4.2). Using the naive method, suppose that initial condition i occurs N_i times. Let $\overline{X}_i$ be the average response with initial condition i:

$$\overline{X}_i = \sum X_{ij}/N_i \quad \text{if } N_i > 0,$$

where X_{ij} is the jth response using initial condition i. The overall average is thus

$$\overline{X} = \sum \overline{X}_i N_i/R \quad \text{if all } N_i > 0.$$

Also, if all N_i are positive,

$$E[\overline{X}] = \sum E[\overline{X}_i] \cdot E[N_i]/R,$$

because $E[\overline{X}_i N_i] = E[\sum_{j=1}^{N_i} X_{ij}] = E[X_{ij}] \cdot E[N_i] = E[\overline{X}_i] \cdot E[N_i]$. From $E[N_i] = P_i R$, we get

$$E[\overline{X}] = \sum E[\overline{X}_i] P_i.$$

This suggests an alternative *poststratified* estimator

$$\hat{X} = \sum \overline{X}_i P_i,$$

defined when all $N_i > 0$ and then (conditionally) unbiased. Comparing $\overline{X}$ and $\hat{X}$, it looks like we have replaced a random variable by its expectation. But this is deceptive. An alternative expression for $\overline{X}$ is $\sum\sum X_{ij}/R$; here N_i disappears, except for being the upper limit of the second sum. Prove or give a counterexample:

$$\text{Var}[\hat{X}] \leq \text{Var}[\overline{X}] \quad (?)$$

The "obvious" estimator of $\text{Var}[\overline{X}_i]$ is the sample variance

$$S_i^2 = \sum (X_{ij} - \overline{X}_i)^2/N_i^2.$$

Show that in general $E[S_i^2] \neq \text{Var}[\overline{X}_i]$. (Hint: The "obvious" estimator pretends that N_i is fixed, but N_i is random.) Do you expect S_i^2 to be biased high or low? See Cochran [(1977), §5A.9] and Kleijnen [(1974), pp. 116–121] for remarks and references. In the above example, N_i is a stopping time (§3.7); give an example where it is not.

2.5. Importance Sampling

To motivate importance sampling, suppose that we wish to estimate

$$\theta = E[C(X)] = \sum_{i=1}^{m} C(X \mid A_i) P(A_i),$$

where the events A_i are generated during a simulation and $C(X \mid A_i)$ is expected cost (or profit) given that A_i occurs.

It often happens that $C(X \mid A_i)$ is large while simultaneously $P(A_i)$ is small for some i. For example, under a reasonable inventory policy, a stockout should be a rare event that may occur when demand is exceptionally heavy; however, a stockout may be so costly that when it does occur it dominates all other costs. If we simulate in a straightforward way, the time required to obtain a reasonable estimate of θ may be prohibitive. The alternative is to sample from a different distribution, applying a correction factor to compensate.

We formulate below an n-period abstract version of the problem. This formulation is only superficially complex. To interpret it concretely, think for example of X_i as the demand in period i, $g(x_1, \ldots, x_n)$ as a cost function, and F as the joint n-period demand distribution function. If successive demands are independent, then F separates; typically g does not separate because inventory is carried over from one period to the next.

Now

$$\theta = \int \cdots \int g(x_1, \ldots, x_n) \, dF(x_1, \ldots, x_n).$$

We consider two cases, slightly abusing notation:

(i) F has density f;
(ii) F is discrete, $P[(x_1, \ldots, x_n)] = f(x_1, \ldots, x_n)$.

Let H be a different distribution function. (In the inventory example, H would put more weight than F on high demands.) Let h be defined analogously to f. Now if $h > 0$ wherever $f > 0$ we have:

Case (i)

$$\theta = \int \cdots \int_{h > 0} \left[\frac{g(x_1, \ldots, x_n) f(x_1, \ldots, x_n)}{h(x_1, \ldots, x_n)} \right] dH(x_1, \ldots, x_n).$$

Case (ii)

$$\theta = \sum_{\substack{h > 0 \\ (x_1, \ldots, x_n)}} \left[\frac{g(x_1, \ldots, x_n) f(x_1, \ldots, x_n)}{h(x_1, \ldots, x_n)} \right] h(x_1, \ldots, x_n).$$

Whereas in a straightforward simulation run, we

(i) sample $(X_1, \ldots, X_n)$ from F;
(ii) output $g(X_1, \ldots, X_n)$,

using importance sampling we

(i') sample $(X_1, \ldots, X_n)$ from H;
(ii') output $g(X_1, \ldots, X_n) f(X_1, \ldots, X_n)/h(X_1, \ldots, X_n)$.

Either way we estimate θ as the average of the output values. This is essentially a Monte Carlo evaluation of the relevant sum or integral, using a slightly more subtle version of the method suggested in Problem 1.9.1. Often g is not given explicitly but is implicitly determined by the simulation program itself, as indicated by Example 2.5.1 below.

To choose an appropriate H, some qualitative knowledge about the problem is needed. This may be available *a priori* or it may be obtained from pilot runs. If θ were known and positive (cf. Problem 2.5.5), taking $h = gf/\theta$ would give a zero-error estimate of θ. Ideally, h should be proportional to gf because then $\int h \, dx = 1$ and $\int gf \, dx = \theta$ jointly imply that $h = gf/\theta$ and hence that the resulting estimator has zero error. Making h mimic gf qualitatively means making h roughly proportional to gf; it is a step in the right direction. All is not rosy, however: disaster can strike. Problem 2.5.7 shows what can go wrong in principle and Section 8.6 shows that this is not merely an abstract possibility.

EXAMPLE 2.5.1 (continuation of Example 1.4.1). Here g is the number of customers served and

$$F(x_1, \ldots, x_n) = \left[\prod_i A_i(x_i)\right]\left[\prod_j B(x_j)\right]\left[\prod_k C_k(x_k)\right][D(x_1)],$$

where $F \leftrightarrow$ joint distribution of everything in sight;

   x_i's $\leftrightarrow$ arrival spacings; A_i's $\leftrightarrow$ spacing distributions;
   x_j's $\leftrightarrow$ service times; $B \leftrightarrow$ service distribution;
   x_k's $\leftrightarrow$ uniform random numbers for generating join/balk decisions;
      $C_k \leftrightarrow$ balking distribution;
   $x_1 \leftrightarrow$ number of tellers present; $D \leftrightarrow$ distribution of number of tellers present.

For the number of tellers present, we use stratified sampling as in Problems 2.4.1 and 2.4.2; per the latter, this number is a constant for most runs. The two-phase sampling procedure mentioned at the end of Problem 2.4.1 might be worthwhile. We assume that all this is already reflected in D above. A natural choice for H has the form

$$H(x_1, \ldots, x_n) = \left[\prod_i A_i(x_i)\right]\left[\prod_j \tilde{B}(x_j)\right]\left[\prod_k C_k(x_k)\right][D(x_1)],$$

where the tilde indicates a modified distribution.

Following the guidelines above, the service times generated by $\tilde{B}$ should be stochastically smaller than those generated by B. Longer service times lead to less customers joining the system, so that, by and large, the output of the simulation will be a decreasing function of any particular service time x_j. The modified service time density is to mimic the old density multiplied by this decreasing function: in other words, short service times should be more probable, and long ones should be less. In addition, we should choose $\tilde{B}$ so that sampling from it is easy and so that $b(x)/\tilde{b}(x)$ is quickly calculated for any x. These qualitative considerations motivate our detailed treatment in Chapter 8. To generate the output we multiply the number of customers served by $f(X_1, \ldots, X_n)/h(X_1, \ldots, X_n)$, canceling the common factors of f and h.

PROBLEM 2.5.1. Formulate a two-period inventory problem, and compute $\text{Var}[\hat{\theta}_H]$ analytically or numerically for various H.

PROBLEM 2.5.2 (Continuation of Problem 2.5.1.) Write a computer program to estimate $\text{Var}[\hat{\theta}_H]$ empirically using Monte Carlo.

PROBLEM 2.5.3. In queueing systems, exceptionally long service times are likely to generate exceptionally long lines and busy periods in their wakes. In planning the capacity of such systems, service-time outliers (corresponding to the tails of the service-time distributions) are therefore of crucial importance. Formulate a queueing model and show how importance sampling can be used advantageously in simulating its behavior.

PROBLEM 2.5.4 [Knuth (1975)]. Show how importance sampling may make the estimation procedure of Problem 1.9.5 more efficient.

PROBLEM 2.5.5. If θ and g are respectively finite and bounded below by a known constant λ, show that without loss of generality we can assume that $\theta > 0$ and $g \geq 0$. Show how to handle the more general case where there is a known constant λ and a known function e such that $g \geq e$ and $\int e \, dF = \lambda$. (Hint: Estimate $\theta - \lambda$.)

PROBLEM 2.5.6. With importance sampling, it may well be difficult to check whether the output is a monotone function of the driving uniform random number stream. Give an example where using importance sampling converts a monotone output to a nonmonotone output. What are the implications for the joint use of importance sampling and any of common random numbers, antithetic variates, and control variates?

PROBLEM 2.5.7. Give a formula for the variance of the importance sampling estimator. Requiring that $f/h \leq k$, a small constant between 1 and 3, say, tends to guard against gross misspecification of h, even though perhaps it eliminates the theoretically optimal h. Why? In multidimensional applications, such as our banking example, trying to enforce this safeguard may eliminate importance sampling as a practical possibility. For example, suppose that $f(x_1, x_2, \ldots, x_n) = f_1(x_1)f_2(x_2) \cdots f_n(x_n)$ and $f_1 = f_2 = \cdots = f_n$ and that, to keep the computational burden reasonable, $h = h_1 h_2 \cdots h_n$ and $h_1 = h_2 = \cdots = h_n$. Show that our safeguard would then require that $f_1(x)/h_1(x) \leq k^{1/n}$ for all x. Because f_1 and h_1 are nonnegative and $\int f_1(x) \, dx = \int h_1(x) \, dx = 1$ (why?), our safeguard severely restricts h_1. Show that, if n is large, it essentially forces $h \equiv f$. (Moral: Because there is no practical way to guard against gross misspecification of h, multidimensional importance sampling is risky.)

PROBLEM 2.5.8 (Gradient estimation). Suppose θ depends on a parameter ξ. We want to estimate simultaneously $\theta(\xi_0)$ and the gradient $\theta'(\xi_0)$ at ξ_0. With this in mind, write $f(x_1, \ldots, x_n)$ more explicitly as $f(x_1, \ldots, x_n; \xi)$ and set $h_{\xi_0}(x_1, \ldots, x_n) = f(x_1, \ldots, x_n; \xi_0)$; likewise, write $g(x_1, \ldots, x_n; \xi)$. Sampling this way from the *original* distribution F_{ξ_0} turns importance sampling upside-down. Why? Interpret

$$L(x_1, \ldots, x_n; \xi) = f(x_1, \ldots, x_n; \xi)/f(x_1, \ldots, x_n; \xi_0)$$

as a likelihood ratio. Importance sampling produces

$$Z(x_1, \ldots, x_n; \xi) = g(x_1, \ldots, x_n; \xi)L(x_1, \ldots, x_n; \xi).$$

Differentiate $Z(x_1, \ldots, x_n; \xi)$ with respect to ξ and then evaluate at ξ_0 to get

$$Z'(x_1, \ldots, x_n; \xi_0) = g'(x_1, \ldots, x_n; \xi_0) + g(x_1, \ldots, x_n; \xi_0)L'(x_1, \ldots, x_n; \xi_0)$$

when ξ is a scalar. Generalize to the case when ξ is a vector. (Hint: This is merely an exercise in notation.) Assuming that we can switch the order of expectation and differentiation, show that

$$\theta'(\xi_0) = E_{\xi_0} Z'(x_1, \ldots, x_n; \xi_0).$$

(Consult an advanced calculus text for conditions that justify this switch. Little more than continuous differentiability of Z in a neighborhood of ξ_0 is needed.) Justify: to estimate $\theta(\xi_0)$, average $g(x_1, \ldots, x_n; \xi_0)$ over runs; to estimate $\theta'(\xi_0)$, average $Z'(x_1, \ldots, x_n; \xi_0)$ over runs.

PROBLEM 2.5.9 (Continuation). In Example 1.4.1 let ξ be the mean service time, $\xi_0 = 2$, k an arbitrary constant, and now let g be the number of customers served minus k/ξ. Show how to compute $Z'(x_1, \ldots, x_n; \xi_0)$. (Hint: Factor $L(x_1, \ldots, x_n; \xi)$ and show that

$$g(x_1, \ldots, x_n; \xi) = \tilde{g}(x_1, \ldots, x_n) - k/\xi,$$

where $\tilde{g}$ does not depend on ξ (since $x_1, \ldots, x_n$ depend only on ξ_0).)

PROBLEM 2.5.10 (Continuation). Combine this estimate with a Robbins–Munro (1951) scheme to search for a local optimum. [Let t be the computer time used. Glynn (1986a) points out for general settings that this scheme converges at rate $O(t^{-1/2})$ while alternatives such as the Kiefer–Wolfowitz algorithm and stochastic quasi-gradient methods are an order of magnitude slower.]

PROBLEM 2.5.11 (Continuation of Problem 2.5.8). Suppose we simulate a finite Markov chain. If the chain's recurrent-transient class structure depends on ξ in a neighborhood of ξ_0, show that Z may not be continuously differentiable there. What restriction on the transition probabilities assures that the gradient estimation scheme works? (Certain other mild technical conditions on Z are needed; see Glynn (1986b).)

PROBLEM 2.5.12 (Counterpart of Problem 2.5.7). Give a formula for the variance of the gradient estimator in Problem 2.5.8. What can you say qualitatively about the size of this variance for Problem 2.5.9?

PROBLEM 2.5.13 (A related restriction). To assure that gradient estimation via likelihood ratios works, we require that the set of possible sample paths is the same for all ξ in a neighborhood of ξ_0. Otherwise, the likelihood ratio can be discontinuous at ξ_0. Why? In a two-point service-time distribution, we can estimate the gradient with respect to the weight on the left point via likelihood ratios but not the gradient with respect to the point's position. Why?

PROBLEM 2.5.14. Compare gradient estimation and common random numbers. Consider the number of parameters, whether they are continuous or discrete, and relevance to noninfinitesimal perturbations.

2.6. Conditional Monte Carlo

Let $\hat{\theta}_1$ be an estimator of the scalar parameter θ. Let Z be a random vector or, more generally, a σ-field. Intuitively, the latter is information generated by the simulation. By assumption, it is not feasible to calculate θ; that is why we are simulating. But suppose that we *can* calculate $\hat{\theta}_2 = E[\hat{\theta}_1 | Z]$. A crude simulation generates $\hat{\theta}_1$ directly. A better way is to use the simulation *only* to generate Z and then to calculate $\hat{\theta}_2$ for that Z. This is called *conditional Monte Carlo*. It exploits the precept that replacing a random variable (here $\hat{\theta}_1$) by its expectation (here $E[\hat{\theta}_1 | Z]$)—whenever legitimate and feasible—is a good idea.

An application of Lemma 2.1.1(i) shows that $E[\hat{\theta}_2] = E[\hat{\theta}_1]$. What we have done is to replace a naive one-stage averaging process $\hat{\theta}_{1R}$ (the mean of the outputs over R simulation runs, say) by a nested two-stage averaging process $\hat{\theta}_{2R}$:

1. For $j = 1, \ldots, R$;
    (a) Generate Z_j.
    (b) Obtain the inner average $\hat{\theta}_2(Z_j) = E[\hat{\theta}_1 | Z_j]$:
        (i) analytically;
        (ii) numerically by a deterministic procedure with negligible error; or
        (iii) by estimating $\hat{\theta}_2(Z_j)$ using a Monte Carlo scheme more efficient than straightforward simulation, calling this estimate $\tilde{\theta}_2(Z_j)$.
2. Estimate the outer average empirically by taking the mean of $\hat{\theta}_2(Z_j)$ or $\tilde{\theta}_2(Z_j)$ over $Z_1, Z_2, \ldots, Z_R$ calling the resulting estimators $\hat{\theta}_{2R}$ and $\tilde{\theta}_{2R}$, respectively.

Remark. One way to implement option (iii) of step 1(b) is itself called conditional Monte Carlo in the literature, an unfortunate confusion of terminology. Granovsky (1981) and Wilson (1984) describe this sophisticated scheme, tracing its history and giving examples. Their work supersedes previous treatments, including that in Hammersley and Handscomb [(1964), Chapter 6]. If $E[\hat{\theta}_2]$ can be estimated readily by numerical integration once $\hat{\theta}_2(z)$ or $\tilde{\theta}_2(z)$ is known on a suitable grid of z-values, then change step 1(a) to generate the grid deterministically and replace step 2 by quadrature.

In what follows we assume that step 1(b) is implemented with option (i) or option (ii). Recall that $\hat{\theta}_2 = E[\hat{\theta}_1 | Z]$.

PROBLEM 2.6.1. Show that

$$\text{Var}[\hat{\theta}_2] \le \text{Var}[\hat{\theta}_1].$$

(Hint: Using the variance-decomposition lemma 2.1.1(ii),

$$\text{Var}[\hat{\theta}_1] = \text{Var}[\hat{\theta}_2] + E[\text{Var}(\hat{\theta}_1 | Z)].$$

Because $\text{Var}[\hat{\theta}_1 | Z]$ is trivially nonnegative for each Z, so is the second term on the right.)

Conditional Monte Carlo is typically used as follows. Let $T_1, T_2, \ldots$ be event epochs (times when the simulation clock activates events) and suppose that $X(T)$ is a well-defined random variable, at least at event epochs. Let

$$\theta = E\left[\sum_{i=1}^{N} X(T_i)\right],$$

where N may be a random horizon. Put

$$\hat{\theta}_3 = \sum_{i=1}^{N} X(T_i).$$

Let S_i be the random state of the system at epoch T_i. Put

$$\hat{\theta}_4 = \sum_{i=1}^{N} E[X(T_i)|S_i].$$

By the definition of a state, also conditioning on S_h for $h < i$ gives the same answer. We usually do not condition on S_j for $j > i$ because:

(i) this anticipates the future;
(ii) the computation may be difficult;
(iii) the variance reduction may be smaller or even zero.

In view of Problem 2.6.1 each summand of $\hat{\theta}_4$ has variance less than or equal to that of the respective term of $\hat{\theta}_3$, but in general this is not enough to assure that $\text{Var}[\hat{\theta}_4] \leq \text{Var}[\hat{\theta}_3]$: the covariance structure among the terms in $\hat{\theta}_4$ differs from that in $\hat{\theta}_3$. Usually these covariance structures are too complex to assess and so we optimistically suppose that of $\hat{\theta}_4$ is not significantly worse. This leads us, heuristically, to prefer $\hat{\theta}_4$ to $\hat{\theta}_3$. We call $\hat{\theta}_4$ an *extended* conditional Monte Carlo estimator. Unlike the *strict* Monte Carlo estimator $\hat{\theta}_2$, one cannot generally prove that $\hat{\theta}_4$ reduces variance; Problem 2.6.3 gives an exception.

In the definitions above, we moved the expectation inside the summation. If N is independent of $X(T_i)$, this is legitimate. In general, check that $\hat{\theta}_4$ estimates θ consistently. Using conditional Monte Carlo, the only function of the simulation is to generate $N, S_1, \ldots, S_N$.

One illustration of this situation is an inventory problem where the epochs are evenly spaced (periodic review), $X(T_i)$ is the cost incurred in period i, and S_i is the inventory position at the beginning of period i. Using conditional Monte Carlo, the *only* function of the simulation is to generate the stock levels at the review points.

For a second illustration, we look again at our banking example of Chapter 1.

EXAMPLE 2.6.1 (continuation of Example 1.4.1). To define the state of the system at arrival epochs, we need to keep track not only of the number i of customers in the system but (in principle) also of the times when the customers being served started service. Fortunately, the probability p_i that a customer balks depends only on the former quantity. The expected value of an indicator variable which is 1 if the customer joins the system and 0 otherwise is $1 - p_i$. What we are after is the expected value of the sum of such indicator variables. Using conditional Monte Carlo to estimate the expected number of customers served, we initialize a counter to 0 and, whenever a customer arrives, increment it by the probability that he does not balk. Thus, if an arrival finds i customers already in the system, we increment by $1 - p_i$. The estimate of the expected number of customers served is the final number on the counter. *Only the estimation method has changed. The system is driven exactly as before.* For example, to determine whether an arrival finding i customers already in the system does in fact balk, we still generate a uniform random number and compare it to p_i.

Carter and Ignall (1975) outline another example. There $X(t)$ is the response time to a fire given that a random fire alarm goes off at t and $S(t)$ is a vector indicating the availability of firefighting equipment at time t and which alarms, if any, have just sounded. Dispatching ambulances dispersed in a district has roughly this structure. The total time to get a patient to the hospital is the time to reach patient (response time) plus the time to transport him to the hospital. If only response time depends on policy or on the counterpart of $S(t)$, then it suffices to consider just response time because the (conditional) expectations of transport time are equal. Thus, transport-time variance does not affect our estimate.

A fourth example is any system with Poisson arrivals. Any instant can be considered a *virtual* arrival instant (cf. §4.9, especially Problems 4.9.9 and 4.9.10). Therefore, we can weight $E[X(T_i)|S_i]$ by a quantity involving the fraction of time that the system is in state S_i. We illustrate this in Problem 4.9.11, which fully exploits the structure of Example 2.6.1.

As a fifth application of conditional Monte Carlo, let $f(x)$ be the cost of a customer waiting time x in a queueing system. Waiting time X is the sum Y of service times and the delay Z waiting for service. Let H be a finite horizon and let N be the number of arrivals up to time H. Assume that N differs negligibly, if at all, from the number of system departures up to time H. We want to estimate

$$C = E\left[\sum_{i=1}^{N} f(X_i)\right],$$

where X_i is the time that customer i spends in the system.

PROBLEM 2.6.2. Show that

$$\sum_{i=1}^{N} E[f(X_i)|Z_i = z_i] = \sum_{i=1}^{N} \int_0^\infty f(y + z_i)\, dG_i(y),$$

where G_i is the conditional distribution of Y_i given z_i. Suppose that G_i is known or can be calculated in a practical way; give examples where this supposition holds and where it does not. Make no change to the simulation itself. Show how to estimate C using conditional Monte Carlo. If the horizon is infinite, we want to estimate average cost per customer. Show that the natural estimator is then a ratio of random variables. Does conditional Monte Carlo still work? Show that it is wrong to replace N by its expectation.

PROBLEM 2.6.3 (continuation). Now suppose that $f(x) = x$. For estimation purposes *only*, replace all service times by their conditional expectations given the state of the system at the start of the respective services. Show how to estimate C using conditional Monte Carlo. [This is considered further in Problem 3.5.1.] Show that for $G/G/s$ queues, where $E[Y_i|Z_i] = E[Y_i]$, the variance reduction is proportional to $\text{Var}[Y_i]$, the service time variance, plus *nonnegative* covariance terms of the form $\text{Cov}(Y_i, Z_j)$. (Hint: Increasing customer i's service time either increases customer j's delay or does not affect it.)

PROBLEM 2.6.4. Let

$$\tilde{\theta} = \lim_{m \to \infty} \frac{1}{m} \sum_{i=1}^{m} E[X(T_i)].$$

Modify $\hat{\theta}_1$ and $\hat{\theta}_2$ accordingly and show that the resulting estimators converge to $\tilde{\theta}$ with probability one. (You may assume weak regularity conditions, but make these explicit.)

If the event epochs are not evenly spaced and the horizon is infinite, then the discrete-time limit $\tilde{\theta}$ may not be the appropriate thing to estimate. Think of $X(t)$ as the cumulative "reward" up to time t. A more relevant quantity may be

$$\gamma = \lim_{t \to \infty} \frac{1}{t} E[X(t)].$$

Now put

$$Y(T_i) = E[X(T_{i+1}^-) - X(T_i^-)|S_i],$$

$$m_j = E[T_{i+1} - T_i|S_i = j],$$

$$r_j = E[Y(T_i)|S_i = j].$$

If it is feasible to calculate m_j and r_j, then conditional Monte Carlo can be used effectively. One case where m_j is easy to calculate occurs when $T_{i+1} - T_i$ depends on S_i but not on S_{i+1}, as in continuous-time Markov chains.

PROBLEM 2.6.5 (Conversion to discrete time). If the underlying process is a Markov renewal process [e.g., see Çinlar (1975)], then

$$\gamma = \sum \pi_j r_j / \sum \pi_k m_k,$$

where π_j is the steady-state probability of being in state j. A naive estimator of γ is

$$\hat{\gamma}_1 = \sum_{i=1}^{n} Y(T_i)/[T_1 + (T_2 - T_1) + \cdots + (T_{n+1} - T_n)].$$

Show how to estimate γ more efficiently by simulating an equivalent discrete-time process. What effect does this have on the number of random numbers generated and on the variance of the estimator of γ? [This has been studied by Hordijk *et al.* (1976) and, more generally, by Fox and Glynn (1986a). See also Problem 3.7.10.]

PROBLEM 2.6.6 (Critical paths revisited). Suppose we have identified a cut set in an acyclic network with random arc lengths such that no two arcs in the set point out of the same node. In estimating expected critical path length (Problem 1.9.8) use conditional Monte Carlo to reduce variance by conditioning on both

(i) the longest path lengths from the start node to all nodes on the same side of the cut set as the start node;
(ii) the longest path lengths from all nodes on the same side of the cut set as the finish node to the finish node.

What complications can arise if two arcs in the cut set point out of the same node? Can you handle them? Do any complications arise if two arcs in the cut set point in to the same node?

2.7. Jackknifing

We begin with the traditional motivation. Let $\hat{\theta}_n$ be a biased estimator, based on n observations, where

$$E[\hat{\theta}_n] = \theta + \sum_{i=1}^{\infty} a_i/n^i.$$

Here the a_i's may be (possibly unknown) functions of θ but not of n. An essential feature is that the bias goes to zero as n gets large. Let $\hat{\theta}_{n-1}$ be an estimator of the same form based on a sample of size $n - 1$. The two samples may overlap. A little algebraic fiddling shows that

$$E[n\hat{\theta}_n - (n-1)\hat{\theta}_{n-1}] = \theta - \frac{a_2}{n(n-1)} - \frac{a_3(2n-1)}{n^2(n-1)} - \cdots,$$

reducing the dominant bias term from $O(1/n)$ to $O(1/n^2)$. This would not have worked if we had simply taken a weighted average of $\hat{\theta}_n$ and $\hat{\theta}_{n-1}$. Numerical analysts call this trick Richardson extrapolation or extrapolation to the limit.

PROBLEM 2.7.1. Verify the above displayed expression.

The reasoning above is shaky, as shown by

PROBLEM 2.7.2. Replace n^i by $(n + k)^i$ in the expression for $E[\hat{\theta}_n]$ and redefine the constants a_i appropriately. Show that $E[(n + k)\hat{\theta}_n - (n + k - 1)\hat{\theta}_{n-1}] = O(1/n^2)$. What does this indicate?

Efron [(1982), §2.6] points out this defect and motivates the estimator $n\hat{\theta}_n - (n-1)\hat{\theta}_{n-1}$ differently. See his monograph for an extensive discussion of the jackknife and related estimators.

Suppose that θ is based on observations $X_1, X_2, \ldots, X_n$. Call $\hat{\theta}_{-i}$ the corresponding estimator based on the same observations, except that X_i is deleted. Now put

$$J_i(\hat{\theta}) = n\hat{\theta} - (n - 1)\hat{\theta}_{-i}$$

and

$$J(\hat{\theta}) = [J_1(\hat{\theta}) + \cdots + J_n(\hat{\theta})]/n.$$

In the literature J_i is called *pseudovalue* i and J is called the *jackknife*. The foregoing development suggests that

$$E[J(\hat{\theta})] = \theta + O(1/n^2).$$

So, under the conditions postulated at the outset, the jackknife may be a good tool. It is handy to reduce bias in the following situations, among others:

1. ratio estimators:
   (a) in regressions with random "independent" variables (cf. Problem 2.3.6);
   (b) the "regenerative" method for simulations (§3.2.2);
2. variance estimation:
   (a) when using control variates with regression-determined weights (§2.3);
   (b) in stochastic processes with serial correlation (§3.3.1).

The jackknife may reduce bias, but may also increase variance: thus there is no guarantee that it reduces mean square error. Its performance is often gauged empirically by method A of Section 1.6, which requires caution as the critique there points out.

For large n, it is customary to regard the pseudovalues as if they were approximately iid—even though they are based on strongly overlapping samples. Perhaps cavalierly, estimate the variance of J in the usual way:

$$\hat{\sigma}_J^2 = \sum_{i=1}^{n} [J_i(\hat{\theta}) - J(\hat{\theta})]^2/(n - 1).$$

So, it might be plausible that

$$\sqrt{n}[J(\hat{\theta}) - \theta]/\hat{\sigma}_J \Rightarrow N(0, 1).$$

Under certain reasonable conditions, this can be shown rigorously.

From a remarkable result of Efron and Stein (1981), it follows immediately that

$$(n - 1)E[\hat{\sigma}_J^2] \geq \text{Var}[J_i(\hat{\theta})],$$

assuming that the X_i's are iid. Summing this inequality over $i = 1, 2, \ldots, n$ and dividing by n^2, we see that $[(n - 1)/n]\hat{\sigma}_J^2$ would be biased high as an estimate of Var $[J(\hat{\theta})]$ if the J_i's were independent. Efron and Stein go on to consider estimators of $\text{Var}[J(\hat{\theta})]$; their results are interesting but inconclusive.

For further details, extensions, and generalizations, see Gray and Schucany (1972), Miller (1974), and Efron (1982).

PROBLEM 2.7.3. Suppose that $\hat{\theta} = (X_1 + \cdots + X_n)/n$. Show that $J(\hat{\theta}) = \hat{\theta}$.

PROBLEM 2.7.4. What happens if we recursively jackknife $J(\hat{\theta})$?

PROBLEM 2.7.5. Let $X_i = (Y_i, Z_i)$ and $\hat{\theta} = (Y_1 + \cdots + Y_n)/(Z_1 + \cdots + Z_n)$. How could you compute the jackknife efficiently?

PROBLEM 2.7.6. Give an estimator of the traffic intensity in a $G/G/s$ queue. Jackknife it. Under what circumstances will the densities of these two estimators be positive for all loads in a neighborhood of one? Under what circumstances is the queue unstable? When is instability predicted with positive probability? Especially in heavy traffic, what does this say about various estimators of average queue length or waiting time?

CHAPTER 3
Output Analysis

3.1. Introduction

In Section 1.6 we asked:

(1) How do we get good estimates of some measure of performance?
(2) How do we get good estimates of the goodness of these estimates?

To answer (1), we must answer (2). To answer (2), we consider the two-way layout below:

	F: finite horizon	I: infinite horizon
A: absolute performance		
R: relative performance		

Our measure of performance, which we estimate using output from a simulation, falls in one of the four empty boxes above. In this chapter we discuss statistical analyses that attempt to quantify the goodness of our estimates. The letters below are keyed to the above layout.

A: All subsequent sections, except Section 3.3.2, are pertinent to this problem. Example: Expected wait under a given policy.
R: In several places in Sections 3.2.2, 3.3.2, and 3.4 we focus directly on relative performance. Example: Expected difference in waits between two given policies.

F: See Section 3.2. The performance measure reflects behavior only up to a certain (possibly random) time. Its dependence, if any, on subsequent behavior is indirect, perhaps via end conditions. Example 1.4.1 illustrates this.

I: See Sections 3.1.1, 3.3, 3.4, 3.5, 3.7, and 3.8. Here we suppose that, as the system runs longer and longer, a limiting (steady-state) distribution independent of the initial state exists for the random variables of interest and that the performance measure depends only on this distribution. This implies that the observations have no long-term trend over time. Loosely speaking, a central limit theorem applies if the observations have finite variances and widely-spaced observations are asymptotically iid. Example: Expected wait in a stationary $M/M/1$ queue where the mean service time is less than the mean spacing between successive arrivals.

To our two-way layout, we add a third factor: fixed sample size versus sequential sampling (§§3.1.2 and 3.3). Problem 1.9.15 previews a subtlety intrinsic to sequential sampling. In practice one does not know when sequential procedures give strictly legitimate confidence intervals. For a given amount of raw data, the more manipulation and checking sequential procedures do, the more suspect they are. Infinite-horizon procedures do more than finite-horizon procedures, as the reader will see.

This chapter discusses statistical inference for the model. The further inference step to the real system, discussed in general terms in Chapter 1, seems beyond systematic treatment without invoking "prior" distributions, eschewed here.

3.1.1. Finite-Horizon Versus Steady-State Performance

A finite horizon can be a time when a system:

(a) reaches a point where plans are reassessed (e.g., a new budget year) or where further extrapolation of initially available data is unreliable;
(b) empties out and, possibly after a delay, starts all over;
(c) accomplishes a specified task;
(d) completes a cycle with respect to exogenous influences (e.g., a telephone exchange running from one low traffic period to the next such as 3 A.M. one day to 3 A.M. the next).

For case (a), rolling the horizon a fixed distance ahead while adapting to new information often works well. Especially in cases (b) and (c), the corresponding time may be random. During this "lifetime" the system may well not reach any kind of steady state. Most real systems fall under the finite-horizon classification exactly for this reason: they do not achieve a steady state, except possibly a degenerate one (consider the system composed of dodo birds and passenger pigeons). To handle case (d) we generally assume that the cycles are stochastically identical *if their respective starting conditions are*; see Problem 2.4.2. This assumption often seems less arbitrary than others but it cannot be checked except by intuition because it asserts something about the future. For cases (b) and (d) one can alternatively view the system as having an infinite horizon provided that its cycles are iid and their average length is not so long that problems associated with case (a) surface. Our banking example 1.4.1 illustrates case (b); trying to recast it by stretching the horizon to infinity seems highly unnatural.

Do we view the system that we are simulating as having a finite or an infinite horizon? The answer strongly affects output analysis. Fairly conventional statistical methods apply to finite-horizon simulations. Serial correlation in any given simulation run is irrelevant and, deliberately, the effect of initial conditions is fully reflected in the analysis. For steady-state simulations, the opposite holds. There we try to estimate infinite-horizon performance from a finite-horizon sample. One reason why this is difficult is that now serial correlation cannot be ignored—far from it. Another is that now we assume that a steady state exists and that we want to estimate steady-state performance; therefore, we do not want the initial conditions to affect our estimate. The proper statistical analysis of steady-state simulations is still an unresolved problem. Thus, even though this situation occurs less frequently in practice, it is by far the most common situation in the literature of simulation output analysis. Before reviewing that literature, we discuss the appropriateness of steady-state simulations to real-world problems.

Any real system has a finite life. When installing a new inventory system with an expected life of 5 years, we may think that this should be treated as a steady-state system. Should we learn, however, that the simulation must run for a simulated 50 years to get accurate steady-state results, then the approach of running ten simulations, each of 5 years, may seem more appropriate. We questioned the stationarity assumption in Section 1.2.2. If the demand is nonstationary, an infinite horizon is (even more) artificial. Shoehorning nonstationary demand into an infinite-horizon model has bad effects. Obvious approaches are to replace a nonstationary demand rate by: (i) an average demand rate; or (ii) the peak demand rate. The former leads to excessive lost sales or backlogging, the latter to excessive inventory and excess warehouse or production capacity. These considerations have counterparts in other contexts.

A related point is that after looking at a precise, single-number performance prediction obtained from a simulation of 50 years, a user may mistakenly conclude that this single number will also accurately predict performance over the next 5 years. Neuts (1977b) and Whitt (1983) forcefully point out that such single numbers do not reflect (possibly large-amplitude) oscillations of the system. Over short time periods such oscillations are notorious. That they can also be an important factor in the "steady state" is less commonly appreciated. Whitt (1983) also shows that the "relaxation" time to (nearly) achieve a steady state can be large. These facts show that exclusive focus on things like mean queue length or mean waiting time, steady state or otherwise, is naive. The ubiquity of this approach in the literature does not justify its automatic adoption.

The key assumption underlying steady-state analysis is that the initial state should have vanishingly small effect on the performance prediction. When studying a specific real system, this assumption may be just the wrong one to make. For example, we may be considering installing a new inventory policy because currently inventories are out of control. Any policy which is installed must start from this state. On the other hand, steady-state simulations may be appropriate when no natural initial condition exists or when many events occur in a short time.

Problems of statistical analysis arise partly because of the requirement that the initial state should have no effect. Because of this, the obvious simulation approach is to make exactly one simulation run, which incidentally should be very long. Thus, there is only one initialization bias to contend with rather than N of them if N independent runs are made. Kelton (1980) and Kelton and Law (1984), though well aware of this argument, nevertheless advocate the multiple-replication approach coupled with a procedure they propose for deleting observations (apparently) not representative of the steady state. We are skeptical about their rationale.

The other source of problems in steady-state statistical analysis is serial correlation. Questions to answer are:

(1) How do we get accurate variance estimates in the face of this correlation? Equivalently, how do we get an acceptably short confidence interval that covers the true parameter with a specified (high) probability?
(2) How many of the early observations should be discarded to reduce the effect of the initial state? How do we know when we have reached steady state? When can we act as if we have reached steady state?

If just one run is available, some people maintain that the answer to the second part of question (2) is *never*. We do not dispute this position. The last part of question (2) has independent interest if we want to modify the way we control the system when it seems to have reached steady state. A good answer would tell us when to switch to a control rule appropriate to steady state. Near the end of Section 3.3.1 we describe a heuristic proposal and cite others, but no definite answer seems possible. Faced with this situation, one could try comparing several composite policies that incorporate heuristic rules for "recognizing" attainment of steady state by simulating their respective performances over a long, but finite, horizon.

If instead we decide to carry out a steady-state analysis as such, question (1) is often more vexing than question (2). Attempting to resolve question (1) usually causes us to run a simulation long enough that the initial state does not greatly affect the final result, thus reducing question (2) to secondary importance. We tackle question (1) first and return to question (2) later, assuming for the time being that no observations are discarded.

3.1.2. Fixed Sample Size Versus Sequential Sampling

The expected performance of a stochastic system is usually presented in terms of a confidence interval, e.g., "We are 90% confident that the expected profit per week is in the interval (5100, 6300)." The width of the confidence interval is loosely referred to as the precision (for a specified probability that the interval covers the true parameter). In any study of a stochastic system, one has the choice of:

(a) Specifying the number of observations beforehand and hoping that the resulting precision is satisfactory.
(b) Specifying the precision beforehand and hoping that the number of observations required is not unacceptably large.

Approach (a) uses a *fixed sample size* while (b) is usually called *sequential sampling*. In (b) the values of the observations in hand determine whether additional observations must be collected.

If a valid confidence interval based on the standard assumptions of normality and independence is to be obtained, the following three statements must be true:

(i) Sufficient observations have been collected so that the central limit theorem allows us to ignore any nonnormality in the data.
(ii) Any lack of independence among samples (steady-state simulations only) is overcome by some mechanism.
(iii) Bias caused by the initial conditions (steady state only) is overcome by some mechanism.

Classical sequential sampling theory is concerned principally with condition (i). Chow and Robbins (1965) prove a fundamental theorem: in the face of nonnormality, confidence intervals computed according to the normality assumption are asymptotically valid as bounds on their lengths are forced to zero provided that the observations are independent; see also Starr (1966).

Because the data must be examined to assess its degree of nonnormality or dependence, one is almost forced to use a sequential procedure. Law and Kelton (1984) test five different fixed sample size procedures (using replication, batch means, autoregressive methods, spectral analysis, and regeneration cycles) commonly proposed for simulations of steady-state performance. They conclude that none can be relied upon to perform uniformly well for all models and sample sizes. The models used in this empirical study were simple queueing models.

3.2. Analysis of Finite-Horizon Performance

Statements about expected performance usually consist of a single number or "point" estimate of expected performance, and a confidence interval or an estimate of mean square error for this point estimate. Call the point estimate $\overline{X}$. Almost always the confidence interval is presented in the form

$$\overline{X} \pm ks(\overline{X}),$$

where $s(\overline{X})$ is an estimate of the standard deviation of $\overline{X}$ and k is a parameter which depends upon how much confidence we want to have in the interval estimate. The value of k is usually determined on the assumption that $\overline{X}$ is normally distributed.

To construct this interval, we have to

(a) determine an $\overline{X}$ which estimates the true performance with little or no bias and, preferably, with small variance;
(b) check that $\overline{X}$ is approximately normally distributed;
(c) find a good way of estimating the standard deviation of $\overline{X}$;
(d) determine suitably two related figures, k, and the number of simulations to perform.

Though for simplicity most of the material in this section is presented in the context of finite-horizon performance, much of it is relevant in steady-state settings as well.

3.2.1. Absolute Performance Estimation

Classical statistical methods apply in this situation with essentially no refinements except for the use of variance reduction techniques (Chapter 2). These techniques do not affect the mechanics for analyzing the output. Let

N = number of simulations performed. (This is usually to be determined, perhaps sequentially.)

X_i = the observation of simulated performance obtained from the ith simulation, $i = 1, 2, \ldots, N$;

$$\overline{X} = \sum_{i=1}^{N} X_i/N;$$

$$s^2 = \sum_{i=1}^{N} (X_i - \overline{X})^2/(N - 1).$$

[Avoid the algebraically equivalent but numerically inaccurate formula $(\sum X_i^2 - \overline{X}^2 N)/(N - 1)$.]

α = probability that the confidence interval statement is true.

If the X_i are iid, then $\overline{X}$ is an unbiased estimate of the expected performance. An unbiased estimate of $\mathrm{Var}[\overline{X}]$ is s^2/N. If in addition the X_i are normally distributed, then a confidence interval for the expected performance is

$$\overline{X} \pm k\sqrt{s^2/N},$$

where k is the ordinate giving a probability $(1 - \alpha)/2$ in each tail of the t-distribution with $N - 1$ degrees of freedom.

Normality Assumption. We rarely know that the X_i are normal. Frequently, we observe that their distribution is decidedly nonnormal. Even then the central limit theorem suggests that $\overline{X}$ probably is approximately normal. As a rule of thumb, it is reasonable to assume that $\overline{X}$ is normal if the following conditions hold:

(a) the X_i are approximately symmetrically distributed about the mean;
(b) there are no outliers more than 3s away from the mean;
(c) $N > 6$;
(d) $\alpha \leq 0.8$.

Consideration (a) is qualitative; it is best investigated using a package such as IDA [see Ling and Roberts (1980)] to examine the histogram either directly or via a normal probability plot. We usually want significantly more than 80% confidence, thus violating condition (d). If any of these conditions is not satisfied, then check normality more rigorously using the test described in Appendix A. If the normality assumption is rejected when applying the test to raw data, there is still hope if the sample is large. Partition the data in groups of 30, say, and then test the hypothesis that the batch means are normally distributed.

Ratio Estimates. As an example, consider estimating the fraction of calls lost by a telephone answering system. On day 1 suppose that 1000 calls out of 3000 were lost while on day 2 no calls were lost out of 2000. Further, suppose the days were independent. Two possible estimates of the fraction lost are

(i) $1000/(3000 + 2000) = \frac{1}{5}$; and
(ii) $(1000/3000 + 0/2000)/2 = \frac{1}{6}$.

Method (i) is appropriate if one asks: "Over a long period of time, what fraction of calls is lost?", while method (ii) is appropriate if the question is: "If we randomly chose a day, what is the expected fraction of calls occurring that day which are lost?". If method (i) is appropriate, consult Section 3.3.2 on regenerative methods.

3.2.2. Relative Performance Estimation

When comparing k policies, often the natural thing to do is to choose the one (say A) with the best empirical performance. More analysis is needed when:

(i) another policy (say B) is already in use, so we will switch to A only if the difference in performance seems significant;
(ii) it is not certain that policy in fact makes a difference.

In case (ii) we need $k(k-1)/2$ *simultaneous* confidence intervals to compare all pairs of policies. Case (i) is perhaps more subtle. It is incorrect to compute a confidence interval for E[performance (A) − performance (B)] as if the identity of A were known in advance. Instead we must in principle compute $k-1$ joint confidence intervals, comparing all policies with B, though intervals not involving the comparison of A and B need not be explicitly constructed. We give details below.

For a given fixed set of random outcomes of turns of a roulette wheel, if a sufficient number of policies (betting rules) are tried, eventually one will be found which appears to be significantly better than the majority for this particular set of outcomes, but which in the long run is neither better nor worse than any other. This pitfall can be avoided as follows. Consider drawing p policies randomly from an urn, all having the same expected performance. We then ask: "What is the probability that the performance differences in a finite number of experiments for these p policies are as 'extreme' as those which we observed for our p real-world policies?". A low probability supports the view that, in the real world, policy makes a difference.

In an honest roulette game, policy makes no difference. Outside casinos, however, we often know beforehand that some of the k policies will perform better than B; the only questions are which is best and by how much. Use the multiple-comparison methods discussed here if the results of your analysis are likely to be vigorously challenged; start with the tentative assumption that policy does not make a difference.

There is a large literature on multiple comparisons, but most of it assumes that empirical performances are independent across policies. However (as discussed in §2.1) properly-designed simulations use common random numbers across policies: independence does *not* hold. With dependent performances there are two valid approaches:

(i) A classical approach uses the Bonferroni inequality:

   P[at least one of c possibly-dependent statements is false]
   $\leq \sum_{s=1}^{c} P$[statement s is false].

   (This is easily proved by induction on c.)
(ii) If the performances are drawn from a multivariate normal distribution, then a joint confidence ellipsoid for all $k(k-1)/2$ policy differences (among other "contrasts") can be constructed using methods of Miller (1981) and Scheffé [(1959), §§3.4, 3.5].

Both methods are conservative. We discuss only method (i), because it does not depend on distributional assumptions, it is easier to explain, and it is usually less conservative than method (ii). Multistage screening heuristics to knock out grossly inferior alternatives seem appealing in practice but are not discussed here.

PROBLEM 3.2.1. Show that if each of c statements taken alone is true with probability $1 - (1 - \alpha)/c$, then

$$P[\text{all } c \text{ statements true}] \geq \alpha.$$

PROBLEM 3.2.2. Show that the following recipe for the construction of c individual confidence intervals produces a joint confidence interval that covers all c parameters with probability at least α. Let

$$c = \text{the number of individual confidence intervals,}$$

$$N = \text{the number of observations of each policy,}$$

$$X_{qr} = \text{output of run } r \text{ for policy } q,$$

$$\bar{X}_i = \sum_r X_{ir}/N,$$

$$s_{ij}^2 = \frac{1}{N-1} \sum_{r=1}^{N} [X_{ir} - X_{jr} - (\bar{X}_i - \bar{X}_j)]^2,$$

$$Z = \sqrt{N}[(\bar{X}_i - \bar{X}_j) - E(\bar{X}_i - \bar{X}_j)]/s_{ij},$$

and let θ_1 and θ_2 be such that

$$P[\theta_1 \leq Z \leq \theta_2] = 1 - (1 - \alpha)/c.$$

Then the individual confidence intervals for $E(\bar{X}_i - \bar{X}_j)$ have the form

$$(\bar{X}_i - \bar{X}_j - \theta_2 s_{ij}/\sqrt{N}, \bar{X}_i - \bar{X}_j - \theta_1 s_{ij}/\sqrt{N}).$$

PROBLEM 3.2.3 (continuation). Suppose that Z has a t-distribution with $N - 1$ degrees of freedom. (Under mild conditions related to those for the central limit theorem, this holds approximately.) Show that it works to set $\theta_2 = -(1 - \alpha)/2c$ percentile of t_{N-1} (a positive number), $\theta_1 = -\theta_2$.

PROBLEM 3.2.4. Why does using common random numbers tend to reduce s_{ij}^2 and hence the length of the confidence intervals?

3.3. Analysis of Steady-State Performance

Suppose we are interested in the average performance of a system after it has been running a very long time. We certainly cannot simulate forever. Our only hope is that the system achieves some approximation to steady-state or long-run behavior during a moderate length of (simulated) time and that extrapolation from this performance to predicted long-run performance is reliable.

Before discussing methods for analyzing long-run simulations, we must carefully define what constitutes an observation. For comparison, consider a problem with finite horizon H. For run i, the *sole* observation has the form

$$X_i = \int_0^H f(Y(t))\, dt + \sum_{j=1}^m g(Y(t_j)),$$

where Y is a stochastic process, f and g are cost functions, the t_j are event times, and $m = \max\{i : t_i \le H\}$.

EXAMPLE 3.3.1. In a queueing situation we might have:

$f =$ cost per unit of waiting time,

$g =$ cost per balking customer,

$Y(t) = [Y_0(t), Y_1(t), Y_2(t), \ldots]$,

$Y_0(t) =$ number of balks since the last event time, i.e., in the half-open interval $(\max\{i : t_i < t\}, t]$,

$$Y_j(t) = \begin{cases} 1, & \text{if customer } j \text{ is waiting at time } t, \\ 0, & \text{otherwise.} \end{cases}$$

For infinite horizons, the definition of an observation depends on the method used for statistical analysis. We consider four ways:

- discretized, equal time spacing (accumulate customer "costs" jointly over each successive fixed interval);
- transaction (accumulate customer "costs" individually over the run);
- regenerative (accumulate "costs" for all customers jointly over successive regeneration cycles);
- asynchronous (accumulate "costs" for all customers jointly between successive state changes).

Fox and Glynn (1986b) analyze asynchronous observations and construct corresponding confidence intervals. This procedure simplifies data collection and, intuitively, facilitates estimation of correlation structure. Fox and Glynn (1986b) conclude that it beats the corresponding one based on equally spaced discretized observations. As pointed out there and in our Section 3.5, any transaction average can be estimated indirectly via a time average with the same mean square error, provided in some cases that specified control variates are used. For other reasons, including the scrambling associated with transaction observations discussed below, Fox and Glynn conclude that this indirect estimation beats direct use of transaction data. Section 3.3.2 discusses the pros and cons of the regenerative method. When regeneration cycles are stochastically short, the regenerative method is probably the best, on balance. When (typically in complex systems) they are long, we believe that using asynchronous observations with the Fox–Glynn method wins. The variance constants needed by that method can be estimated using batch means

(Section 3.3.1), spectral analysis (Section 3.3.3), or autoregressive methods (Section 3.3.4); see, however, Problem 3.6.10. We now discuss the four observation types in the order indicated above.

For discretized observations with spacing Δ between successive observations, a (temporally aggregated) observation is

$$X_i = \left[\int_{i\Delta}^{(i+1)\Delta} f(Y(t))\, dt + \sum_{j=c}^{d} g(Y(t_j)) \right] \bigg/ \Delta,$$

where

$$c = \min\{ j: t_j \geq i\Delta \},$$
$$d = \max\{ j: t_j < (i+1)\Delta \}.$$

The integral could be approximated as $f(Y(i\Delta))\Delta$, but this can produce bias even when the process is aperiodic.

We can choose to observe a transaction-based process provided that overall cost functions f and g separate as

$$f(Y(t)) = f_1(Y_1(t)) + f_2(Y_2(t)) + \cdots$$

and

$$g(Y(t_j)) = g_1(Y_1(t_j)) + g_2(Y_2(t_j)) + \cdots,$$

where Y_i is a process representing only transaction i and f_i and g_i are cost functions for transaction i. Think of a transaction as a customer or job. The observation of transaction i is

$$X_i = \int_0^T f_i(Y_i(t))\, dt + \sum_{j=1}^{e} g_i(Y_i(t_j)),$$

where T is the time when the simulation terminates (the run length) and

$$e = \max\{ j: t_j \leq T \}.$$

EXAMPLE 3.3.2. We want to estimate mean waiting time in a system. Let

$$Y_i(t) = \begin{cases} 1, & \text{if customer } i \text{ is in the system at time } t, \\ 0, & \text{otherwise,} \end{cases}$$

$$f_i(1) = 1,$$
$$f_i(0) = 0,$$
$$g_i \equiv 0.$$

In Section 3.4 we discuss proper analysis of transaction-based observations. There is no really suitable way to associate a time with a transaction-based observation, though perhaps either the arrival time or the departure time is less unnatural than other choices. In general, transactions can overtake other transactions: first in does not imply first out. Generally, a time series is amenable to analysis if and only if the serial correlation of two arbitrary terms X_i and X_j falls off rapidly as the *index* spacing $|i - j|$ increases, showing no sporadic peaks. Usually, discretized observations equally spaced *in time* behave like this, but—especially in non-FIFO systems—transaction-based observations often do not. Even when a series of transaction-based observations is covariance stationary, moving from the time domain to the transaction domain can produce a messy covariance structure. It scrambles past, present, and future and cuts connections between time spacing and index spacing. This may make it awkward to estimate the variance of the average of trans-action-based observations; more on this later. Testing output analysis methods on first-in, first-out systems is not stringent enough.

✳   For the *regenerative* method detailed in Section 3.3.2, we define an observation differently. The system regenerates at times $\tau_1, \tau_2, \tau_3, \ldots,$ where $\tau_1 = 0$. At a regeneration point, only the current state matters for predicting the future—the past is irrelevant. In other words, at regeneration points the system has no memory. For example, arrival instants to an empty system are regeneration points for the standard textbook queues. (One can invent situations where this is not true: suppose, for example, that service times increase stochastically with arrival spacings.) Cycle i is the interval $[\tau_i, \tau_{i+1})$. For cycle i, the *sole* observation is (X_i, T_i) where

$$T_i = \tau_{i+1} - \tau_i,$$

$$X_i = \int_{\tau_i}^{\tau_{i+1}} f(Y(t))\, dt + \sum_{j=a}^{b} g(Y(t_j)),$$

$$a = \min\{j: t_j \geq \tau_i\},$$

$$b = \max\{j: t_j < \tau_{i+1}\}.$$

Often X_i simply sums transaction costs over a cycle.

For equally spaced and transaction observations, the point estimator with N observations is the empirical mean

$$\overline{X} = \frac{1}{N} \sum_{i=1}^{N} X_i$$

and the confidence interval has the form

$$(\overline{X} - s, \overline{X} + s),$$

where s depends on the method. For example, a naive method uses $s = (\theta s_X)/\sqrt{N}$ where s_X^2 is the sample variance of the X_i and θ is an appropriate percentage point of the t-distribution with $N - 1$ degrees of freedom.

As an estimate of $\text{Var}[\bar{X}]$, s_X^2/N is often badly biased because it neglects correlation among the X_i's. We will discuss three other estimates of $\text{Var}[\bar{X}]$ using batch means, and spectral, and autoregressive approaches. These estimates are biased, but generally less than s_X^2/N.

If the integrals in the definition of X_i are computed exactly, then $\bar{X}$ is an unbiased estimate of expected performance. With equally spaced observations, even at high sampling frequency (Δ small) $\bar{X}$ does not necessarily have low variance. The reason is that serial correlation between successive observations generally increases as Δ decreases. Grassmann (1981) and references therein consider this phenomenon further.

The observations must usually be made *synchronously* as above or (typically not as good) at artificial clock chimes generated by a Poisson process (see §4.9, especially Problems 4.9.9, 4.9.10, and 4.9.20). Otherwise, it is likely that $\bar{X}$ is grossly biased except for Markovian systems and the following class of (possibly non-Markovian) systems:

> Let $X(t)$ be the number of customers in the system at time t. In any queue in equilibrium with Poisson arrivals where $X(\cdot)$ changes by unit jumps, observations can be taken at arrival or departure instants without injecting bias. Here the queue lengths at a random point in time, left by a departure, and seen by an arrival are stochastically identical.

See Cooper [(1981), pp. 185–188] or Kleinrock [(1975), p. 176 and p. 236 (Problem 5.6)]. The statistics associated with synchronous and asynchronous observations have different distributions. Halfin (1982) demonstrates certain asymptotic optimality properties of $\bar{X}$ as an estimator.

For the regenerative method our estimator is not $\bar{X}$. We estimate

$$r = E[X_1]/E[T_1]$$

(see §3.7 on renewal theory) by

$$R = \sum_{i=1}^{N} X_i / \sum_{j=1}^{N} T_j,$$

but generally

$$E[R] \neq r.$$

It turns out that more bias is introduced in constructing a confidence interval for r because to get the length of the interval we again estimate r by R. So the regenerative method gets around bias due to influence of initial conditions and serial correlation, but generally creates other bias.

An asynchronous observation free of spacing parameters like Δ is

$$X_i = \int_{S_i}^{S_{i+1}} f(Y(t))\, dt + g(Y(S_i)),$$

where S_k is the kth event time with $S_0 = 0$. Set

$$\alpha_i = S_{i+1} - S_i,$$

$$\bar{X}_n = (1/n) \sum_{i=0}^{n-1} X_i,$$

$$\bar{\alpha}_n = (1/n) \sum_{i=0}^{n-1} \alpha_i,$$

$$r = EX_i/E\alpha_i,$$

$$r_n = \bar{X}_n/\bar{\alpha}_n.$$

Under weak regularity conditions, Fox and Glynn (1986b) prove that

$$r_n \to r \qquad \text{with probability one}$$

and

$$n^{1/2}(r_n - r) \Rightarrow \sigma N(0, 1)/E\alpha_0,$$

where r is the expected steady-state average reward and σ^2 is a variance constant specified in Problem 3.6.15. This brief discussion barely gives the flavor of the Fox–Glynn paper. Consult it for several other results and proofs.

Schruben (1983b) presents an output-analysis method based on "standardized time series" (STS) keyed to equally spaced observations, though it may work with transaction observations. Section 3.8 discusses STS.

At the end of Section 3.3.1 we discuss initialization bias. That discussion applies to output analysis using batch means, spectral analysis, or autoregressive methods applied to equally spaced, transaction, or asynchronous observations. It is irrelevant to regenerative methods (Section 3.3.2). Section 3.8.2 discusses transients in the context of STS.

3.3.1. Batch Means

We have a sequence of possibly correlated observations $X_1, X_2, \ldots, X_m$, e.g., of inventory level in a simulated series of days. Define

$$\bar{X} = \sum_{i=1}^{m} X_i/m,$$

$$s_X^2 = \sum_{i=1}^{m} (X_i - \bar{X})^2/(m - 1),$$

$\quad b = $ a batch size specified by the statistical analyst. (We give guidelines for choosing b.)

$n = m/b$. (We assume that m is a multiple of b to make n an integer.)

$$Y_i = \sum_{1+(i-1)b}^{ib} X_i/b, \text{ i.e., the average of the } i\text{th batch of observa-}$$
tions of size b,

$$\bar{Y} = \sum_{i=1}^{n} Y_i/n,$$

$$s_Y^2 = \sum_{i=1}^{n} (Y_i - \bar{Y})/(n-1).$$

We are interested in $\mu \equiv \lim_{i \to \infty} E(X_i)$ and in $\text{Var}[\bar{X}]$. The batch means method estimates μ with the statistic $\bar{X}$. The definitions imply that $\bar{Y} = \bar{X}$. An unbiased estimate of $\text{Var}[\bar{X}]$ if the X_i are independent is s_X^2/m. More generally:

$$\text{Var}[\bar{X}] = \text{Var}[X_i]/m + (2/m^2) \sum_{i=1}^{m-1} \sum_{j=i+1}^{m} \text{Cov}(X_i, X_j).$$

To assume independence is to disregard the covariance term. Usually the X_i are positively correlated, so disregarding the covariance term underestimates $\text{Var}[\bar{X}]$.

Another unbiased estimate under the independence assumption is s_Y^2/n. If the X_i are correlated, then both estimates are probably biased. Which is more biased? Generally it is s_X^2/m, because the correlation of the Y_i tends to be less than that of the X_i.

The batch means method in its fixed sample-size form follows.

(1) Form the n batched observations Y_i.
(2) Calculate $\bar{Y}$ and s_Y^2.
(3) Use $\bar{Y}$ as a point estimate of μ.
(4) Use $\bar{Y} \pm k s_Y/\sqrt{n}$ as the interval estimate of μ, where k is the appropriate ordinate from the t-distribution with $n-1$ degrees of freedom.

Conway's (1963) example uses transaction-based observations. From a simulation of $m = 9,300$ jobs through a jobshop, the sample variance of the shop time was $s_X^2 = 5602.136$. The estimate of $\text{Var}[\bar{X}]$ based on this is $s_X^2/m = 0.6024$. When the same data were batched into groups of size $b = 100$ (so $n = 93$), the sample variance of the Y_i was $s_Y^2 = 469.98$. The batched data give an estimate of $\text{Var}[\bar{X}]$ as $s_Y^2/n = 5.054$. Apparently, naive use of the unbatched data seriously underestimates $\text{Var}[\bar{X}]$ in this example. Maybe batch means overestimates $\text{Var}[\bar{X}]$ but this seems less likely intuitively.

There is no safe way to choose a batch size b without looking at the data. We must choose b sufficiently large that the correlation among the X_i is small while at the same time maintaining a reasonable value of n. This drives us to use

A Sequential Batch Means Procedure. The sequential procedure we describe here is due to Law and Carson (1979). It is very *ad hoc* but has been more extensively tested than most other batch means based procedures, passing with moderate success. It has no rigorous support.

The batch means method tries to reduce serial correlation by batching observations. Thus, the procedure monitors the apparent serial correlation and (in step 5 below) calls for a larger batch size (and hence further sampling) if the apparent serial correlation is "large." Standard estimates of serial correlation are seriously biased when based upon a small number of observations. The procedure therefore partitions (in step 2) the observations into a large number of small batches when estimating serial correlation. If this estimated serial correlation is either:

(i) positive but moderately small and (step 3) halving the number of batches does not increase estimated serial correlation; or
(ii) the estimate in step 2 is negative;

then it is assumed that the true serial correlation among a small number of large batches is negligible and (in step 4) a confidence interval based on these large batches is constructed. If this interval is short enough, the procedure accepts it and stops; otherwise, it collects more observations and returns to step 2.

The specific implementation suggested by Law and Carson follows. Although they recommend values of certain parameters, do not use these slavishly (see below).

(1) Set $i \leftarrow 1$, $n_0 \leftarrow 600$, $n_1 \leftarrow 800$. Collect n_1 observations. [Comment: n_0 is needed for step 5.]
(2) Partition the n_i observations into 400 batches, each of size $n_i/400$. If the estimated serial correlation in these batches is greater than 0.4, go to (5). If it is negative, go to (4).
(3) Partition the n_i observations in 200 batches, each of size $n_i/200$. If the estimated serial correlation among these 200 batches is greater than that among the 400 batches, go to (5).
(4) Partition the n_i observations into 40 batches, each of size $n_i/40$. Construct a *nominal* p-percent confidence interval assuming the 40 batches are independent and ignoring that the batches were constructed sequentially. If the interval is acceptably small relative to the current value of $\overline{X}$ (say half-width/$\overline{X} < \gamma$), stop; otherwise go to (5).
(5) Set $i \leftarrow i + 1$, $n_i \leftarrow 2n_{i-2}$. Collect $n_i - n_{i-1}$ additional observations and go to (2).

Law and Carson suggest using jackknifing (§2.7) to compute the estimates of serial correlation. With $X_1, X_2, \ldots, X_n$ the sequence of batch means, let

$$\bar{X} = \sum_{i=1}^{n} X_i/n,$$

$$\rho = \sum_{j=1}^{n-1} (X_j - \bar{X})(X_{j+1} - \bar{X}) \bigg/ \sum_{j=1}^{n} (X_j - \bar{X})^2,$$

$$k = n/2 \quad (n \text{ is assumed even}),$$

$$\bar{X}_1 = \sum_{j=1}^{k} X_j/k,$$

$$\rho_1 = \sum_{j=1}^{k-1} (X_j - \bar{X}_1)(X_{j+1} - \bar{X}_1) \bigg/ \sum_{j=1}^{k} (X_j - \bar{X}_1)^2,$$

$$\bar{X}_2 = \sum_{j=k+1}^{n} X_j/k,$$

$$\rho_2 = \sum_{j=k+1}^{n-1} (X_j - \bar{X}_2)(X_{j+1} - \bar{X}_2) \bigg/ \sum_{j=k+1}^{n} (X_j - \bar{X}_2)^2.$$

The jackknifed estimate of serial correlation is

$$\hat{\rho} = 2\rho - (\rho_1 + \rho_2)/2.$$

Jackknifing easily generalizes to more than two groups. There is little empirical data supporting any specific number of groups.

Schmeiser's (1982) results seem to indicate that changing the 40 in step 4 to 30 would be an improvement.

Law and Carson [see also Law and Kelton (1982)] applied their sequential procedure to fourteen different queueing and inventory systems for which the steady-state expected performance is known exactly. The goal of each simulation was to produce a 90% confidence interval for which the ratio of the confidence interval half width to the point estimate was less than a specified tolerance γ; thus, $p = 90$ in step 4. This experiment was repeated 100 times for each system, and the number of times that the approximate interval included the known expected performance was tabulated. For true 90% confidence intervals the expected number of times this would occur is 90 out of 100. A brief summary of performance appears below.

Fraction of time approximate 90% confidence interval included true value		
Confidence interval width ratio, γ	0.075	0.15
Best coverage	0.98	0.98
Average coverage	0.883	0.872
Worst coverage	0.84	0.76

Before attempting to extrapolate this performance to other systems, recall the caveat about extrapolation from Section 1.6. One might expect similar coverage for queueing and inventory systems, but extrapolation to other systems is based mainly on optimism. The arbitrary constants in the method may be a cause for some skepticism in this regard. Law and Carson do not indicate to what extent these constants were "tuned" to the systems tested.

Initialization Bias. Batch means, spectral, and autoregressive methods have to contend with initialization bias because for most systems the steady state is approached only asymptotically, if at all. Which observations belong to the transient phase (and should therefore be discarded) is an ill-defined question. All the suggested rules of thumb are merely heuristics. They drop all observations up to a certain point but differ on the specification of that point. Fewer observations are thrown away using one long run than with several shorter ones. The effective sample size is the number of retained observations. All the statistical methods for analyzing the retained observations ignore the fact that the sample was obtained by selection from a larger set of observations; this adds another heuristic element to these procedures.

When using a single run, Conway (1963) suggests the following rule for dropping observations: discard initial observations until the first one left is neither the maximum nor the minimum of the remaining observations. Though there is little theoretical support for this, no obviously better rule has been suggested since Conway's paper. If the batch means method is used, then apparently Conway intended that this rule be applied to the batch means and not to the original observations. Gafarian *et al.* (1978) show that if Conway's rule (in fact, any of about half a dozen other rules) is applied to individual observations, it will not work well.

The current state of a nontrivial system fluctuates over time, making recognition of equilibrium difficult. Batching tends to smooth high-frequency fluctuations, perhaps revealing more clearly a trend—if any—in that run. Other techniques, such as moving averages or low-pass filters, also tend to smooth. Recent diverse proposals to delimit the transient period are given in the *Proceedings of the Winter Simulation Conference* [(1981), pp. 113–120]. Whether this can be done reasonably by observing a single run is controversial. Section 3.8.2 briefly describes Schruben's (1986) way to handle initialization bias.

For certain inventory systems which follow a base-stock ordering policy, one can show that steady state is achieved in a finite length of time. Consider the following simple version. Each period an order is placed to bring the amount on hand plus the amount on order up to a constant S. It takes exactly L periods for an order to arrive and become on-hand inventory. It is easy to show that, if initially the amount on hand plus the amount on order is less than S, then after L periods the system state distribution is the steady-state distribution. Similar arguments can be applied to inventory systems which follow the so-called (s, S) ordering policy.

3.3.2. Regenerative Methods

The principal difficulty with batch means and similar methods is that the observations generally are dependent. Sometimes we can get around this by a different kind of batching, this time with (astutely-selected) random batch sizes. Many systems have regeneration or renewal points. At such a point future behavior is independent of past behavior. An example is the arrival of a job to an empty queueing system: given that the system is empty, the waiting time of the new arrival and all future arrivals is usually independent of how the system reached the empty state. Thus, if each cycle between successive renewal points corresponds to an observation, then (voilà!) one has independent observations and bias due to initial conditions disappears. Restrain your delight, however: we will show shortly that these gains are not free.

EXAMPLE 3.3.3. Assemblies arrive every 40 seconds at a station in an assembly line. The time required to process an assembly is a random variable with mean 30 seconds. If a particular outcome of this random variable is greater than 40 seconds, then incoming assemblies accumulate until the team can catch up. Plotting the number of assemblies in process or waiting gives a graph like Figure 3.3.1. The arrows identify the regeneration points. The time between successive arrows is called a busy cycle. What happens during a specific busy cycle is independent of what happened during other busy cycles.

PROBLEM 3.3.1. Why is it not true in general that the instants when the system empties are also renewal points? When *is* this true?

Cox and Smith [(1961), p. 136] sketch the germ of the regenerative method for simulation, though they do not push their development to the point of constructing confidence intervals. Donald Iglehart and his collaborators are mainly responsible for taking this method to its current state.

Bias of Regenerative Method Estimates. A price paid for independence under the regenerative method is that typical performance estimates are biased. Bias was also a fundamental problem with methods like batch means. The problem is still confronting us: it is simply less apparent.

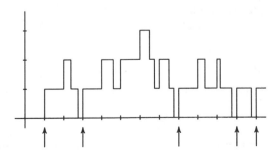

Figure 3.3.1. Regeneration points for a system.

Bias arises under the regenerative method because our estimators are almost always ratios of random variables. In general, the expectation of a ratio is not equal to the ratio of the expectations. To illustrate this, take an oversimplified version of the assembly line queueing problem just described. For the ith busy cycle there are two data of interest: W_i, the total wait incurred at the station during the ith cycle, and T_i, the length of the ith cycle. We are interested in the expected wait per unit time, i.e., in $E[W_i]/E[T_i]$ as Problem 3.7.8 shows. Based on observing N cycles, we have the estimator

$$\sum_{i=1}^{N} W_i \Big/ \sum_{i=1}^{N} T_i,$$

which can be rewritten as

$$\left(\sum_{i=1}^{N} W_i/N\right)\Big/ \sum_{i=1}^{N} (T_i/N).$$

Under reasonable assumptions, this approaches $E[W_i]/E[T_i]$ with probability one; for any finite N, however,

$$E\left[\sum_{i=1}^{N} W_i \Big/ \sum_{i=1}^{N} T_i\right] \neq E[W_i]/E[T_i],$$

except in trivial cases.

PROBLEM 3.3.2. Construct a simple example illustrating this assertion for the case $N = 1$.

The limit above was taken over cycle-completion epochs. If instead we take the limit in continuous time, we get the same answer. This plausible but subtle result follows from renewal theory, as the solution to Problem 3.7.8 shows.

Estimating Formulas for the Regenerative Method. Regenerative estimates are almost always ratios of random variables. For example, in an inventory simulation we may be interested in the ratio (total holding and shortage costs during a cycle)/(cycle length); in a queueing simulation we may be interested in the ratio (total waiting time during a cycle)/(number of jobs processed during a cycle).

We use X_i and Y_i to denote the numerator and the denominator, respectively, of the ratio associated with the ith cycle. We are interested in an estimate of $E(X_i)/E(Y_i)$ and in a confidence interval for that estimate.

Let N be the number of cycles simulated, and define $\bar{X}$, $\bar{Y}$, s_X^2, and s_Y^2 in the usual way. Let

$$\bar{Z} = \bar{X}/\bar{Y},$$

$$s_{XY}^2 = \sum_{i=1}^{N} (X_i - \bar{X})(Y_i - \bar{Y})/(N - 1),$$

$$s^2 = s_X^2 - 2\bar{Z}s_{XY}^2 + \bar{Z}^2 s_Y^2,$$

$k =$ appropriate ordinate from the normal distribution.

Then $\bar{Z}$ is the (biased) point estimate of $E(X_i)/E(Y_i)$ and a confidence interval for $E(X_i)/E(Y_i)$ is

$$\bar{Z} \pm ks/\bar{Y}\sqrt{N},$$

which we justify in Section 3.7.

Iglehart (1978) describes alternative ways of constructing confidence intervals. Based on empirical comparisons, he recommends for small sample sizes only a jackknifed version of the above confidence interval:

$$\tilde{Z} \pm k\tilde{s}/\bar{Y}\sqrt{N},$$

where $\tilde{Z}$ is $\bar{Z}$ jackknifed and

$$\tilde{s}^2 = s_X^2 - 2\tilde{Z}s_{XY}^2 + (\tilde{Z})^2 s_Y^2.$$

Computing $\tilde{Z} \pm k\tilde{s}/\bar{Y}\sqrt{N}$ requires only slightly more computation than $\bar{Z} \pm ks/\bar{Y}\sqrt{N}$ provided that the jackknife's structure is exploited. Once again recall the caveat of Section 1.6 regarding extrapolation. Do not take as a fact that the jackknifed estimator is best for systems beyond those Iglehart studies.

Sequential Procedures for the Regenerative Method. Using a sequential procedure for the regenerative method is particularly recommended because the point estimate has an unknown bias and because the normality assumption is particularly questionable when working with a ratio of random variables. The general form of such a procedure follows.

(1) Take an initial sample.
(2) Construct a confidence interval. If its length is greater than desired, go to (5).
(3) [Optional]: Test the point estimator for bias and for approximate normality. If it does not pass, go to (5).
(4) Use the confidence interval just calculated and stop.
(5) Collect further samples and go to (2).

A conventional sequential sampling procedure is obtained if step (3) is deleted. With step (3) included, no analog of the Chow–Robbins (1965) theorem applies and so the (nominal) coverage probability of the confidence interval has less theoretical support. Even so, this step seems worthwhile intuitively.

Specific implementations of steps 1, 3, and 5 follow.

Step 1: Choose an initial batch size of n cycles. Let h_i be any suitable point estimate calculated from the ith batch that is (presumably) less biased than the classical estimate $\bar{X}/\bar{Y}$ calculated from that batch. Fishman (1977) recommends the Tin estimator (described along with several others in

Iglehart [(1978), p. 58]), but in view of Iglehart's empirical comparisons (in another setting) the jackknife estimator provisionally seems better. Let

$$\bar{Z}_i = \sum_{q=1}^{i} h_q/i,$$

$$s_i^2 = \sum_{q=1}^{i} (h_q - \bar{Z}_i)^2/(i-1).$$

Here $\bar{Z}_i$ is the average of i jackknife estimators $h_1, \ldots, h_i$ each of which in turn is the average of n pseudovalues. Fishman (1977) suggests collecting N batches of n cycles each where N is the smallest odd integer such that $ks_N/\sqrt{N} < \delta/2$. Here, as above, k is an appropriate ordinate from the normal distribution, and δ is the desired width of the confidence interval.

Step 3: Fishman suggests calculating the classical point estimator Z_c based on all the Nn observations. Accept Z_c as unbiased if

$$|Z_c - \bar{Z}_N|/\bar{Z}_N < a,$$

where a is a suitable tolerance. To test for normality, apply the test described in Appendix A to the h_i.

Step 5: Should either test fail, increase the batch size n and return to step (2). Law and Kelton suggest simply doubling n.

Law and Kelton (1982) perform extensive empirical tests of this sequential procedure and find it slightly more accurate than the batch means method in general. The method has no rigorous justification and so the standard caveat about extrapolation beyond the cases tested applies. Heidelberger and Lewis (1981) give a more elaborate, but not necessarily better, alternative method.

Lavenberg and Sauer (1977) show that with step 3 deleted the (nominal) coverage probabilities are asymptotically exact when the length tolerance in step 2 approaches zero. Their empirical studies indicate that coverage is satisfactory if stopping occurs when the length of the confidence interval divided by the current value of $\bar{Z}$ is less than 0.05, giving a relative length of at most 5%.

Pros and Cons of the Regenerative Method. Advantages:

(1) There is no initial transient problem.
(2) It produces independent observations.
(3) It is simple to understand.
(4) No prior parameters are needed, except for the sequential version with its step 3 included.
(5) It is asymptotically exact, with the same proviso.

Disadvantages:

(1) The cycle lengths are unknown beforehand. If we simulate for some (possibly random) number of cycles, so is the simulation length—making it hard to plan a simulation: even using fixed sample sizes rather than sequential sampling, the computer time needed is a random variable. If we terminate at a "fixed" time rather than after some number of cycles, then: (i) the number of cycles is random; and (ii) on average, the last cycle is longer than the first (see Problem 3.7.6). Note that (i) means it is difficult to control the precision of our estimates and (ii) means that additional bias may be introduced, depending how we handle the last cycle; if we simulate to the end of the last cycle, then we have to contend with a random simulation length with no prior bound. For further discussion, see Section 3.7.

(2) The cycle lengths may be embarrassingly long. Indeed just to observe a few cycles may require a very long simulation. If a system consists of several unrelated components, the expected time between successive visits to a given state of the entire system may be very long even though each sub-system reaches steady-state behavior quickly. A number of authors try to alleviate this difficulty by using regeneration *sets*; for example, see Gunther and Wolff (1980) and the references therein. When long cycle lengths indicate slow approach to steady state, *any* reasonable way to estimate steady-state performance accurately requires (perhaps impracticably) long run lengths. When long cycle lengths and fast convergence to steady state coexist, the regenerative method requires longer run lengths than others to achieve a given statistical precision. We have no statistical test for such coexistence.

(3) Effective synchronization when estimating performance via pairwise comparisons of cycles generated using common random numbers or antithetically is difficult. One can use paired comparisons effectively only if the expected time interval between *simultaneous* renewals of the systems in question is finite. A condition generally necessary for this is that the simulation operate in discrete time. When the discrete-time conversion method of Problem 2.6.5 is applicable, Heidelberger and Iglehart (1979) study the use of common random numbers for estimating relative performance. One expects the average time between simultaneous renewals to be an order of magnitude larger than that for individual renewals, aggravating the problem of observing a reasonable number of cycles in a reasonable length of time.

(4) The relevant mean and variance estimators are biased, though consistent.

(5) Because of (4) and the involvement of ratio estimators, confidence intervals based on (asymptotically valid) normality are more suspect for finite sample sizes than is usual in classical applied statistics.

(6) Detecting cycle boundaries sometimes may require a lot of checking. Getting cycle observations occasionally requires evaluation of a complex formula.

Recent Work on Regenerative Methods. The research literature is active. Glynn (1982a, b, c) shows that "virtually any well-behaved simulation has a regenerative-type structure," contrary to some folklore. He shows how to find and exploit such structure, but it is too soon to assess whether his approach significantly alleviates the disadvantages listed earlier.

Somewhat arbitrarily, we cite just five other papers: Iglehart and Shedler (1983), Iglehart and Stone (1982), Iglehart and Lewis (1979), and Glynn (1982d, e). Probably in a few years someone will write a monograph covering this literature and intervening developments, necessarily involving more sophisticated mathematics than we assume the average reader of this book knows.

3.3.3. Spectral Analysis Methods

With the usual notation, let

$$\overline{X} = \sum_{t=1}^{n} X_t/n,$$

$$R_s = E[(X_t - \mu)(X_{t-s} - \mu)] \quad \text{for } s = 0, 1, \ldots, n,$$

$$\overline{R}_s = \sum_{t=s+1}^{n} (X_t - \overline{X})(X_{t-s} - \overline{X})/(n - s).$$

Then under the assumption that $\{X_t\}$ is covariance stationary (i.e., R_s does not depend on t)

$$\text{Var}[\overline{X}] = \left[R_0 + 2 \sum_{s=1}^{n-1} (1 - s/n)R_s \right]/n.$$

PROBLEM 3.3.3. Prove this.

An obvious temptation is to replace R_s by $\overline{R}_s$ in the above formula and use it to estimate $\text{Var}[\overline{X}]$. Drawback: When the X_t are dependent, then $\overline{R}_s$ is generally a very biased estimate of R_s, with particularly high mean square error for s close to n. In fact, the truncated estimator

$$V_m = \left[\overline{R}_0 + 2 \sum_{s=1}^{m} (1 - s/n)\overline{R}_s \right]/n$$

is generally a better estimate of $\text{Var}[\overline{X}]$ for $m \ll n$ than it is when $m = n - 1$.

A plot of the expected value of this expression as a function of m is typically similar to Figure 3.3.2. Even though the successive X_i are usually positively correlated, the $\overline{R}_s$ generally are negative for some large s. Intuitively, if a steady state exists, then any movement away from it must sooner or later be followed by a movement in the opposite direction: crudely, what goes up must come down. One normally expects the magnitudes of the serial correlations to be damped as the lag s increases, though $|\overline{R}_s|$ generally goes down

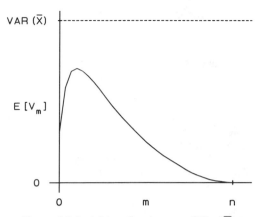

Figure 3.3.2. A biased estimator of Var($\overline{X}$).

significantly slower than $|R_s|$. At very long lags, empirical serial correlation usually is spurious. As s increases, $\overline{R}_s$ is subject to more sampling fluctuation because it is the average of fewer terms.

PROBLEM 3.3.4. Show that $E[V_n] = 0$ and hence that not all $\overline{R}_s$ are positive. If some $\overline{R}_s$ in the set $\{\overline{R}_0, \ldots, \overline{R}_n\}$ is positive, show that there is also a negative element. (Hint: Show more. Rearrange $n^2 V_n$ as

$$\sum \sum (X_i - \overline{X})(X_j - \overline{X}) = 0.)$$

This problem supports the claim that what goes up must come down.

Because the $\overline{R}_s$ have high mean square errors, consider

$$\tilde{V}_m = \left\{ \overline{R}_0 + 2 \sum_{s=1}^{m} w_m(s)[1 - s/n]\overline{R}_s \right\}/n.$$

Except for the weighting function $w_m(s)$, this is V_m. One suggested weighting is the so-called Tukey–Hanning *window*:

$$w_m(s) = [1 + \cos(\pi s/m)]/2.$$

We sketch how this window arises. The *spectral density f* is the Fourier transform of the covariance sequence:

$$f(\lambda) = \frac{1}{2\pi} \sum_{s=-\infty}^{\infty} R_s \cos(\lambda s), \qquad -\pi \le \lambda \le \pi.$$

Comparing the expressions for $f(0)$ and V_m, we see that V_m is a natural (truncated) estimate of $2\pi f(0)/n$. It turns out to be a bad estimate; see below.

Proviso. This transform to the frequency domain appears natural only when using discretized observations equally spaced in time.

We get an unweighted estimate $\hat{f}(\lambda)$ of $f(\lambda)$ by modifying V_m in the obvious way.

PROBLEM 3.3.5. How? [Hint: Multiply $\bar{R}_s$ by $\cos(\lambda s)$ and the resulting expression by $n/2\pi$.]

PROBLEM 3.3.6. Smooth $\hat{f}$ as $\tilde{f}(\lambda) = \hat{f}(\lambda)/2 + \hat{f}(\lambda - \pi/m)/4 + \hat{f}(\lambda + \pi/m)/4$. Show that the Tukey–Hanning window drops out. [Hint: $e^{i\theta} = \cos\theta + i\sin\theta$. Replacing $\cos(\lambda s)$ in the expression for $\hat{f}(\lambda)$ by $e^{i\lambda s}$ and then taking real parts changes nothing.] Typically f peaks sharply at zero; when this happens, show that $f(0) > \tilde{f}(0)$. [So-called "pre-whitening" compensated by "recoloring" may attenuate this bias. Most books on spectral analysis of time series discuss this technique.] For readers who know what "aliasing" means: discuss the relation between observation spacing and bias in $\tilde{f}(\lambda)$.

Brillinger (1981) and Hannan [(1970), pp. 277–278], for example, show that such smoothing converts a bad estimator $\hat{f}(\lambda)$ of $f(\lambda)$ to a better one.

PROBLEM 3.3.7. Suggest a more elaborate smoothing that accounts for more than just the immediately "neighboring" frequencies. Under what circumstances would it pay to use a *fast Fourier transform* (FFT)? Estimate the work relative to a straightforward evaluation. (Strang [(1986), §5.5] and Press *et al.* [(1986), Chapter 12] neatly unravel the FFT.)

From Brillinger [(1981), Theorem 4.4.1] or Hannan [(1970), Theorem 11, p. 221], for example,

$$\sqrt{n}(\bar{X} - \mu) \Rightarrow N(0, 2\pi f(0)).$$

Under reasonable conditions (see below) $\tilde{f}(\lambda)$ is a consistent estimator of $f(\lambda)$. Hence $n\tilde{V}_m$ is a consistent estimator of $n \operatorname{Var}[\bar{X}] = 2\pi f(0)$. The continuous mapping theorem [Billingsley (1968)] then gives

$$\bar{X} \Rightarrow N(\mu, \tilde{V}_m).$$

Using this asymptotic distribution to construct confidence intervals is rather crude for moderate sample sizes. We discuss a (partly heuristic) improvement below.

Most books on spectral analysis of time series show that asymptotically $\tilde{V}_m$, properly normalized, is a weighted sum of independent chi-square variates. They argue that the distribution of this sum is well approximated by a chi-square distribution with $8n/3m$ degrees of freedom. However, we know no reference that quantifies the error in this approximation. For consistent estimation as $n \to \infty$ of $f(\lambda)$ and hence $\operatorname{Var}[\bar{X}]$, we want $m \to \infty$ but $m/n \to 0$. There is no clear guidance on the choice of m. Taking m proportional to $\log n$ or $\sqrt{n}$ works. Using $m \ll n/4$ seems common practice.

In elementary statistics we learn that if $\bar{X}$ has a normal distribution and if a consistent estimator s^2 of its variance:

(i) is independent of $\bar{X}$, and
(ii) has a chi-square distribution,

then $(\overline{X} - \mu)/s$ has a t-distribution. In our case, $\tilde{V}_m$ satisfies (i) because it is a weighted sum of $m + 1$ $\overline{R}_i$'s and each $\overline{R}_i$ is independent of $\overline{X}$, because adding the same constant to each X_j would leave all $\overline{R}_i$'s unchanged. Bearing in mind that $\tilde{V}_m$ also satisfies (ii), asymptotically and approximately, and the continuous mapping theorem (see below), it is plausible that $(\overline{X} - \mu)/\sqrt{\tilde{V}_m}$ has approximately a t-distribution with $8n/3m$ degrees of freedom for n and m large but $m = o(n)$. The corollary to the continuous mapping theorem, proved in Billingsley [(1968), pp. 30–31] for example, that we are using is:

$$\text{if } h \text{ is continuous and } Z_n \Rightarrow Z,$$

$$\text{then } h(Z_n) \Rightarrow h(Z).$$

To construct a confidence interval, one can pretend (perhaps cavalierly) that this approximate t-distribution is exact. Law and Kelton (1979) replace any negative $\overline{R}_s$ by 0, though for typical cases this seems to us to have a good rationale only for small and moderate s. With this change, they find that the confidence intervals obtained are just as accurate as those given by the simple batch means method. Duket and Pritsker (1978), on the other hand, find spectral methods unsatisfactory. Wahba (1980) and Heidelberger and Welch (1981a) aptly criticize spectral-window approaches. They present alternatives based on fits to the logarithm of the periodogram. Heidelberger and Welch (1981a, b) propose a regression fit, invoking a large number of asymptotic approximations. They calculate their periodogram using batch means as input and recommend a heuristic sequential procedure that stops when a confidence interval is acceptably short. Heidelberger and Welch (1982) combine their approach with Schruben's model for initialization bias to get a heuristic, composite, sequential procedure for running simulations. Because the indicated coverage probabilities are only approximate, they checked their procedure empirically on a number of examples and got good results. Despite this, we believe that spectral methods need further study before they can be widely used with confidence. For sophisticated users, they may eventually dominate batch means methods but it seems premature to make a definite comparison now.

3.3.4. Autoregressive Methods

By assuming more about the underlying stochastic process, we obtain a method that looks less sophisticated than spectral approaches and requires less computations. Because we suggest that unsophisticated users not try either approach, their relative sophistication becomes unimportant. Except in trivial simulations, the work needed to do the output analysis by any of the methods in this book is small compared to the work need to generate the observations. So we believe that the relative amount of computation required by spectral and autoregressive methods is unimportant too. The key disadvantage of autoregressive methods is that (strong) assumptions are made which often do not hold. The circumstances under which these assumptions hold (only) to a first approximation and when autoregressive methods are then robust have not been well explored.

To apply the method, one looks for an approximation to $\text{Var}[\overline{X}]$ that depends only on certain parameters $p, b_1, \ldots, b_p$ that have to be estimated from the data. The idea is to transform the original dependent observations X_s to new observations

$$Y_t = \sum_{s=0}^{p} b_s(X_{t-s} - \mu), \qquad t = p + 1, \ldots, n,$$

$$\mu = E[X_s], \qquad b_0 = 1,$$

where the Y_t are *supposedly* iid. Here p is the order of the representation, assumed known until further notice, and the b_s are constants. We see that p is roughly analogous to the parameter m of the spectral approach. For finite p, Priestley [(1981), §7.8] indicates the *approximate* correspondence between autogressive and spectral estimation. Here, the (assumed) linearity in the representation of Y_t has no counterpart in the spectral approach. If p were infinite, there would be a correspondence via Wold's theorem (e.g., see Priestley [(1981), p. 222]; this seems academic. The iid condition, the obvious fact that $E[Y_t] = 0$, and trivial manipulation of $E[Y_t Y_{t+j}]$ give

$$\sum_{r=0}^{p} \sum_{s=0}^{p} b_r b_s R_{j+r-s} = 0, \qquad j = 1, 2, \ldots, p.$$

PROBLEM 3.3.8. Prove this.

PROBLEM 3.3.9. Show that

$$\sum_{r=0}^{p} b_r R_{r-h} = 0, \qquad h \neq 0,$$

satisfies the equation displayed above. [Remark: This also follows immediately from the (superfluous) condition $E[Y_t X_s] = 0$, $t > s$.] Assuming nothing more about the b_i's, show that

$$R_r = -\sum_{s=1}^{p} b_s R_{r-s}, \qquad r \neq 0.$$

What does this recursion imply about efficient computation of the R_s? How can it be used to estimate the b_s? (Hint: Vary p.)

PROBLEM 3.3.10. Show that

$$\sum_{s=0}^{p} b_s R_s = \sigma_Y^2.$$

(Hint: $\sigma_Y^2 = E[Y_t^2]$. Write this as a double summation and use the previous problem to show that all but one of the sums vanish.)

Let $B = b_0 + \cdots + b_p$. Assuming that a steady state exists, as n gets large

$$\overline{Y} \approx B(\overline{X} - \mu),$$

$$\text{Var}[\overline{Y}] \approx B^2 \, \text{Var}[\overline{X}],$$

$$\text{Var}[\overline{X}] \approx \sigma_Y^2/nB^2.$$

PROBLEM 3.3.11. Justify these three relations. (Hint: Interchange summations somewhere, divide by n, and argue that

$$(1/n)\sum_{t=1}^{n} X_{t-s} \approx \overline{X}, \qquad s = 0, 1, \ldots, p.)$$

When $\overline{R}_s$ replaces R_s, we get the estimates $\hat{b}_s$, $\hat{\sigma}_Y^2$, $\hat{B}$, and $\hat{\sigma}_X^2$. Because the Y_t are assumed to be iid, the central limit theorem gives

$$\sqrt{n}\,\overline{Y}/\sigma_Y \Rightarrow N(0, 1).$$

From this it is natural to conjecture that $\sqrt{n}(\overline{X} - \mu)/\hat{\sigma}_X$ has a t-distribution asymptotically. Fishman [(1978), p. 252] outlines an argument, supported by literature citations, that this conjecture is correct:

$$\sqrt{n}(\overline{X} - \mu)/\hat{\sigma}_X \Rightarrow t(d),$$

where d is estimated by

$$\hat{d} = n\hat{B}/2 \sum_{j=0}^{p} (p - 2j)\hat{b}_j.$$

PROBLEM 3.3.12. Find a confidence interval for $\overline{X}$. Do you expect $\hat{d}$ to be positive? Why? Can $\hat{d}$ be negative? Why? Suggest what to do if $\hat{d}$ is negative.

PROBLEM 3.3.13. Does $E[Y_t Y_s] = 0$, $t > s$ imply $E[Y_t X_s] = 0$, $t > s$?

Now drop the assumption that p is known. When the tentative order of the autoregressive representation is i, replace $\hat{\sigma}_Y^2$ by $\hat{\sigma}_{iY}^2$ for clarity. Denote by $F(c, d)$ the F-distribution with (c, d) degrees of freedom. Hannan [(1970), p. 336] shows that

If the correct autoregressive order is at most q, then the statistic

$$[(n - q)/(q - i)][(\hat{\sigma}_{iY}^2/\hat{\sigma}_{qY}^2) - 1] \Rightarrow F(q - i, n - q)$$

as n increases, $i < q$.

To select p, successively test this statistic for $i = 0, 1, 2, \ldots$, stopping when i is accepted. The final i is p.

Remark. An analogous procedure in ordinary stepwise multiple regression tests successive ratios of empirical residual variances. There the regression coefficients are estimated by least squares and the test statistics have corresponding F-distributions.

The approximations involved in using this autoregressive method follow.

(1) The desired autoregressive representation is assumed to exist when in fact it may not.
(2) The correct order p must be estimated from the data.
(3) The expression for σ_X^2 based on σ_Y^2 is approximate for finite n.
(4) The determination of the b_s is based on replacing R_s by an estimate $\bar{R}_s$ which is biased for finite n.
(5) The sequential nature of the procedure means the distribution of the test statistic can only be "guesstimated."

Given these approximations, the method requires extensive empirical testing before it can be used with confidence.

In an empirical study (using rather small samples) based on the simple single-server queue and on a computer time-sharing model, Law and Kelton (1979) find that the autoregressive method gives no more accurate confidence intervals than the simple batch means method with the number of batches equal to 5. Our usual caveat about extrapolation applies.

Caution. Autoregressive methods, while perhaps sometimes appropriate with discretized observations, appear unlikely to be so with transaction-based observations. In the former case, the regression coefficients ought to vary with the gaps between observations. Though we hesitate to replace "unlikely" with "never," anyone with a sequence of transaction-based observations should carefully check results in the autoregression literature for explicit or implicit assumptions about the nature of the observations before postulating that they are generated autoregressively. Combining these remarks with our categorical condemnation of using conventional spectral methods with transaction-based observations, we conclude that such observations generally can be handled properly only by batch means methods—pending development of modified spectral or autoregressive methods.

3.3.5. Recommendations

If you have decided that a steady-state simulation is appropriate, you must then choose which output analysis method to use. Because fixed sample sizes give random statistical precision, always use a sequential method. If you can't afford the computation cost, don't undertake the simulation. If the regeneration cycles tend to be short, choose a regnerative method. Otherwise use asynchronous observations and choose a batch-means method or a spectral-analysis method; if you don't understand the latter, use the former.

No completely satisfactory method for analyzing the output of steady-state simulations has yet been devised. Certainly, no consensus has been reached regarding the relative merits of existing methods. The above recommendations are lukewarm, but we have nothing better to suggest—other than perhaps to reassess whether a steady-state simulation should be carried out.

To readers not expert in time-series analysis, spectral and autoregressive methods may seem nonintuitive. Keep them out of reach of unsophisticated users. As Woolsey and Swanson [(1975), p. 169] say: "People would rather live with a problem they cannot solve than accept a solution they cannot understand." In the mathematical programming context, this is perhaps an overstatement to the extent that the user accepts on faith the claimed optimality of a solution with respect to a model that he does understand. There at least he can check that the solution is feasible, makes sense in the context of the model, and is at least as good as he can "eyeball." No analog exists for analysis of simulation output. The unsophisticated user has no way to check that a statistical package works correctly and no way to properly evaluate any (perhaps hidden) assumptions or approximations that it makes. When using or writing your own statistical analysis routines, stick to methods that you have a good "feel" for.

3.4. Analysis of Transaction-Based Performance

If despite our recommendation to use asynchronous observations you want to use transaction-based observations, we suggest:

First, batch the observations as in Section 3.3.1. The performance of successive transactions, e.g., completion times of successive jobs in a queue, is usually so highly correlated that simple variances calculated from individual transactions are meaningless.

Second, maintain the identity of a transaction from one policy to another when possible. Use a distinct random number stream for each transaction type and, when feasible, generate all attributes for a transaction as it enters the system. One attribute should be an ID number. When doing comparisons, the nth batch in policy A should be composed of the same transactions (those with the same ID numbers) as the nth batch in policy B. This generally does not happen, for example, if a batch is constructed by taking transactions with successive completions: transactions may leave the system in different orders for different policies. Likewise, batch via ID numbers when using antithetic variates.

PROBLEM 3.4.1. Review the three synchronization steps stated in the introduction to Chapter 2. Give a queueing network example in which it is not feasible to implement step 3. Indicate to what extent implementing step 2 by giving each node its own service-time generator and one-step transition generator synchronizes the simulation. In this example, does it seem harder to synchronize the simulation with transaction-based observations than with discretized or regenerative-based observations? What combination of observation type and analysis method would you recommend? Justify your answer.

PROBLEM 3.4.2. Combine the sequential batch means procedure of Section 3.3.1 with a method that tries to attenuate initialization bias by looking at a sequence of batch means. Now suppose that we are estimating relative performance by batch means and that the sign of the initialization bias is the same in each case. Why does this make bias attenuation easier?

PROBLEM 3.4.3. Explain why there is in general no unbiased way to handle transactions still in the system when the simulation run ends. (Hint: See Problem 3.7.6.)

3.5. Indirect Estimation via $r = \lambda s$

Call the expected steady-state time-average reward r; whether based on equally spaced, asynchronous, or regenerative observations, we get the same value. Likewise, call the expected steady-state transaction-average reward s. According to Heyman and Sobel [(1982), §11.3], $r = \lambda s$ under weak regularity conditions. This generalizes the familiar $L = \lambda W$ where L is the average number in system, W is the average time in system, and λ is the arrival rate. Glynn and Whitt (1986a,b) give (apparently) the last word on generalizing $L = \lambda W$, including cumulative input processes that may not be representable as integrals or sums and central-limit-theorem versions of $r = \lambda s$. Use bars to denote sample averages. Abbreviate mean square error as **mse**. There are two cases:

(i) λ unknown:
  Glynn and Whitt (1987) show in a precise sense that

$$\bar{r} \approx \bar{\lambda}\bar{s}$$

  and hence that

$$\mathbf{mse}(\bar{r}) \approx \mathbf{mse}(\bar{\lambda}\bar{s}),$$
$$\mathbf{mse}(\bar{s}) \approx \mathbf{mse}(\bar{r}/\bar{\lambda}).$$

(ii) λ known:
  Let

$$\alpha = -\operatorname{Cov}(\bar{r}, \bar{\lambda})/\operatorname{Var} \bar{\lambda},$$
$$\beta = -\lambda \operatorname{Cov}(\bar{r}, \bar{\lambda})/\operatorname{Var} \bar{\lambda},$$

with $\hat{\alpha}$ and $\hat{\beta}$ being consistent estimators of α and β respectively. Glynn and Whitt (1987) show in a precise sense that

$$\bar{r} + \hat{\alpha}(\bar{\lambda} - \lambda) \approx \lambda\bar{s} + \hat{\beta}(\bar{\lambda} - \lambda),$$

and hence that

$$\mathbf{mse}[\bar{r} + \hat{\alpha}(\bar{\lambda} - \lambda)] \approx \mathbf{mse}[\lambda\hat{s} + \hat{\beta}(\bar{\lambda} - \lambda)],$$
$$\mathbf{mse}[\bar{s} + \lambda^{-1}\hat{\beta}(\bar{\lambda} - \lambda)] \approx \mathbf{mse}[\lambda^{-1}\bar{r} + \lambda^{-1}\hat{\alpha}(\bar{\lambda} - \lambda)].$$

Thus, in each case $\bar{\lambda} - \lambda$ is a control variate and the formulas for α and β follow from the answer to (the simple) Problem 2.3.3. Roughly speaking, the contribution of Glynn and Whitt is to show that this control variate is equally effective for $\bar{r}$ and $\lambda\bar{s}$. To motivate this control variate, we write

$$\bar{\lambda}\bar{s} = \lambda\bar{s} + (\bar{\lambda} - \lambda)\bar{s}$$
$$= \lambda\bar{s} + (\bar{\lambda} - \lambda)s + (\bar{\lambda} - \lambda)(\bar{s} - s)$$
$$\approx \lambda\bar{s} + s(\bar{\lambda} - \lambda)$$

and now recognize, in view of Problem 2.3.3, that s is not in general the optimal control-variate weight. By expanding nonlinear control functions in a Taylor series, Glynn and Whitt (1985) go on to show that, asymptotically, only the linear term matters.

These results supersede those in pioneering papers: Law (1975), Carson (1978), (1980), and Carson and Law (1980).

PROBLEM 3.5.1. Can conditional Monte Carlo (Section 2.6) be applied with equal ease to time-based and transaction-based observations?

PROBLEM 3.5.2. Discuss the compatibility of the estimators mentioned in this section with common random numbers. Relate your discussion to Problem 2.1.6.

3.6. Problems

PROBLEM 3.6.1. You have collected a sequence of observations of the state of a system by means of simulation. The mean of these observations is used to estimate the mean steady state of the system. If you were told that the individual observations are highly negatively correlated, would you have more or less confidence in your estimate?

PROBLEM 3.6.2. You have run a simulation for 2000 simulated days. The confidence interval calculated for the long run expected profit per month is 1475 ± 40 for a certain fixed coverage probability. You would like to have a confidence interval of width ± 25. How many additional days should be simulated? [Apply (perhaps cavalierly) an implicit variance estimate from the first confidence interval to the construction of the second.]

PROBLEM 3.6.3. You want to use simulation to estimate the expected difference between the time-in-system for the $M/M/1$, SPT (job with shortest processing time first) queue and the same queue with the FIFO discipline. You have available a statistical analysis routine which requires you to group the output statistics into batches of 40 jobs each. The statistic associated with each batch is the total time-in-system for all jobs in the batch. Describe in detail how you would do the batching in each case.

PROBLEM 3.6.4. Let X_{ij} be the performance in the ith simulation of policy j. Define

$$\bar{X}_{rs} = \sum_{i=1}^{N} (X_{ir} - X_{is})/N,$$

$$V_{rs} = \sum_{i=1}^{N} (X_{ir} - X_{is} - \bar{X}_{rs})^2/(N - 1).$$

X_{ir} and X_{is} were based on the same random events. For a particular set of simulations:

$$N = 10,$$
$$\bar{X}_{12} = 11,$$
$$\bar{X}_{13} = 15,$$
$$V_{12} = 9,$$
$$V_{13} = 16.$$

Make a statement about the likelihood that the expected performance of policy 1 is the best of the three. What assumptions underly your statement?

PROBLEM 3.6.5. The number of jobs completed on the ith day in a shop is denoted by C_i. These statistics have been collected for 21 days. The following computations have been made:

$$\bar{C} = \sum_{i=1}^{21} C_i/21 = 45,$$

$$\sum_{i=1}^{21} (\bar{C} - C_i)^2/20 = 36,$$

$$\sum_{i=2}^{21} (\bar{C} - C_i)(\bar{C} - C_{i-1})/20 = 5.$$

Estimate Var$[\bar{C}]$, stating the assumptions you are making.

PROBLEM 3.6.6. Airplanes are processed through a maintenance facility in two steps. The first step includes inspection and standard maintenance, and takes the same time for all planes. In the second step problems uncovered in the first step are fixed. Both steps can only process one plane at a time.

At present a FIFO discipline is used at each step. Because the first step reveals for all practical purposes the processing time in the second step, it is proposed to use a shortest processing time queue discipline in the second step. This will require that a more extensive information system be installed in the first step. However, having a lower average number of airplanes in the system is a substantial benefit.

You are to develop an accurate estimate of the long-run reduction in the average number of airplanes in the maintenance facility under the new procedure, and a confidence interval for this reduction. The arduous task of plowing through historical data and analyzing it has already been performed. It is reasonable to assume that arrivals to the facility occur in a Poisson stream at rate 1.2 planes per week, although in reality arrivals are slightly more regular than Poisson. The first step takes 5 days for all planes. The time for the second step is uniformly distributed over the continuous interval (2.5, 8.5) days.

Design a simulation experiment to accomplish this.

PROBLEM 3.6.7. You have a sequence of 30 observations X_i of the state of a system. $\bar{X}$ is used to estimate the mean steady state. Under which situation:

$$\frac{1}{29} \sum_{i=1}^{29} X_i X_{i+1} < \bar{X}^2 \quad \text{or} \quad \frac{1}{29} \sum_{i=1}^{29} X_i X_{i+1} > \bar{X}^2,$$

would you have greater confidence in the estimate?

PROBLEM 3.6.8. The Head Quarters, a local hair salon, is thinking of going the franchise route. They want you to predict the expected revenue per customer based on 40 days of historical data. They feel that each day is statistically independent of every other. For each day i, $i = 1$ to 40, the statistics R_i, total revenue, and H_i, the number of customers, are available. Assume (rightly or wrongly) that demand is stationary. Specify concisely how you would predict likely revenue per customer.

PROBLEM 3.6.9. As a result of analyzing a complex simulation of an even more complex real system, the following four confidence statements were made:

Statement	Probability of truth
(a) The communications processor will not be overloaded	0.98
(b) Temporary storage will not overflow	0.99
(c) Expected revenue per day is in excess of $35,000	0.94
(d) Ten operators will be sufficient to handle all complaint calls without delays	0.97

What can you say about the probability that all four of the above statements are true?

PROBLEM 3.6.10. List the estimators discussed in this chapter which are biased, including estimators of variance. Which of these estimators are consistent in the sense that the bias goes to zero as the number of observations goes up? For the estimators not provably consistent, suggest modifications that make them consistent. (Note: Comparison of estimators by way of inconsistent estimators of their respective mean square errors would be treacherous indeed. Hint: Key parameters of the estimation procedure, such as batch size, to the sample size.)

PROBLEM 3.6.11. Suggest sequential versions of spectral and autoregressive methods.

PROBLEM 3.6.12. In an arbitrary queueing system, let $g(x)$ be the cost to a customer who spends time x there. To estimate $E[g(X)]$ using transaction-based observations, simply take the sample average of the values $g(X_i)$, where X_i is the time that customer i spends in the system. Show how to estimate $E[g(X)]$ using regenerative-based observations consisting of pairs: (cost for all customers served during a cycle, the number of customers served during that cycle); construct a confidence interval for $E[g(X)]$ using these observations. If g is linear, show how to use queue-length statistics from equally-spaced observations to (indirectly) estimate $E[g(X)]$. If g is nonlinear, can equally-spaced observations be used to estimate $E[g(X)]$? (Hint: Suppose customer n arrives at A_n and leaves at B_n. Put

$$f_n(t) = \begin{cases} g'(t - A_n), & A_n \le t \le B_n, \\ 0, & \text{otherwise,} \end{cases}$$

and call observation i

$$h_i = \int_{i\Delta}^{(i+1)\Delta} \sum f_n(t) \, dt / \Delta.$$

Let H be the average of the h_i's and let λ be the arrival rate. Use Heyman and Stidham (1980) to justify the estimator H/λ.) We do not assume that g is differentiable. Though g' appears above, it is merely a notational convenience. The integral can be evaluated easily in terms of g only.

PROBLEM 3.6.13. Using the notation and results of Problem 1.9.16:

(i) How could you use $\{X^k\}$ to estimate the average value of f on R?
(ii) How would you estimate the goodness of that estimate?
(iii) How is (ii) related to the problem of estimating steady-state performance?

PROBLEM 3.6.14. Every second you win or lose a dollar, each with probability $1/2$. Still measuring in seconds, at times $n \cdot 10^{100} + u_n$ where n runs through the positive integers and u_n is a standard uniform variate you win 10^{10000} dollars. Argue that this example defeats all "reasonable" rules for attenuating initialization bias. Can you find a less pathological example?

PROBLEM 3.6.15. For asynchronous observations, show that the variance constant in the asymptotic normal distribution is

$$\sigma^2 = \text{Var}(X_0 - \alpha_0 r) + 2 \sum_{k=1}^{\infty} \text{Cov}[(X_0 - \alpha_0 r), (X_k - \alpha_k r)].$$

(Hint: You may assume that $n^{1/2}(\bar{X}_n - \alpha_0 r) \Rightarrow \sigma N(0, 1)$.) Give at least three different ways to estimate σ^2 consistently. Simplify the above expression for σ^2 in the following two cases:

(i) when the process is regenerative;
(ii) when the process is stationary with $S_n = n$.

To what does the latter case correspond?

PROBLEM 3.6.16. With what types of observations does spectral analysis seem most compatible? least compatible? Repeat for autoregressive methods and standardized time series.

PROBLEM 3.6.17 (Discrete-time conversion revisited). With asynchronous observations, suppose that Y is a Markov renewal process and $Y(S_0)$, $Y(S_1)$, $\ldots$ is an imbedded Markov chain. Show that

$$E[X_i | Y(S_0), Y(S_1), \ldots] = f(Y(S_i)) E[H(Y(S_i)) | Y(S_i), Y(S_{i+1})] + g(Y(S_i)),$$

where $H(Y(S_i))$ is the holding time in $Y(S_i)$. Relate this to discrete-time conversion.

PROBLEM 3.6.18 (Continuation). When Y is a continuous-time Markov chain and you have found its generator, show that the conditional expected holding time $E[H(Y_i(S_i))|Y(S_i)]$ is trivial to compute. Give several queueing networks with highly structured generators where $E[H(Y(S_i))|Y(S_i)]$ depends on $Y(S_i)$ in a highly structured way. What implications does such a structure have for efficient simulation via a row generator?

PROBLEM 3.6.19 (Continuation). When Y is a $M/M/s$ queue with arrival rate λ and service rate μ, S_i is an arrival or departure epoch, $Y(S_i)$ is the number of customers in the system at S_i^+, show that

$$E[H(Y(S_i))|Y(S_i)] = 1/[\lambda + \mu \min(s, Y(S_i))],$$

and generalize to the case where the queue has finite capacity.

PROBLEM 3.6.20 (Continuation of Problem 3.6.18 for finite horizons). Uniformize Y and condition on the number of jumps and the sequence of states visited up to the horizon. Consider both fixed horizons and random horizons, such as hitting a given set. Compute the corresponding conditional expectations. Is there necessarily a variance reduction? Assess the work. Can you radically reduce the work, perhaps by introducing slight bias? (Reference: Fox and Glynn (1986d).)

PROBLEM 3.6.21 (Overlapping batches). Meketon and Schmeiser (1984) show that the variance estimator V_m of Section 3.3.3, with no window $w_m(s)$, closely corresponds to the estimator they derive for overlapping batches. Does the discussion in Section 3.3.3 suggest that the estimator $\tilde{V}_m$, with a window, beats the estimator based on overlapping batches?

PROBLEM 3.6.22 (Spaced batches). Let n be the number of observations. Alternate long blocks of length $\lfloor n^{3/4} \rfloor$ and short (spacer) blocks of length $\lfloor n^{1/4} \rfloor$. Under mild conditions, the proof of Theorem 27.5 in Billingsley (1986) shows that the long blocks are asymptotically iid as $n \to \infty$ and that the short blocks have negligible effect on the asymptotic distribution $N(0, 1)$ of the normalized sum. Does this suggest a good batching strategy? (Hint: Discard the spacers.) Previous proposals in the literature involve spacers whose total relative size is not asymptotically negligible.

3.7. Renewal Theory Primer

The reader interested in a deeper understanding of the renewal theory underpinnings of the regenerative method may find this section helpful. It probably contains more material than is strictly necessary; skip judiciously if you prefer.

Let $X_1, X_2, X_3, \ldots$ be a sequence of random variables, e.g., cycle lengths in a regenerative simulation. Put

$$S_n = X_1 + X_2 + \cdots + X_n, \qquad S_0 = 0,$$

and define $N(t)$ by

$$N(t) = k \quad \text{if and only if} \quad S_k \leq t \quad \text{and} \quad S_{k+1} > t.$$

Now assume that the X_i are jointly independent with common distribution F, where $F(0-) = 0$ and $F(0+) < 1$. The iid assumption on the X_i is strong. In particular, it implies that the sequence of X_i is stationary in the sense that the distribution of X_i does not depend on the history and, hence, not on the clock time S_{i-1}. With the above assumptions, the sequence of X_i is a renewal process and N is the associated counting process. In the special case where the X_i are exponential variables, N is Poisson.

Examine the iid condition critically in each concrete situation. When it is appropriate, powerful results are available, some cited below. Attempts have been made to weaken the iid hypothesis slightly and still retain enough structure so that many of the standard results remain intact. These alternative hypotheses are difficult to justify or to check in real situations.

Our brief exposition below is based mainly on Crane and Lemoine (1977), Feller (1971), and Karlin and Taylor (1975), Consult these references for proofs, elaborations, sidelights, and many additional results.

Let

$$\mu = E[X_i],$$

$$\sigma^2 = \text{Var}[X_i],$$

$$a(t) = E[N(t)].$$

Proposition 3.7.1. *Asymptotically, $N(t)$ is normally distributed with expectation t/μ and variance $t\sigma^2/\mu^3$.*

This proposition allows us to make inferences about mean and variance of interevent times even though our data may be in the form of mean and variance of number of events in an interval. Brown *et al.* (1981) study several unbiased (simulation) estimators of $a(t)$, "which compete favorably with the naive estimator, $N(t)$."

The next proposition is useful in determining $E[S_{N(t)+1}]$. Let M be an integer-valued random variable that is a function of $X_1, X_2, \dots$. It is a *stopping time* if for $m = 0, 1, 2, \dots$ the event $M = m$ is independent of X_{m+1}, $X_{m+2}, \dots$. Thus, if M is a stopping time, then S_M does not depend on any X_i not explicitly in the sum $X_1 + \cdots + X_M$. To say the same thing slightly differently, the tentative "last" element in the sum *can* determine whether there is another one to be added or not; it *cannot* determine whether it itself is in or not. From the definitions, $N(t)$ is not a stopping time but $N(t) + 1$ is because

$$\{N(t) = n\} \Leftrightarrow \{S_n \leq t, S_{n+1} > t\}$$

and

$$\{N(t) + 1 = n\} \Leftrightarrow \{S_{n-1} \leq t, S_n > t\}.$$

Proposition 3.7.2 (Wald's Equation). *If the X_i are iid and M is a stopping time, then*

$$E\left(\sum_{i=1}^{M} X_i\right) = E(X_1)E(M).$$

PROOF. Define

$$Y_m = \begin{cases} 1, & \text{if } M \geq m, \\ 0, & \text{if } M < m, \end{cases}$$

giving

$$\sum_{m=1}^{M} X_m = \sum_{m=1}^{\infty} X_m Y_m$$

and

$$E\left(\sum_{m=1}^{M} X_m\right) = E\left(\sum_{m=1}^{\infty} X_m Y_m\right) = \sum_{m=1}^{\infty} E(X_m Y_m).$$

Now Y_m does not depend on X_m because $Y_m = 1$ if and only if we continue sampling after successively observing $X_1, \ldots, X_{m-1}$; so

$$E\left(\sum_{m=1}^{M} X_m\right) = \sum_{m=1}^{\infty} E(X_m)E(Y_m) = E(X_1) \sum_{m=1}^{\infty} E(Y_m)$$

$$= E(X_1) \sum_{m=1}^{\infty} P[M \geq m] = E(X_1)E(M),$$

because M is nonnegative. $\square$

Proposition 3.7.3. *If F is not concentrated on numbers of the form $0, c, 2c, 3c, \ldots$, then*

$$-1 \leq a(t) - t/\mu \to (\sigma^2 - \mu^2)/2\mu^2.$$

If the above condition does not hold, then interpret the arrow rigorously in the $(C, 1)$ sense and intuitively to mean roughly that $a(t) - t/\mu$ fluctuates about $(\sigma^2 - \mu^2)/2\mu^2$, but not too wildly. In the displayed expression, the inequality follows from $t \leq S_{N(t)+1}$ and Wald's equation.

The proposition below uses the concept of directly Riemann integrable (DRI) functions. Formal definitions are given in Feller (1971) and Karlin and Taylor (1975). Monotone functions that are integrable in the ordinary sense are DRI. In fact, it is fair to say that functions that are integrable in the usual sense but not DRI are pathological.

Proposition 3.7.4 (Key Renewal Theorem). *Let G be the distribution function of a positive random variable with mean α. Suppose that b is DRI and that B is the solution of the equation*

$$B(t) = b(t) + \int_0^t B(t - x) \, dG(x). \qquad (*)$$

(i) *If G is not concentrated on numbers of the form* 0, c, 2c, 3c, . . . , *then*

$$B(t) \rightarrow \frac{1}{\alpha} \int_0^\infty b(x) \, dx;$$

(ii) *otherwise,*

$$\lim_{n \to \infty} B(k + nc) = \frac{c}{\alpha} \sum_{n=0}^\infty b(k + nc).$$

The equation (∗) often arises by conditioning on X_1. At X_1 the process "renews." One "renewal" equation is

$$a(t) = \int_0^t [1 + a(t - x)] \, dF(x).$$

In this case, $a(t) \rightarrow \infty$ and so the key renewal theorem adds nothing. The problems below illustrate nontrivial applications.

Proposition 3.7.5. *If in* (∗) $G = F$, *then* $B(t) = b(t) + \int_0^t b(t - x) \, da(x)$.

We now give three important definitions:

$$\gamma_t = S_{N(t)+1} - t \quad \text{(excess life, residual life)},$$
$$\delta_t = t - S_{N(t)} \quad \text{(age)},$$
$$\beta_t = \gamma_t + \delta_t \quad \text{(total life)}.$$

PROBLEM 3.7.1. Draw a picture that illustrates these concepts.

PROBLEM 3.7.2. Put $R_z(t) = P[\gamma_t > z]$. Find a renewal equation of the form (∗) that $R_z(t)$ satisfies. [Hint: Consider the equation

$$R_z(t) = \int_0^\infty P[\gamma_t > z | X_1 = x] \, dF(x),$$

which gives $R_z(t) = 1 - F(t + z) + \int_0^t R_z(t - x) \, dF(x)$.]

PROBLEM 3.7.3. Show that, if F is continuous,

$$R_z(t) \rightarrow \frac{1}{\mu} \int_z^\infty [1 - F(y)] \, dy$$

for fixed z. Show that the density of this asymptotic distribution is $[1 - F(z)]/\mu$ and that its mean is $(\mu/2) + \sigma^2/2\mu$.

PROBLEM 3.7.4. Show that

$$\lim_{t \to \infty} P[\beta_t > z] = \frac{1}{\mu} \int_z^\infty y \, dF(y).$$

PROBLEM 3.7.5. Find explicit expressions for $E[\gamma_t]$, $E[\delta_t]$, and $E[\beta_t]$ when N is stationary Poisson.

PROBLEM 3.7.6. Find the mean of the asymptotic distribution of total life. Compare it to μ. Show that in steady state β_t is stochastically larger than X_1 or give a counterexample.

Intuitively, the length of the renewal interval covering a *fixed* point is likely to be longer than average. So, upon reflection, the answer to Problem 3.7.6 is not surprising. Similarly, the expected length of every member of the set of spacings that collectively cover a fixed interval is longer than the unconditional expected spacing length. Successive bus headways, say, sampled in this way are length biased.

By coupling the renewal sequence $X_1, X_2, \ldots$ with an associated sequence $Y_1, Y_2, \ldots$, we obtain a structure that significantly extends the range of application of renewal theory. Assume that $(X_1, Y_1), (X_2, Y_2), (X_3, Y_3), \ldots$ is an iid sequence; however, X_i and Y_i may be dependent. In a queueing system, X_i could be the length of busy period i and Y_i the number of customers served during that period. Now set

$$C(t) = \sum_{i=1}^{N(t)+1} Y_i.$$

Think of $C(t)$ as the "cost" (freely interpreted) up to the end of the current renewal interval.

PROBLEM 3.7.7. Show that

$$E[C(t)] = [a(t) + 1]E[Y_1]. \tag{$**$}$$

(Hint: Show that $N(t) + 1$ is a stopping time for $C(t)$ and apply Wald's equation.) Show that

$$E[C(t)] = E[Y_1] + \int_0^t E[C(t - x)] \, dF(x).$$

Give another proof of $(**)$ using Proposition 3.7.5.

PROBLEM 3.7.8. According to the *elementary renewal theorem*:

$$a(t)/t \to 1/\mu.$$

Use this and $(**)$ to prove

$$E[C(t)/t] \to E[Y_1]/\mu. \tag{$***$}$$

Show that

$$E[Y_{N(t)+1}]/t \to 0;$$

this says that the cost of the last cycle is asymptotically negligible, a nontrivial fact. Let

$$C_1(t) = \sum_{i=1}^{N(t)} Y_i$$

and

$$C_2(t) = \text{cost up to time } t.$$

Show that (∗∗∗) holds with C replaced by C_1 or C_2. {Reference: Ross [(1970), pp. 52–54]}.

To motivate (∗∗∗) heuristically, observe that

$$C(t)/t = \frac{\sum_{i=1}^{N(t)+1} Y_i/[N(t)+1]}{[S_{N(t)+1} - (S_{N(t)+1} - t)]/[N(t)+1]}.$$

Intuitively, it seems plausible that with probability one

$$\sum_{i=1}^{N(t)+1} Y_i/[N(t)+1] \rightarrow E[Y_1],$$

$$S_{N(t)+1}/[N(t)+1] \rightarrow \mu,$$

$$(S_{N(t)+1} - t)/[N(t)+1] \rightarrow 0,$$

but this does *not* follow immediately from the strong law of large numbers. The result that $E[C(t)]/t \rightarrow E[Y_1]/\mu$ is more subtle than at first appears.

We now construct a confidence interval for the expected average long-run cost. Put

$$r = E[Y_1]/\mu,$$

$$D_j = Y_j - rX_j,$$

and note that $E[D_j] = 0$.

Set

$$\bar{Y}_n = (Y_1 + \cdots + Y_n)/n,$$

$$\bar{X}_n = (X_1 + \cdots + X_n)/n,$$

$$\bar{D}_n = (D_1 + \cdots + D_n)/n,$$

$$\delta^2 = \text{Var}[D_1] = E[D_1^2].$$

Assume that $\delta^2 < \infty$. Noting that $\bar{D}_n$ is the average of n iid random variables, the central limit theorem gives

$$\sqrt{n}\,\bar{D}_n/\delta \Rightarrow N(0, 1).$$

With $\hat{r}_n = \bar{Y}_n/\bar{X}_n$, this is equivalent to

$$\lim_{n \to \infty} P[\sqrt{n}\,[\hat{r}_n - r]/(\delta/\bar{X}_n) \le k] = \Phi(k),$$

where Φ is the standard normal cdf. Up to now this is not operational because δ^2 is unknown. In addition, $\hat{r}_n$ is biased and so we may wish to use another estimator that appears to have less bias. Now let $\tilde{r}_n$ and $\tilde{s}_n$ be such that with probability one

$$\sqrt{n}(\tilde{r}_n - \hat{r}_n) \rightarrow 0,$$

$$\tilde{s}_n \rightarrow \delta.$$

Then it follows from the continuous mapping theorem [Billingsley (1968)] that

$$\lim_{n \to \infty} P[\sqrt{n}[\tilde{r}_n - r]/(\tilde{s}_n/\overline{X}_n) \le k] = \Phi(k).$$

PROBLEM 3.7.9. Show how to find *asymptotically* valid confidence intervals for r in terms of $\tilde{r}_n$ and $\tilde{s}_n$.

PROBLEM 3.7.10. Does using the discrete-time conversion method of Problem 2.6.5 always reduce $\mathrm{Var}[\overline{D}_n]$? When does that method tend to shorten confidence intervals constructed as in Problem 3.7.9 without adversely affecting the coverage probabilities?

Note that

$$\delta^2 = \mathrm{Var}[Y_1] - 2r\,\mathrm{Cov}(Y_1, X_1) + r^2\,\mathrm{Var}[X_1]$$

and set

$$\tilde{s}_n^2 = \tilde{s}_y^2 - 2\tilde{r}_n \tilde{s}_{yx} + (\tilde{r}_n)^2 \tilde{s}_x^2,$$

where y is short for Y_1 and x is short for X_1. For example, we may choose $\tilde{r}_n = \hat{r}_n$ and

$$\tilde{s}_y^2 = \hat{s}_y^2 = \frac{1}{n-1} \sum_{j=1}^{n} (Y_j - \overline{Y}_n)^2,$$

$$\tilde{s}_x^2 = \hat{s}_x^2 = \frac{1}{n-1} \sum_{j=1}^{n} (X_j - \overline{X}_n)^2,$$

$$\tilde{s}_{xy} = \hat{s}_{xy} = \frac{1}{n-1} \sum_{j=1}^{n} (Y_j - \overline{Y}_n)(X_j - \overline{X}_n).$$

Another possibility is to choose $\tilde{r}_n$, $\tilde{s}_y$, $\tilde{s}_x$, and $\tilde{s}_{xy}$ equal to jackknifed versions of $\hat{r}_n$, $\hat{s}_y$, $\hat{s}_x$, and $\hat{s}_{xy}$, respectively. With any of these choices, we have an operational method for constructing confidence intervals.

PROBLEM 3.7.11. Which of the above estimators are biased? Show that $\tilde{s}_n^2$ is biased, no matter what combination of the above estimators is used.

The bias in $\tilde{r}_n$ and $\tilde{s}_n^2$ and the correlation of these estimators does not affect the *asymptotic* validity of the confidence intervals constructed in Problem 3.7.9, but it does suggest that in practice such intervals should be regarded skeptically unless n is extremely large.

Implicit in the above confidence intervals is a sampling plan where the number n of regeneration cycles is fixed and the time $t = S_n$ is random. It may be more convenient to fix t and then sample $N(t)$ cycles, where $N(t)$ is random. From Billingsley [(1968), p. 146] we get

$$\sqrt{N(t)}\, \overline{D}_{N(t)}/\delta \Rightarrow N(0, 1)$$

and so the entire foregoing development carries over with n replaced by $N(t)$.

However, an additional bias is thereby introduced. The cycle covering t is stochastically longer than the average (see Problem 3.7.6) and it has been neglected. If we take "asymptotically" literally, all is still well. If there is some flexibility in the computer time budget, continue to the end of the cycle covering t. Meketon and Heidelberger (1982) show that (under mild regularity conditions) the bias of the corresponding point estimate of r is reduced from $O(1/t)$ using $N(t)$ cycles to $O(1/t^2)$ using $N(t) + 1$ cycles. It can be shown that

$$\sqrt{N(t) + 1}\,\bar{D}_{N(t)+1}/\delta \Rightarrow N(0, 1)$$

and, presumably, the convergence rate is faster with $N(t) + 1$ than with $N(t)$. However, lower bias does not necessarily imply lower mean square error. Meketon and Heidelberger show that here bias reduction is costless: the mean square error in either case is $\delta^2/tE[X_1] + O(1/t^2)$. Analysis of sequential experiments with time rather than the number of cycles as an adjustable parameter is an interesting, open problem.

If $\bar{D}_\tau$ is the estimator based on the cycles completed with work τ and w is the expected work per cycle, then

$$\tau^{1/2}\bar{D}_\tau/\delta \Rightarrow w^{1/2}N(0, 1)$$

by the elementary renewal theorem and a converging-together argument.

3.8. Standardized Time Series

Since the issue of relative merits of STS and the other methods for steady-state output analysis discussed in Section 3.3 has not yet been entirely resolved in the literature, we sketch here the key ideas of STS largely following Schruben (1986b). Standardization aims to reduce time series of diverse origins to a single tractable form to which limit theorems can be applied to construct confidence intervals that are asymptotically valid under mild assumptions. As Schruben points out, standardizing a time series closely parallels standardizing an ordinary scalar statistic depending on iid observations, except that STS requires index scaling detailed below. It can be widely applied, for example, in situations where the conditions required for valid application of autoregressive analysis seem not to hold or where the regenerative method is infeasible due to (perhaps infinitely) long cycle lengths. With some complex models, establishing asymptotic validity of any method may be hard. The required "mixing" condition often resists rigorous checking, through intuitively it typically holds. It may be unclear which method is least risky. We have no definitive advice.

Glynn and Iglehart (1985, 1988) show that, asymptotically, any method which consistently estimates the appropriate steady-state variance constant (see Problem 3.6.10) beats STS in a sense they make precise. Mild additional restrictions beyond those needed for the validity of STS assure consistent variance estimation; but see Problem 3.8.2. On the other hand, without contradicting the preceding statement, there are situations where consistent variance estimation holds but STS fails. Glynn and Iglehart (1986) show that the regenerative method works for some systems where the functional central limit theorem leading to the Brownian bridge in Step 5 below does not apply.

When STS works and consistent variance estimation holds for alternative methods, STS may be competitive for small to moderate sample sizes. Perhaps the sample-size crossover point where alternative methods win is often very large.

Duersch and Schruben (1986), Heidelberger and Welch (1983), and Nozari (1986) document implementations of STS. In this brief treatment, we deal only with concept of standardization and not with the mechanics of any particular implementation. See Problem 3.8.1.

3.8.1. Steady State

Consider b independent runs or b (supposedly) independent batches with n observations each. Let j index runs or batches. The mean of the first k observations of $\{X_{ij}: i = 1, 2, \ldots, n\}$ is

$$\bar{X}_{kj} = (1/k) \sum_{i=1}^{k} X_{ij} \qquad \text{(run } j \text{ or batch } j\text{)},$$

with overall mean

$$\bar{\bar{X}}_n = (1/b) \sum_{j=1}^{b} \bar{X}_{nj}.$$

Our estimator of the unknown steady-state mean μ is $\bar{\bar{X}}_n$. The steady-state variance constant is

$$\sigma^2 = \lim_{n \to \infty} n \operatorname{Var} \bar{X}_{nj},$$

introduced here only to clarify the distribution of the ratio in $(*)$ below. Using STS, we construct asymptotically valid confidence intervals for μ using

$$([2b - 1]bn)^{1/2}(\bar{\bar{X}}_n - \mu)/\sigma Q \Rightarrow t_{2b-1} \qquad (*)$$

as $n \to \infty$ where Q is defined below and t_{2b-1} is the t distribution with $2b - 1$ degrees of freedom. Step 6 below (optionally) estimates σ, which cancels out of $(*)$ because Q has σ in its denominator. To prove $(*)$, it suffices to show that

(i) $(bn)^{1/2}(\bar{\bar{X}}_n - \mu)/\sigma \Rightarrow N(0, 1)$ (standard normal);
(ii) $Q^2 \Rightarrow \chi^2_{2b-1}$ (chi-square with $2b - 1$ degrees of freedom);
(iii) the variates $N(0, 1)$ and χ^2_{2b-1} above are independent.

Of course, (i) is merely the central limit theorem, though for dependent variables (e.g., see Billingsley [(1986), Theorem 27.5]) when inter-batch correlation is admitted. From (i)–(iii), we get $(*)$ via the continuous mapping theorem [Billingsley (1968), p. 30], since $(2b - 1)^{1/2}N(0, 1)/\chi_{2b-1}$ is t_{2b-1}. We sketch below the argument for (i)–(iii) without making explicit the (mild) conditions needed; Schruben (1983b) details the latter. For steady state, $EX_{ij} = \mu$ for all i, j.

First we standardize and then find Q.

Step 1 (*Center*): Let

$$S_{nj}(k) = \bar{X}_{nj} - \bar{X}_{kj}, \qquad k = 1, \ldots, n,$$
$$S_0 = 0.$$

Step 2 (*Scale magnitude*): Set

$$\tilde{S}_{nj}(k) = kS_{nj}(k)/n^{1/2}\sigma.$$

Step 3 (*Scale index*): By piecewise-constant interpolation, we now get our STS defined in continuous time as

$$T_{nj}(t) = \tilde{S}_{nj}(\lfloor nt \rfloor), \qquad 0 \le t \le 1.$$

Step 4 (*Compute area under T_n*): Set

$$A_{nj} = \sum_{k=1}^{n} T_{nj}(k/n).$$

[Other functionals of T_n work.]

Step 5 (*Note limit theorems*): Schruben (1983b) shows that

$$\begin{pmatrix} T_{nj} \\ n^{1/2}(\bar{X}_{nj} - \mu)/\sigma \end{pmatrix} \Rightarrow \begin{pmatrix} W^0 \\ N(0, 1) \end{pmatrix} \qquad \text{as } n \to \infty,$$

where W^0 is a Brownian bridge (e.g., see Billingsley [(1968), pp. 64–65] independent of the standard normal variate $N(0, 1)$. With

$$V_n = 12/(n^3 - n),$$

Schruben also shows that as $n \to \infty$

$$V_n A_{nj}^2 \Rightarrow \chi_1^2,$$

chi-square with one degree of freedom. One argues that this holds for the limiting Brownian bridge and then applies the continuous mapping theorem. To use (∗), no familiarity with Brownian bridges is needed.

Step 6 (*Pool estimators*): Set

$$\tilde{A}_{nb} = \sum_{j=1}^{n} V_n A_{nj}^2,$$

$$\hat{\sigma}_{nb}^2 = \frac{1}{b-1} \sum_{j=1}^{b} (\bar{X}_{nj} - \bar{\bar{X}}_n)^2,$$

$$\tilde{\sigma}_{nb}^2 = (b-1)\hat{\sigma}_{nb}^2/\sigma^2,$$

$$Q^2 = \tilde{A}_{nb} + \tilde{\sigma}_{nb}^2.$$

Step 7 (*Apply independence and limit theorems, in particular continuous mapping*):

$$\tilde{A}_{nb} \Rightarrow \chi_b^2 \quad \text{since } \sum_{i=1}^{b} \chi_1^2 = \chi_b^2,$$

$$\tilde{\sigma}_{nb}^2 \Rightarrow \chi_{b-1}^2 \quad \text{(standard statistics)},$$

$$Q^2 \Rightarrow \chi_{2b-1}^2 \quad \text{since } \chi_b^2 + \chi_{b-1}^2 = \chi_{2b-1}^2.$$

This justifies (ii) above and step 5 justifies (i) and (iii).

3.8.2. Transients

In steady state, assumed up to now, $EX_{ij} = \mu$ for all i, j. In considering initialization bias now, we can plot $T_{nj}(t)$ as a function of t to reveal departures from this assumption. Schruben (1982) quantifies this consideration. As he remarks, "the computed significance level for the test is a guide and should not be substituted for intuition and common sense." This is especially important when his test is applied to time series where his hypotheses are in doubt (see Problem 3.6.14) or when his test is (heuristically) implemented sequentially or with just one run. Schruben's (1986b) procedure, different from his (1982) suggestion, uses (∗) on the series examined sequentially in reverse chronological order. When the confidence intervals become unstable as we move forward in the reversed series, we take this to signal nonstationarity. The remaining observations are then discarded. A confidence interval is constructed using the retained observations if there are enough of them; otherwise, the sample size (run length) is increased starting from the former final observation in chronological order. And so on.

Examining reversed series to detect nonstationarity can be coupled with any steady-state confidence-interval procedure. Using (∗) is just one possibility.

PROBLEM 3.8.1. Detail a specific implementation. Show how to compute the required statistics recursively. Explain why checking confidence-interval stability amounts to sequential testing of the stationarity hypothesis.

PROBLEM 3.8.2. Pooling estimators as in Step 6 above increases the degrees of freedom in (∗), tending to shorten confidence intervals. Step 7 implicitly assumes that $\tilde{A}_{nb}$ and $\tilde{\sigma}_{nb}^2$ are independent. Does this condition hold? If not, does it appear to be a reasonable heuristic? Suggest a multiple-run scheme where $\tilde{A}_{nb}$ is pooled with an estimator of σ^2 based solely on the other runs. Is pooling then rigorously justified? In answering this question, assume that σ^2 is consistently estimated. When pooling, STS is more robust than which alternatives? Would you pool?

PROBLEM 3.8.3. Compare the limit distributions of the normalized overall mean of the observations for STS (see Step 5 above) and for asynchronous observations (see §3.3). Conclude that STS, at least in its current form, is incompatible with asynchronous observations. Does this affect the attractiveness of STS?

CHAPTER 4

Rational Choice of Input Distributions

Simulation incorporates the randomness of the real world. How do we decide upon the nature of this randomness? How do we specify distributions for our random variables? To make these decisions and specifications, we start from the data available. We have:

(a) either lots of data or "little" data;
(b) from either the distribution of interest or a "related" distribution.

Having little data is bad; having little data on the wrong distribution is worse. "Little" data means any combination of

(L1) small sample;
(L2) summary statistics only, e.g., mean, variance, min, max, median, mode;
(L3) indirect qualitative information, based for example on interviews with informed people or on experience in related situations.

Wrong but "related" distributions are sometimes the only sources of information available:

(R1) Wrong amount of aggregation. An inventory simulation might need daily demand data, but our client has only monthly demand data.
(R2) Wrong distribution in time. Almost always our data are historical, but we really want next month's demand distribution—or, worse, next year's.
(R3) Wrong distribution in space. We want to simulate the New York City Fire Department but we have data only from Los Angeles.
(R4) Censored distributions. We want the demand data but have only the sales data. Sales understate demand when there are stockouts. (See Problem 4.9.21.)

(R5) Insufficient distribution resolution. In a message switching system that used the U.S. national phone system, the length of a typical phone call was of the same order of magnitude as the resolution accuracy of the phone system for timing calls. Hence, data on call durations was inaccurate.

With "little" or "wrong, but related" data, be skeptical about any distributions specified. Sensitivity analysis is particularly called for in such bold-inference situations. Sections 4.1 to 4.5 discuss what qualitative information might lead us to conjecture that a normal, lognormal, exponential, Poisson, or Weibull distribution is appropriate. Because such conjectures are tentative, check sensitivity both to the parameters of the distribution conjectured and to its form. Do sensitivity analyses using paired simulation runs with common random numbers.

The goal of a sensitivity analysis is to show that output depends only weakly on which of a set of plausible distributions is used. A theoretical distribution usually has only one or two parameters, which can be varied continuously. This makes sensitivity analysis easy, provided one considers only the (limited) shapes that the theoretical distribution can take. Fitting theoretical distributions to data is popular, but in Section 4.8 we argue that this is not necessarily wise. Section 4.6 discusses a quasi-empirical distribution that blends a continuous, piecewise-linear distribution which closely mimics the data (interpolating the usual empirical distribution) with an exponential tail. For empirical comparisons, see Section 8.1.

Quite often (wrongly, we believe) sensitivity studies consist solely in varying the mean and variance of the input distributions, simply by using transformations of the form $Y = a + bX$. With *rare* exceptions (like the mean wait in a stationary $M/G/1$ queue), these two parameters do not suffice to determine the expectation of the output distribution, let alone the distribution itself. Especially in the tails of the input distributions, the data give only meager support to the assumed forms and there are significant qualitative differences in the behavior, for example, of the exponential, Weibull, and normal. Usually a meaningful robustness investigation must check sensitivity of the performance measure to the form (e.g., skewness) of the input distributions. Compared to focusing only on the mean and variance, this extended sensitivity analysis may be more difficult to explain to the user and may require more effort to carry out. But, for credibility, it is needed.

Depending on the random number generator, there may be a positive probability of generating the numbers 0 and ∞ from some "continuous" distributions. When transforming uniform random numbers, underflow or overflow errors can occur. We mention the overflow problem here because distributions with nonfinite support are artifices; to avoid statistical anomalies we may wish to reject transformed numbers that are more than a few standard deviations away from their expected values. Fox and Glynn (1986c) study this situation. If the uniform variates are transformed monotonely, this can be done by proper choice of positive tolerances ε_1 and ε_2 and rejection of uniform variates outside the interval $[\varepsilon_1, 1 - \varepsilon_2]$. To avoid synchronization problems (Chapter 2), rather than directly reject uniform random numbers outside this interval, we can (equivalently) transform each uniform random number as follows:

$$U \leftarrow \varepsilon_1 + (1 - \varepsilon_1 - \varepsilon_2)U.$$

In Chapter 2, we saw that with few exceptions only inversion is compatible with synchronization. Therefore, consider the speed of computation of the inverse when choosing a distribution to postulate. If the exact inverse cannot be quickly calculated, alleviate the problem by approximating a given F^{-1} by an easier-to-compute function Q^{-1}, bounding $\sup\{|F^{-1}(x) - Q^{-1}(x)|: x$ and $F^{-1}(x)$ can be represented as single-precision floating point numbers in the computer at hand}. Accuracy to six decimal digits over this range probably suffices for most applications, but values of x near one and zero may cause problems. In fact, for distributions with infinite tails, to avoid a vacuous criterion bound $\sup\{|F^{-1}(x) - Q^{-1}(x)|: \varepsilon_1 \leq x \leq 1 - \varepsilon_2\}$ consistent with our clipping recommendation above. Even if the relative error in the tails is small, the absolute error may be large. Check errors in the first few moments. Another reasonable criterion is:

$$\Delta(F, Q) = \tfrac{1}{2} \sum |f_i - q_i| \qquad \text{(discrete case)},$$

$$\Delta(F, Q) = \tfrac{1}{2} \int |f(x) - q(x)| \, dx \quad \text{(continuous case)}.$$

This is a special case of a criterion advocated by Devroye (1980):

$$\Delta(F, Q) = \sup\left\{ \left| \int_A dF - \int_A dQ \right| : A \text{ is a Borel set} \right\}.$$

The first and second integrals are the probabilities of A under F and Q, respectively. The restriction to Borel sets is a mathematical nicety to ensure that these probabilities exist. To show the correspondence between the two definitions, note that the area above q but below f equals the area above f but below q because the area under f and the area under q are both one. Let A be the (possibly disconnected) set where f is above q. If A is Borel, it is a worst case; if not, take the sup over Borel subsets of A. Devroye (1982) gives readily-computed upper bounds, and in some cases explicit expressions, for Δ.

4.1. Addition and the Normal Distribution

When should we expect to find the normal distribution? The *central limit theorem* [e.g., see Feller (1971)] states that under fairly mild conditions the average of independent, but not necessarily identically distributed, random variables is asymptotically normally distributed. Let $X_1, X_2, \ldots$ be independent random variables with

$$E[X_k] = 0, \qquad E[X_k^2] = \sigma_k^2,$$

$$s_n^2 = \sigma_1^2 + \cdots + \sigma_n^2, \qquad s_n \to \infty \text{ as } n \to \infty,$$

$$Y_n = (X_1 + \cdots + X_n)/s_n.$$

Define the distribution of Y_n as F_n and call the standard normal distribution N. We quote two key theorems from Feller [(1971), §§16.5, 16.7].

Theorem 4.1.1. *If for $k = 1, 2, \ldots$*

$$E[|X_k|^3] < \lambda_k, \qquad r_n = \lambda_1 + \lambda_2 + \cdots + \lambda_n,$$

then for all n and x

$$|F_n(x) - N(x)| \leq 6r_n/s_n^3.$$

If the X_i's are iid, then 6 can be replaced by 3 in this bound.

Theorem 4.1.2. *Suppose that there exists a nonempty interval $[-a, a]$ in which the characteristic functions of the X_k's are analytic and that*

$$E[|X_n|^3] \leq M\sigma_n^2,$$

where M is independent of n. If x and n vary so that $s_n \to \infty$ and $x = o(s_n^{1/3})$, then

$$\frac{1 - F_n(x)}{1 - N(x)} \to 1,$$

with an error $O(x^3/s_n)$.

The first of these theorems gives a remarkable bound on the absolute difference between F_n and N. It refines the central limit theorem but says little about the quality of the normal approximation in the tails, a gap filled by the second theorem. The analyticity condition is mild, virtually always satisfied in practice.

PROBLEM 4.1.1. If the condition $x = o(s_n^{1/3})$ were dropped, give a counterexample to the conclusion of Theorem 4.1.2. (Hint: Consider the symmetric binomial distribution.)

If the normal distribution is used to approximate the distribution F of a nonnegative random variable, truncate the normal at $F(0)$. If it is used to approximate a discrete distribution, truncate the normal if appropriate and then round any random number generated from it to the nearest value assigned positive mass by F. We consider the quality of the normal approximation to the Poisson below.

In most applications the bound in Theorem 4.1.1 is extremely conservative. The normal approximation is often invoked (with a prayer) even when that bound seems worthless.

PROBLEM 4.1.2. Show how to use Theorem 4.1.1 to bound the error of the normal approximation to the Poisson with parameter α. (Hint: The Poisson distribution with parameter α is the n-fold convolution of Poisson distributions with parameter α/n. Show that $\lambda_k = \alpha/n + O(\alpha^3/n^3)$, $r_n/s_n^2 \to 1$, and hence that $|F_\alpha(x) - N_\alpha(x)| \le 3/\sqrt{\alpha}$, where F_α is the Poisson distribution with parameter α and N_α is the normal distribution with mean and variance α.) For what values of α is the error less than 0.001? (Answer: According to the bound, $\alpha \ge 9,000,000 \Rightarrow$ error < 0.001. This condition on α is sufficient but far from necessary.) [Devroye (1981a) and Ahrens and Dieter (1982b) give exact algorithms for generating Poisson variates that run in $O(1)$ average time; with high probability they exit with a truncated normal variate.]

For large α, generating random numbers from the Poisson in a straightforward way is harder than generating from the approximating normal.

In Appendix A we give a goodness-of-fit test for the normal distribution. Apply this test (or any other) only when there are *a priori* grounds to conjecture that the distribution in question can be well approximated by a normal distribution (cf. §4.8). The only clearly prudent *a priori* ground is a belief that the central limit theorem applies, though some might invoke the normal if it occurred in a "wrong but related" situation like those mentioned earlier.

Likewise, postulating (even tentatively) a multivariate normal distribution is prudent only when one believes that the multivariate central limit theorem [e.g., see Feller (1971), p. 260] applies. The multivariate normal is the only multivariate distribution for which the covariance matrix completely captures the dependency structure. A discrete multivariate distribution requires a large amount of data to estimate its parameters. A continuous multivariate distribution, other than the multivariate normal, requires more: a (not completely natural) specification of the form of the dependency. Illustrating this difficulty, Schmeiser and Lal (1982) specify various bivariate gamma distributions, justify their choices, and show how to generate random variates from them. Section 5.3.2 shows how to generate random variates from a multivariate normal.

4.2. Multiplication and the Lognormal

If $\log X$ has the normal distribution, then X has the lognormal distribution. A rationale for the lognormal is akin to that for the normal. Let

$$X = Z_1 \cdot Z_2 \cdots Z_n,$$

where the Z_i's are iid positive random variables. Then

$$\log X = \log Z_1 + \log Z_2 + \cdots + \log Z_n$$

and so, under mild conditions, the asymptotic distribution of $(\log X)/n$ is normal; hence, the asymptotic distribution of X is lognormal, by the continuous mapping theorem [e.g., see Billingsley (1968), pp.. 30–31]. Section 5.3.5 shows how to generate lognormal variates and relates the lognormal to proportional growth models.

4.3. Memorylessness and the Exponential

A distribution F with density f has *failure rate* $f(t)dt/(1 - F(t))$ at t: the probability of failure in the infinitesimal interval $(t, t + dt)$ given survival up to t. Constant failure rate, connoting lack of memory and complete randomness, characterizes the exponential distribution. Epstein (1960) and Harris (1976) give tests for constant failure rate.

Lack of memory simplifies modeling, analysis, and simulation: given the current state, the past is stochastically irrelevant. The corresponding state space reduction often is crucial for tractability. In a general queue, the number of customers in a system does *not* constitute a state—contrary to common abuse in the literature. Erlang's method of stages approximates a nonexponential distribution by a sum of exponential distributions and augments the state space to keep track of the current summand number (stage), thus inducing memorylessness. Kleinrock [(1975), Chapter 4] and Cooper [(1981), §4.8] give modern treatments. Neuts [(1981), Chapter 2] generalizes and unifies this area by introducing phase-type distributions and their matrix-theoretic treatment. This greatly widens the scope of applicability of continuous-time Markov chains, which often have large state spaces. It remains to exorcise the curse of dimensionality. With deterministic numerical approaches, state aggregation is the only possible cure; closely bounding the resulting error is hard. Heyman and Sobel [(1984), §6.4] give the flavor of the difficulty in a related setting. Deterministic numerical approaches, when feasible, sometimes can be imbedded easily in a dynamic-programming framework, greatly simplifying optimization.

With simulation, we can sometimes exorcise the curse without aggregation (beyond that already implicit in the model) because we do not have to keep track of a state vector. The key is to construct a row generator for the transition matrix of the Markov chain that takes an order of magnitude less space than the state vector while producing rows quickly as needed. This may well be feasible provided that the matrix is sparse and structured. We believe that this proviso must hold in reasonable models. Fox and Glynn (1986d) compare deterministic and simulation approaches. When an optimal policy for a model with state aggregation can be disaggregated to apply to the original problem, simulation can be used to check the behavior of the disaggregated policy. Simulation cannot establish its optimality, but in general nothing can do that.

PROBLEM 4.3.1. Show that Erlang's method gives a distribution with increasing failure rate and coefficient of variation less than one.

PROBLEM 4.3.2. Show how to approximate a distribution with decreasing failure rate and coefficient of variation greater than one by a finite mixture of exponential distributions. [This mixture is a *hyperexponential* distribution.] Specify a two-step process to generate a variate from this mixture. Show that from the completion of the first step to the next event time the system is memoryless.

PROBLEM 4.3.3. Combine the methods of Problems 4.3.1 and 4.3.2 to approximate an arbitrary distribution. How might this approximation be used to check a simulation program? Can you see any other use for it in simulations?

4.4. Superposition, the Poisson, and the Exponential

In Section 4.9 we indicate how the superposition of renewal processes could yield a Poisson process. We also give axioms for a (possibly nonstationary) Poisson process. These could be used to rationalize directly postulating this process in many cases. One consequence of these axioms is that the successive spacings are independent and exponentially distributed and at any instant the time to the next arrival is exponentially distributed. So rationalizing a Poisson process indirectly rationalizes exponentially-distributed interarrival times. This is a basis for a statistical test for the Poisson. It is not enough to check whether the arrival spacings are exponentially distributed. One must also check whether they are independent.

4.5. Minimization and the Weibull Distribution

Galambos (1978) thoroughly goes over the extensive asymptotic theory of extreme values. We recommend his book for those wishing to study this fascinating subject in depth. Here we quote just one result [see Galambos (1978), p. 56].

Theorem 4.5.1. Let $\alpha(F)=\inf\{x: F(x)>0\}> -\infty$. Put $F^*(x)=F(\alpha(F)-1/x)$, $x < 0$. Suppose that $t \to -\infty \Rightarrow F^*(tx)/F^*(t) \to x^{-\gamma}$, for some positive constant γ. Let $W_n = \min(X_1, X_2, \ldots, X_n)$ where the X_i's are iid with distribution F. Then there is a sequence $d_n > 0$ such that

$$\lim_{n \to \infty} P[W_n < \alpha(F) + d_n x] = \begin{cases} 1 - \exp(-x^\gamma), & x > 0, \\ 0, & x \le 0. \end{cases}$$

The normalizing constants d_n can be chosen as

$$d_n = \sup\{x: F(x) \le 1/n\} - \alpha(F).$$

The limit distribution above is called the Weibull distribution with shape parameter γ. For $\gamma > 1$ (<1), the Weibull distribution has increasing (decreasing) failure rate. At the frontier $\gamma = 1$, we get the exponential distribution. Checking the hypotheses of the theorem may be difficult. In many circumstances F is not known, but sometimes the Weibull distribution (with shape, scale, and location parameters to be estimated) can be justified directly.

Galambos [(1978), pp. 189–191] gives one such example where the iid condition in the theorem has been significantly weakened. Dannenbring (1977) suggests the use of the Weibull distribution for estimating the value of the minimum $\alpha(F)$ for large combinatorial problems in a Monte Carlo setting. To estimate the distribution of W_n, we generate blocks $(X_{1i}, X_{2i}, \ldots, X_{ni})$ for $i = 1, 2, \ldots, n$ say, and calculate the corresponding W_{ni}'s. We discuss variate generation from the Weibull distribution in Section 5.3.14 and from other extreme value distributions in Section 5.3.15.

4.6. A Mixed Empirical and Exponential Distribution

If we have a respectable number of observations, say 25, and no insight *a priori* regarding the form of the distribution, then an obvious approach uses the empirical distribution. Probably it poorly fits the true underlying distribution in the right tail. Our proposal below compromises between simplicity and accuracy. There is some intuitive and theoretical support (see Theorem 4.7.1) for fitting an exponential distribution to the right tail of the distribution. An overall fit using, say, a gamma or a Weibull distribution would be smoother and in this sense probably would more closely mimic the (generally unknown) true distribution. However, with a reasonable number of breakpoints in the empirical distribution (i.e., with enough data), differences in smoothness would appear negligible.

We order the observations so that $X_1 \le X_2 \le \cdots \le X_n$ and then fit a piecewise-linear cdf to the first $n - k$ observations and a shifted exponential to the right of observation X_{n-k} with mean chosen so that the means of the overall fitted distribution and the sample are the same. To the left of X_{n-k} we use linear interpolation between breakpoints of the empirical cdf; the latter jumps by $1/n$ at each breakpoint. Section 5.2.4 gives formulas for the cdf and variance of this quasi-empirical distribution and a simple, fast variate-generating technique using inversion. Linear interpolation instead of splines, say, is used to expedite variate generation.

A reasonable value of k might be the integer in $\{1, 2, 3, 4, 5\}$ that gives the best fit to the empirical variance. If the number of breakpoints is reasonably large, any k in this range probably works well. Only when the number of breakpoints is large is the empirical variance likely to be an accurate estimator of the true variance. In that case, the empirical variance and variance of the fitted distribution match well. If the number of observations is small, a rough match suffices.

We smooth the empirical cdf (by linear interpolation) because we believe that the distribution generating the data has a density. According to the Kolmogorov–Smirnov theorem [e.g., see Feller (1971), §I.12], the unsmoothed empirical cdf converges uniformly to the true cdf with probability one. The same holds for our smoothed cdf if we let $k \to 0$ as $n \to \infty$. At any point the density of our quasi-empirical distribution is itself a random variable. Devroye (1981b) pointed out to us that at no point does it converge in probability to the true density at that point. This is somewhat disconcerting, but we still believe that using our quasi-empirical distribution is often a lesser evil than other alternatives for reasons discussed in Section 4.8 and below.

No matter how much data there is, we would coalesce the breakpoints in a quasi-empirical distribution to form another quasi-empirical distribution with at most a few dozen breakpoints. This simultaneously weakens the impact of the above pathology concerning the convergence of the densities, saves memory, and facilitates insight. In line with our remarks in the introduction to this chapter, we would truncate the exponential tail to avoid statistical anomalies.

To recap: this quasi-empirical distribution is an easy-to-use, often-rational choice.

PROBLEM 4.6.1. Suppose that we want to fit a *theoretical* distribution G by a distribution that looks like the above quasi-empirical distribution. The natural thing to do is to define $X_1, X_2, \ldots, X_{n-k}$ by

$$G(X_i) = i/n, \qquad i \leq n - k,$$

to set

$$X_i = 0, \qquad i > n - k + 1$$

and to (try to) define X_{n-k+1} so that $X_{n-k+1} > X_{n-k}$ and the mean of the quasi-empirical distribution equals the mean of G. (The piecewise linear interpolation may shift the mean to the right and the exponential tail must be chosen to offset this effect.) Give a G such that for $n = 100$ and $k = 1$ this is impossible. Show that for any fixed n we can choose k so that it *is* possible. (Hint: Try $k = n$.) In your example, what is the smallest k that works? [The pathology mentioned above concerning convergence of densities does not apply here because we are fitting a theoretical distribution, not data.]

PROBLEM 4.6.2. For sensitivity analysis, we can replace the exponential tail by Weibull tails of various shapes. Generalize the definition of F and the algorithm for generating random numbers from it.

Devroye [(1986), §14.7] analyzes the appropriateness of certain other quasi-empirical distributions for simulation using sophisticated results from the theory of density estimation. Generating variates from these alternative distributions by inversion would be hard.

4.7. Extreme Values and Spacings

What motivates the exponential tail in our quasi-empirical distribution (§4.6)? As the sample size n goes to infinity, the successive spacings between the k largest observations (k fixed) asymptotically become independent and exponentially distributed for a wide class of distributions that includes the gamma and Weibull families. A more precise statement is

Theorem 4.7.1 [Weissman (1978a, b)]. *Suppose that*

$$\frac{d}{dx}\left|\frac{1 - F(x)}{f(x)}\right| \to 0, \quad \text{where } f = F'.$$

Let $b_n = F^{-1}(1 - 1/n)$, $a_n = 1/nf(b_n)$, $X_1 \leq X_2 \leq \cdots \leq X_n$ *be an ordered sample from* F, *and* $D_{ni} = X_{n+1-i} - X_{n-i}$, $i = 1, 2, \ldots, n - 1$. *Then for* k *fixed,* $D_{n1}, D_{n2}, \ldots, D_{nk}$ *are asymptotically independent and, for* $i \leq k$, D_{ni}/a_n *is asymptotically exponentially distributed with mean* $1/i$.

It is important to note that X_{n-k} is random. The hypotheses of the theorem do *not* imply that, to the right of a large *fixed* point, F has an approximately exponential tail.

PROBLEM 4.7.1. Find a_n when F is the Weibull distribution with $F(x) = 1 - \exp(-x^\alpha)$. (Answer: $a_n = [\log n]^{1/\alpha}/\alpha \log n$.)

PROBLEM 4.7.2. From the answer to Problem 4.7.1 it follows that $a_n \equiv 1$ when $\alpha = 1$, i.e., when F is exponential. In this case, prove without using the theorem that D_{ni} is *exactly* exponentially distributed with mean $1/i$. (Hint: If each of k operating light bulbs has life exponentially distributed with mean one, then the overall failure rate is k.) This result goes back at least to Renyi (1953).

Exponential tails are consistent with theoretical distributions that have asymptotically flat failure rate curves, such as the gamma and hyperexponential families. If the failure rate goes to infinity, as for a Weibull with shape parameter greater than one, approximating with exponential tails would at first sight seem grossly wrong. Surprisingly, even here, using our procedure of Section 4.6 that assumes exponential tails to the right of X_{n-k} seems reasonable, in view of Theorem 4.7.1, whenever a_n is insensitive to n over a wide range. For example, take the Weibull with $\alpha = 2$: we see that, although $a_n \to 0$, the rate of convergence is extremely slow; for instance, $a_{500} \approx 0.201$ and $a_{1000} \approx 0.190$.

4.8. When Not to Use a Theoretical Distribution

Much ado has been made in the literature about the putative benefits of fitting theoretical distributions (e.g., gamma, beta), specified by a few parameters, to data for use as input to a simulation.

These benefits may be real in cases where using a theoretical distribution can be justified on *a priori* grounds, illustrated in the preceding sections, or on the basis of goodness-of-fit tests, such as chi-square and Kolmogorov–Smirnov (discussed in §6.6). A well-known defect of these tests, often not properly acknowledged, is that they have notoriously low power. That is, the probability of rejecting a fit, even though the data do not in fact come from the theoretical distribution fitted, is small except in those rare cases where there is a huge amount of data available. There is a *big* difference between fitting data and fitting the underlying distribution from which the data come.

The power is even less if goodness-of-fit tests are applied to a gamut of candidate theoretical distributions to select the best fit out of the set. A common practice is to form a short list of candidates based on an informal look at the data, perhaps using interactive graphics packages, and then to use the distribution that fits best according to goodness-of-fit tests. This is fine so far. It is a mistake, however, in this situation to go further and attach any credibility to (wrong) statements like "the distribution that fit best was not rejected at the 90% level." Statistics associated with tandem goodness-of-fit tests have unknown distributions. An additional reason for caution is that an informal look at the data to select candidates for fitting affects the distribution of these goodness-of-fit statistics in ways impossible to analyze mathematically.

It is statistically dishonest to evaluate a policy using the data that was used to develop the policy. It is only slightly less dishonest to evaluate a policy using a theoretical distribution fitted to the data used to develop it. One can legitimately develop a policy with, say, half the data, chosen in an unbiased way, and then evaluate it with the other half. Interchanging the roles of the two halves can provide a worthwhile cross-validation comparison. The breakpoints in our quasi-empirical distribution correspond to parameters and there are a lot of them. In general, the more parameters there are to adjust, the more important cross-validation is.

PROBLEM 4.8.1. With a really huge amount of data concerning the real system, the need for variate generation may disappear. Why?

Estimating the tails of distributions accurately from a limited amount of data is impossible. This is a disconcerting, unavoidable drawback of every approach. There is no consensus on what to do. Galambos [(1978), p. 90], gives a striking example of the intrinsic difficulty. Motivated heuristically by Theorem 4.7.1, our quasi-empirical distribution of Section 4.6 assumes that these tails are shifted exponentials, with parameters chosen so that the mean of the fitted distribution equals the sample mean. Though we have not manufactured this assumption out of the air, we do make it with fingers crossed. In any application we would carry out a sensitivity analysis of the tail shape.

Apart from the cases mentioned in the second paragraph of this section, we see only disadvantages in using theoretical fits to observations from the true distribution. Specifically:

1. Unless the theoretical family used for fitting has reliable qualitative support, fitting loses and distorts information; this distortion does not vanish as the sample size goes to infinity. Do not neglect the risk that the

theoretical distribution does not generate the data no matter what parameters are chosen. God does not usually tell us from what distribution the data come. Going the other direction, a theoretical distribution could be approximated by a piecewise linear fit, possibly with an exponential tail, to expedite variate generation. Rather than this approximate method, why not use the quasi-empirical distribution directly? Even if there is little or no data and the input distribution simply reflects a subjective assessment, a cdf that starts piecewise linear and ends with an exponential tail seems at least as reasonable as something chosen, say, from the beta family.

2. Parameter estimation is often difficult or nonrobust, involving the solution of nonlinear equations. Outliers may result in grossly wrong estimates.

3. It is often harder (see Chapter 5) to generate random numbers from a theoretical distribution than from the quasi-empirical distribution (§4.6). Unlike the latter, the former generally cannot be generated efficiently and easily by inversion, which is by far the most effective method for use with variance reduction techniques (Chapter 2). If postulating a particular distribution causes difficulties, why do it?

PROBLEM 4.8.2. Lilliefors (1967, 1969) estimates by Monte Carlo the critical values and power for the Kolmogorov–Smirnov test for the normal and exponential distributions with unknown parameters. Write your own computer program to do this, using any variance reduction techniques that seem worthwhile.

The views expressed in this section go counter to widely and easily-accepted mythology. Those who feel an almost-irresistible urge to fit data to hypothesized distributions should reread our discussion of model validation in Section 1.2. The fit may be spurious. An hypothesized distribution is a model that should be validated. The more sensitive the output is to this distribution, the more stringent the validation checks ought to be. Only then can we gain genuine insight. If you are not convinced, perhaps it is because one man's myth is another man's dogma. Fox (1981) concludes that the notion that fitting "standard" distributions to data is necessarily good is both a dogma and a myth.

4.9. Nonstationary Poisson Processes

We tailor our discussion of nonstationary Poisson processes to the simulation context. In particular, several problems show how to exploit Poisson processes to strengthen conditional Monte Carlo.

Let N be a stochastic (e.g., arrival) process. We suppose that:

(i) N is a nondecreasing, right-continuous, step function (piecewise constant);

(ii) on all sample paths all jumps (occurring at "arrival" epochs) of N have integral height;

(iii) the set of sample paths where a jump size is greater than one (e.g., two arrivals simultaneously) has probability zero (though such paths may exist);

(iv) $N(t)$ is the number of jumps in the half-open interval $(0, t]$;

(v) for any $t, s \geq 0$, the number $N(t + s) - N(t)$ of jumps in $(t, t + s]$ is independent of the history $\{N(u): u \leq t\}$ up to and including t.

These axioms, taken from Çinlar (1975), characterize a (possibly *nonstationary*) Poisson process. They are more intuitive and easier to check than those usually given.

PROBLEM 4.9.1. $N(0) = ?$

If $N(t + s) - N(t)$ is independent of t, then the process is *stationary*. (Sometimes we say nonhomogeneous and homogeneous instead of nonstationary and stationary, respectively.) Returning to the general case, call the integrated rate function

$$a(t) = E[N(t)]$$

and the instantaneous rate function $\lambda(t) = a'(t^+)$. From now on we assume that a is strictly increasing and continuous, except where weaker conditions are explicitly stated. It would be easy to superpose a process that has jumps at fixed times, corresponding to scheduled arrivals say.

Using the following result, going back at least to Parzen [(1962), p. 26], we can transpose results for stationary processes to nonstationary processes and vice versa. What the proposition says is that transforming the time scale by a converts a nonstationary process to a stationary process. To go the other direction, transform the time scale by a^{-1}.

Proposition 4.9.1. *In a nonstationary Poisson process with random arrival times* $T_1, T_2, \ldots$, *set*

$$X_i = a(T_i),$$

$$\tilde{N}(t) = number\ of\ X\text{'}s\ in\ (0, t],$$

$$\tilde{a}(t) = E[\tilde{N}(t)].$$

Then $d\tilde{a}(t)/dt \equiv 1$. *In other words,* $\tilde{N}$ *is a stationary Poisson process with rate 1.*

PROOF. Graph $a(t)$ versus t with the X_i's, t, and $t + \Delta$ on the vertical axis and the T_i's, $a^{-1}(t)$, and $a^{-1}(t + \Delta)$ on the horizontal axis. This shows that

$$E[\tilde{N}(t + \Delta) - \tilde{N}(t)|\tilde{N}(u): u \leq t]$$
$$= E[N(a^{-1}(t + \Delta)) - N(a^{-1}(t))|N(a^{-1}(u)): a^{-1}(u) \leq a^{-1}(t)]$$
$$= E[N(a^{-1}(t + \Delta) - N(a^{-1}(t))]$$
$$= a(a^{-1}(t + \Delta)) - a(a^{-1}(t))$$
$$= (t + \Delta) - t = \Delta. \qquad \square$$

We exploit Proposition 4.9.1 in Section 5.3.18 to generate a nonstationary Poisson process by *monotonely* transforming the time-scale of a stationary Poisson process via a^{-1}. The latter can itself be generated *monotonely* by generating its exponentially-distributed spacings by inversion. This sets the stage for efficient variance reduction, as described in Chapter 2 (cf. Theorem 2.1.1). The method is reasonably efficient if, for example, a (and hence a^{-1}) is piecewise linear and a fast method (cf. §5.2.1) is used to find the right piece of a^{-1}. In Section 5.3.18 we give another method, due to Lewis and Shedler (1979a), for generating a nonstationary Poisson process based on "thinning" another (e.g., stationary) Poisson process with an everywhere-higher rate function. With a careful implementation, the process (event times) generated will be a monotone function of the driving uniform random number stream. As with the time-scale transformation method, the events can be generated one at a time as needed. Lewis and Shedler (1979a) adapt the method to generate two-dimensional Poisson processes.

Proposition 4.9.2. *Let N be a nonstationary Poisson process and a be continuous. Set $b(t, s) = a(t + s) - a(t)$. Then, for any $t, s \geq 0$,*

$$P[N(t + s) - N(t) = k] = [b(t, s)]^k [\exp(-b(t, s))]/k!$$

for $k = 0, 1, 2, \ldots$.

PROBLEM 4.9.2. Prove this proposition. (Hint: Combine Proposition 4.9.1 and

$$T_i \in (t, t + s] \Leftrightarrow X_i \in (a(t), a(t + s)].$$

You may assume that Proposition 4.9.2 holds in the stationary case where $a(t) = \lambda t$ for a positive constant λ; e.g., see Çinlar (1975) or Karlin and Taylor (1975). Without loss of generality, take $\lambda = 1$.)

PROBLEM 4.9.3. Let $T_1, T_2, \ldots$ be the successive arrival times of a nonstationary Poisson process. Under the hypotheses of the above proposition, show that

$$P[T_{n+1} - T_n > t \,|\, T_1, \ldots, T_n] = \exp(-b(t, T_n))$$

for $n = 1, 2, \ldots$. (Hint: Combine

$$\{T_{n+1} - T_n > t\} \Leftrightarrow \{N(T_n + t) - N(T_n) = 0\}$$

and Proposition 4.9.2 with $k = 0$.)

We obtain a *compound* Poisson process by deleting axiom (ii) above and replacing axiom (iii) by

(iii′) the magnitudes of the successive jumps are identically distributed random variables independent of each other and of the jump times.

So if successive arrival spacings are exponentially distributed (per Problem 4.9.3) and arrivals occur in iid batches, the arrival process is compound Poisson.

PROBLEM 4.9.4. How would you simulate such a process?

Suppose again that (iii) holds and that arrival i is of type j with probability p_j independent of i and the other arrivals.

PROBLEM 4.9.5. How would you simulate such a process?

PROBLEM 4.9.6. What can you say about the process consisting only of the type j arrivals? Can you simulate it without simulating the entire original process? If so, how?

A different situation occurs when independent Poisson arrivals streams are superposed. For example, stream j could consist of type j arrivals; the superposed stream would then be the overall stream where all arrivals of all types are recorded without regard to type. Here, type could refer to size, sex, priority, origin, etc.

Proposition 4.9.3. *Let $N_1, N_2, \ldots, N_k$ be independent (possibly nonstationary) Poisson processes with respective integrated rate functions $a_1, a_2, \ldots, a_k$.*

Set $a(t) = a_1(t) + a_2(t) + \cdots + a_k(t)$. If a_i is continuous for $i = 1, 2, \ldots, k$, then the superposition of the arrival processes, denoted symbolically by $N = N_1 + N_2 + \cdots + N_k$, is a Poisson process with integrated rate function a.

PROBLEM 4.9.7. Prove this proposition when $a_i(t) = \lambda_i t, i = 1, \ldots k$. (Hint: Use generating functions.)

PROBLEM 4.9.8. Give a general proof.

Proposition 4.9.3 holds for all t and all k. The assumption that the component streams are Poisson may sometimes be too strong. However, if no small subset of the component streams "dominates" the overall stream, then we can relax the condition that the component streams be Poisson provided that we are willing to settle for an asymptotic result. Under mild conditions, the overall stream is effectively Poisson for large t and k. Roughly, these conditions amount to supposing that the renewal epochs (see §3.7) of each individual stream are not regularly spaced and are extremely rare so that the superposition is not greatly affected by a small subset of these streams. Check that the superposition has two properties, asymptotically:

 (i) its spacings are exponentially distributed with the same mean;
(ii) these spacings are independent.

For the case when the component streams are renewal processes, see Khintchine (1969) for a precise theorem and a rigorous proof.

Feller's (1971) heuristic argument for (i) follows. Let the interarrival times of stream i have distribution F_i and mean μ_i. With n such streams, put

$$1/\alpha = 1/\mu_1 + 1/\mu_2 + \cdots + 1/\mu_n.$$

Our assumptions imply that, for fixed y, $F_k(y)$ and $1/\mu_k$ are small—uniformly in k. Let $W_k(t, z)$ be the probability that the wait until the next arrival in stream k is less that z, given that the process has been running for time t. By the result of Problem 3.7.3,

$$W_k(t, z) \to W_k(z) = \frac{1}{\mu_k} \int_0^z [1 - F_k(y)]\, dy$$

$$\approx z/\mu_k,$$

since $F_k(y)$ is small in the fixed interval $(0, z)$. So the probability that the time to the next arrival of the overall stream is greater than z is

$$\prod_{i=1}^n (1 - W_i(t, z)) \approx \prod_{i=1}^n (1 - z/\mu_i) \approx e^{-z/\alpha}.$$

In the case of equal μ_i's the last approximation is a familiar result from elementary calculus.

More generally, the component streams can be "point" processes [e.g., see Çinlar (1972)]. Çinlar surveys results that give conditions under which the superposition of such processes converges to a (possibly nonstationary) Poisson process and the rate of this convergence. Under mild conditions, the "difference" Δ_n between a Poisson process and the superposition of n point processes is of order $1/n$. The implicit constant in this rate depends on other parameters of the problem. Albin (1982) reports queueing simulation experiments where for n fixed Δ_n increases significantly with the traffic intensity and the deviation from one of the squared coefficient of variation of the component process arrival spacings.

PROBLEM 4.9.9 (A Poisson arrival and a random observer see stochastically identical states). In a system with (possibly nonstationary) Poisson arrivals, let $\delta(s, t) = 1$ if the system is in state s at time t and $=0$ otherwise. Here and below we could replace s by any measurable set. Assume that the left-hand limit implicit in $\delta(s, t^-)$ exists with probability one, with or without any of the conditioning below. Intuitively, this limit is one if the state seen by an arrival about to enter the system at time t is s, not counting himself. Let B_t be the set of all arrival epochs strictly prior to t. Assume that the system is *nonanticipative*:

$$P[\delta(s, t^-) = 1 \,|\, B_t \text{ and arrival at } t] = P[\delta(s, t^-) = 1 \,|\, B_t].$$

Remark. A queue in which departures are fed back directly or indirectly to the arrival stream, as occurs at some nodes of certain networks of queues, is generally anticipative. For an important class of such networks, Melamed (1979) shows that the resulting merged arrival stream cannot be Poisson (even) in steady state. Most queueing theorists conjecture that this is so in general. So in the context of Poisson arrivals, the assumption that the queue is nonanticipative is usually superfluous; but see Wolff's (1982) Example 2 of an exception.

Prove that for Poisson arrivals

$$P[\delta(s, t^-) = 1 | \text{arrival at } t] = P[\delta(s, t^-) = 1].$$

(Hint: There are four ingredients:

(i) for Poisson arrivals, the conditional distribution G of B_t given an arrival at t equals the unconditional distribution F of B_t;

(ii) nonanticipation;

(iii) conditioning/unconditioning;

(iv) $P(X | Y) = \int P[X | Y, Z] \, dG(Z | Y).$

Check that

$$P[\delta(s, t^-) = 1]$$

$$= \int P[\delta(s, t^-) = 1 | B_t] \, dF(B_t) \qquad \text{using (iii)}$$

$$= \int P[\delta(s, t^-) = 1 | B_t \text{ and arrival at } t] \qquad \text{using (ii)}$$

$$\cdot \, dG(B_t | \text{arrival at } t) \qquad \text{using (i)}$$

$$= P[\delta(s, t^-) = 1 | \text{arrival at } t],$$

the last equality using (iv) with $X = \{\delta(s, t^-) = 1\}$, $Y = \{\text{arrival at } t\}$, and $Z = B_t$.)
Show by example that this is not necessarily true if the arrivals are not Poisson. On the positive side, see Melamed (1982) for a literature review and generalizations. Our argument follows Strauch's (1970) pattern. The above result applies at an arbitrary but fixed time, which suffices for Problem 4.9.10 below. Wolff (1982) proves the corresponding limit theorem, needed for Problem 4.9.20. Under a nonanticipation condition similar to ours, he shows that *Poisson arrivals see time averages* in a precise sense that corresponds to the intuitive meaning of this phrase.

PROBLEM 4.9.10 (continuation). Now we combine the result of Problem 4.9.9 with conditional Monte Carlo, using the notation and assumptions of Section 2.6 and Problem 4.9.9. The estimator $\hat{\theta}_4$ of Section 2.6 is a sum, the ith summand corresponding to epoch T_i. Here we assume that these epochs correspond to Poisson arrivals. Instead of considering the system only at these epochs, think of every instant of time as a virtual, or fictitious, arrival epoch—regardless of whether a real arrival occurs then. By Problem 4.9.9, instead of summing over the real arrival epochs as in $\hat{\theta}_4$, we can integrate over the virtual arrival epochs, i.e., in continuous time. The expected number of real arrivals in an infinitesimal interval $(t, t + dt)$ is $a(t + dt) - a(t)$. Let $S(t)$ be the state at time t.

The above considerations motivate a fifth estimator,

$$\hat{\theta}_5 = \int_0^T E[X(t) | S(t)] \, da(t).$$

Interpret $\hat{\theta}_5$ intuitively. [Optional: Indicate how it generalizes Wolff's (1982) equation (12).] Let Z be (the σ-field generated by) the uncountable set of random variables $\{S(t) : 0 \leq t < T\}$. In $\hat{\theta}_4$ let S_i be the state found by (Poisson) arrival i. Taking the limit of Riemann–Stieltjes sums, show (heuristically) that generally

$$\hat{\theta}_5 \neq E[\hat{\theta}_4 | Z],$$

but that

$$E[\hat{\theta}_5] = \theta.$$

(The cognoscenti familiar with measure-theoretic probability should try to give a rigorous proof.)

PROBLEM 4.9.11 (continuation of Problem 4.9.10 and Example 2.6.1). For every simulation run

(i) during every interval in which the number in system remains constant and in which the arrival rate is constant, calculate

$P[\text{a hypothetical new arrival would not balk} | \text{number in system}]$
$\cdot (\text{interval length}) \cdot (\text{arrival rate}),$

(ii) sum over all such intervals.

Show that the expectation of this sum equals the expected number served. (For empirical results, see §8.7.)

PROBLEM 4.9.12. Suppose that it is feasible to estimate the integrated arrival rate function a from data. How would you test statistically that the arrival process is (possibly nonstationary) Poisson? (Hint: One way is to transform the data as in Proposition 4.9.1 and then to use a chi-square test. To test for a stationary Poisson process, use the raw data directly.) If the arrival process is cyclic, then the intervals used in a goodness-of-fit test should be "desynchronized" with it. Why? If we postulate that the process is cyclic and that a is piecewise linear with specified breakpoints, how would you estimate its slope in each piece? The data used to check goodness of fit should be distinct from the data used to estimate a. Why?

PROBLEM 4.9.13. Show that it is incorrect to generate the next event time in a nonstationary Poisson process with instantaneous rate function $\lambda(t)$ $[= a'(t^+)]$ by adding an exponential variate with mean $1/\lambda(s)$ to the current clock time s.

PROBLEM 4.9.14. Suppose that in a stationary Poisson process exactly k arrivals occur in a fixed interval $[S, T]$. Given this, show that these k arrivals are independently and uniformly distributed in that interval. Without the stationarity condition, show that the conditional joint distribution of these arrivals is the same as that of the order statistics of k iid random variables with distribution

$$F(t) = [a(t) - a(S)]/[a(T) - a(S)], \qquad S \leq t \leq T.$$

What is the relationship of this problem to Problem 1.9.12? Show how to simulate a Poisson process using the (well-known) result in this problem, assuming that an efficient method exists to generate k. Chouinard and McDonald (1982) give a rank test for nonstationary Poisson processes, related to this result, in which the particular forms of F and a are irrelevant.

PROBLEM 4.9.15. Lewis and Shedler (1979b) consider the instantaneous rate function

$$\lambda(t) = \exp\{\alpha_0 + \alpha_1 t + \alpha_2 t^2\}.$$

Review the motivation for this form in that paper and the references therein. How would you validate this model and estimate its parameters? Study its robustness numerically.

Show that the Lewis–Shedler algorithm does not transform uniform random numbers monotonely.

PROBLEM 4.9.16. Consider a rate function $\lambda(t) = \lambda(t + T)$ for all t where T is the smallest period for which this holds. Let a Poisson process with this rate function generate arrivals to an arbitrary queueing system. Show that any time point which is an integral multiple of T and at which the system is empty is a regeneration point. Under what moment conditions can the regenerative method for constructing confidence intervals be applied? An alternative to constructing such confidence intervals is to construct confidence intervals based on a finite-horizon simulation where the horizon is a small multiple of T. When does this alternative seem preferable to constructing confidence intervals for steady-state behavior?

PROBLEM 4.9.17 (continuation). Give *easily-checked* conditions for applying the regenerative method. [This is an open problem. Harrison and Lemoine (1977) give related results for single-server queues. They show that, if

$$\bar{\lambda}E(S_1) < 1,$$

where

$$\bar{\lambda} = \int_\sigma^T \lambda(t)\, dt/T$$

and S_1 is a service time, the expected spacing between successive regeneration points is finite. Interpreting $\bar{\lambda}E(S_1)$ as the "traffic intensity," this result is not surprising; however, it is not obvious.]

PROBLEM 4.9.18. Consider a single-server queue with nonstationary Poisson arrivals, general iid service times, and FIFO service order. Use Problem 4.9.10 to improve the estimators in Problems 2.6.2 and 2.6.3.

PROBLEM 4.9.19. A number of results in Section 3.7 involve a limit as a *nonrandom* point in time goes to infinity. Instead consider a limit as n goes to infinity of the nth event epoch generated by some *random* process. If the process is stationary Poisson, show that the respective limits for each result are equal. Under what other circumstances are these limits equal?

PROBLEM 4.9.20. Let an arrival stream be generated by a stationary Poisson process with instantaneous rate λ. Call the time at which the nth regeneration cycle ends T_n. Based on observing the system over $[0, T_n]$, let A_n be the number of arrivals and let B_n be the number of arrivals that see the system in state s just before they arrive. We want to estimate the stationary probability p that an arrival finds the system in state s. A naive estimator is

$$\hat{p}_1 = B_n/A_n.$$

Using $\hat{p}_1$, mimic the procedure in Section 3.7 to obtain a confidence interval I_1 for p. Using the notation of Problem 4.9.9 and the result of that problem, another estimator of p is the proportion of time that a *hypothetical* arrival would have seen the system in state s:

$$\hat{p}_2 = \int_0^{T_n} \delta(s, t)\, dt/T_n.$$

Show that this has the form of a standard regenerative estimator and use the procedure of Section 3.7 to construct a confidence interval I_2 for p using $\hat{p}_2$. Let

$$D_1 = B_n - pA_n$$

and

$$D_2 = \int_0^{T_n} \lambda \delta(s, t)\, dt - p\lambda T_n.$$

Show that $E[D_1] = E[D_2] = 0$ and that $D_2 = E[D_1 | Z]$ where Z is (the σ-field generated by) the uncountable set of random variables $\{T_n, \delta(s, t): 0 \le t \le T_n\}$. Show that

$$\text{Var}[D_2] \le \text{Var}[D_1].$$

Argue, heuristically, that the length of I_2 tends to be shorter than the length of I_1. Reconsider your argument in view of Cooper's [(1981), pp. 291–295] example.

PROBLEM 4.9.21. We observe

$N = $ the number of customers entering a system during an interval
of length t,
$F = $ fraction time that the system was full during that interval,

and know that, when the system is full, arriving customers do not enter but go away. Assume that arrivals are Poisson at an unknown rate λ. Discuss the estimator

$$\hat{\lambda} = N/t(1 - F).$$

Is it unbiased?

PROBLEM 4.9.22. Prove: With Poisson arrivals, homogeneous or not, the number that come in any fixed period t is more likely to be even than to be odd. We view 0 as even. (Hint: Show that for a homogeneous Poisson with rate 1

$P[\text{even number of arrivals in } t]$
$= e^{-t}[1 + t^2/2! + t^4/4! + \cdots]$
$= e^{-t}[1 + t + t^2/2! + t^3/3! + \cdots + 1 - t + t^2/2! - t^3/3! + \cdots]/2$
$= e^{-t}[e^t + e^{-t}]/2 > \frac{1}{2}.$

For the general case, use Proposition 4.9.1.)

PROBLEM 4.9.23. Under what nontrivial conditions can you deduce the limiting distribution of $(X_1 + \cdots + X_n) \bmod t$?

PROBLEM 4.9.24 (Gradient estimation). We use the notation and results of Problem 2.5.8. Let g be independent of ξ, taken here to be the intensity of a Poisson process. Fix N, the number of Poisson events (say arrivals). Let the corresponding (random) amount of simulated time be T. [Don't confuse this with the computer time to simulate.] Let the X_i's corresponding to successive arrival spacings be $X_{i_1}, \ldots, X_{i_N}$. Show that the joint density of these spacings is

$$p(X_{i_1}, \ldots, X_{i_N}; \xi) = \xi^N \exp(-\xi[X_{i_1} + \cdots + X_{i_N}])$$
$$= \xi^N \exp(-\xi T),$$

and that

$$\frac{dp(X_{i_1}, \ldots, X_{i_N}; \xi)}{d\xi} = \left[\frac{N}{\xi} - T \right] \xi^N \exp(-\xi T).$$

Conclude that

$$Z'(X_1, \ldots, X_n; \xi_0) = \left[\frac{N}{\xi_0} - T \right] g$$

since $g' = 0$ and the other factors in the likelihood ratio cancel. Problem 2.5.8 then gives

$$\frac{d}{d\xi} E_\xi g = E_\xi \left[\left(\frac{N}{\xi} - T \right) g \right] \qquad (*)$$

assuming that we can switch the order of expectation and differentiation.

PROBLEM 4.9.25 (Continuation). Now fix T and let N be random. Show that the joint density of the successive spacings is

$$q(X_{i_1}, \ldots, X_{i_N}; \xi) = \frac{(\xi T)^N e^{-\xi T}}{N!} \cdot \frac{N!}{T^N}$$

$$= \xi^N e^{-\xi T}$$

as before. Conclude that $(*)$ above holds under the same condition. [Reiman and Weiss (1986), under their (mild) "amiability" condition, prove $(*)$ directly even when both N and T are random. A heuristic argument observes that likelihood ratios are martingales; so when $(*)$ holds for fixed N, it "should" hold when N is a stopping time (cf. Billingsley [(1986), pp. 495–496]). With Poisson arrivals, we can therefore let N be the number of customers served during a busy period say.] Generalize $(*)$ to nonstationary Poisson processes.

CHAPTER 5
Nonuniform Random Numbers

5.1. Introduction

In a typical simulation one needs a large number of random variates with the "proper" statistical properties. Chapter 4 provides background for deciding what is proper. We now consider the problem of generating these random variates artificially. All the methods to be presented for generating random variables transform uniformly distributed random variates. It might therefore seem logical to discuss first the generation of the latter. However, most computer languages have built-in functions for producing random variables uniform over the interval (0, 1), while only a few provide generators for nonuniform variates. Only when we become concerned about the finer points of true randomness do we wish to look more closely at generators of uniform random variates. Chapter 6 recommends methods for constructing such generators.

Chapter 5 has two major parts:

(i) Section 5.2: Fundamental methods of generating random variates of arbitrary distribution, also called nonuniform random numbers.
(ii) Sections 5.3 and 5.4: Handbook for variate generation from specific distributions, continuous and discrete, respectively.

Section 5.5 (problems) and Section 5.6 (timings) finish the chapter.

We assess variate-generation speed by summing the time required to generate 100,000 variates with fixed distribution parameters. If a generator requires significant set-up time, it is noted. Generators may rate differently depending on whether they are implemented in a high-level language or in assembly language. In our experiments, the generators were coded only in

Fortran except that in each case both Fortran and assembly versions of the uniform generator were used. When variate-generation speed is a bottleneck, we do not disparage assembly-coded generators but we do not give empirical comparisons of such implementations because of their necessarily local nature. In each case the generator is written as a subroutine or function subprogram in which all parameters are passed; this is typical in practice. The variates are summed and the result printed so that smart optimizing compilers do not optimize away the call to the subprogram. Portable Fortran codes are listed in Appendix L. We summarize their computational performance in Table 5.6.1 at the end of this chapter.

In most complex simulations, logical and statistical operations take at least as much time as nonuniform random number generation. For a desired statistical precision, reducing the number of runs via variance reduction methods—not efficiency of variate generation—is the overriding consideration in such cases. As discussed in Chapter 2, inversion is most compatible with variance reduction techniques that depend on inducing correlation. Sometimes we can have our cake and eat it too: Section 8.1 shows that using inversion does not necessarily sacrifice variate-generation speed, especially if one is willing to be a bit flexible about the input distributions selected.

In this chapter we stress inversion and simplicity, sometimes at the expense of variate-generation speed. Devroye's (1986) impressive, encyclopedic book also considers inversion but concentrates mainly on other methods. He discusses algorithms that are uniformly fast for all parameter values in given families of distributions. Like us, he stresses simplicity and is willing to pay a modest price in variate-generation speed for it. Like him, we believe that simple methods are usually more reliable in conception and implementation and are likely to survive longer. Correctness of a variate generator is particularly hard to judge empirically because statistical tests may not be strong enough to catch sneaky bugs and the underlying uniform-variate generator that drives it is necessarily imperfect. While we favor using mathematical subroutine libraries, because of the special difficulty of adequately testing variate-generator packages we prefer those uncluttered by fancy tricks and intricate detail. We and Devroye both cite more complicated methods without discussing them in detail.

The introduction to Chapter 4 points out advantages of truncating extreme tails and shows that, with inversion, truncation is easily implemented without loss of synchronization. For several standard distributions our inversion routines are approximate. They are accurate except perhaps in the extreme tails. This exception is irrelevant when these tails are truncated.

5.2. General Methods

We consider methods which fall in one of the following categories, where C indicates relevance to continuous distributions, D to discrete, and N indicates a method illustrated by application to the normal distribution:

(a) inversion [C, D];
(b) tabular approximation of the inverse [C];
(c) rejection [C, N];
(d) composition [C, N, D];
(e) alias method [D];
(f) functional approximations of the inverse [C, N];
(g) others [C, N].

5.2.1. Inversion

As discussed in Section 1.5, inversion is a method for generating variates of any distribution. When the inverse cdf F^{-1} is easy to compute, efficient implementation of inversion is trivial. For cases when the inverse cdf is hard to compute, see Sections 5.2.2 and 5.2.9 for "tabular" and "functional" approximations, respectively. The methods of those sections apply to continuous distributions. Yuen (1982) cleverly adapts hashing onto a table of previously-generated F-values to find an initial interval containing a root of $F(X) = U$ and then inverts by regula falsi; her work is not detailed here solely to avoid preemption of her publication. We discuss below computation of the inverse cdf for general discrete distributions on a finite number of points. Important special discrete distributions, some on an infinite number of points, are treated in Section 5.4.

If a distribution is concentrated on a few points, it is easy to tabulate the cdf and then generate variates by sequential linear search.

PROBLEM 5.2.1. Give details. How would you determine in which direction to scan a cdf table? Show how to replace initial complete cdf tabulation by cdf calculation at certain points as needed.

If a discrete distribution is concentrated on more than a few points, then in general we begin as before by tabulating the cdf; this may require significant set-up time, but this seems an unavoidable price to pay for fast marginal variate generation speed. Suppose that the cdf F is concentrated on N points: $1, 2, \ldots, N$.

PROBLEM 5.2.2. Show that linear search takes $O(N)$ worst-case time, and give an example where linear search takes $O(N)$ average time. Show that ordering the values according to their respective probabilities speeds variate generation but destroys monotonicity.

PROBLEM 5.2.3. Show that the binary search method below runs in $O(\log N)$ worst-case time.

1. Set $L \leftarrow 0, R \leftarrow N$.
2. Generate U.

3. Do the following while $L < R - 1$:
   (a) set $X \leftarrow \lfloor (L + R)/2 \rfloor$;
   (b) if $U > F(X)$, set $L \leftarrow X$; otherwise, set $R \leftarrow X$. [Comment: We ensure throughout
       that $F(L) < U \leq F(R)$.]
4. Return R and stop.

Can the average time be less than the worst-case time?

Fox [(1978b), pp. 704–705] outlines a bucket scheme to replace pure
binary search for handling tables. Start by using m buckets, each of width
one, numbered $1, 2, \ldots, m$. Set $p_i = \min\{j: F(j) > (i - 1)/m\}$ and $q_i =
\min\{j: F(j) \geq i/m\}$. Now $i - 1 < mU \leq i \Rightarrow F(p_i - 1) < U \leq F(q_i)$. In
bucket i store the indices p_i and q_i. This yields the following algorithm:

1. Generate U.
2. Set $K \leftarrow \lfloor mU \rfloor + 1$ [Comment: K is uniformly distributed on the
   integers $1, 2, \ldots, m$.]
3. Search the values $p_K, p_K + 1, \ldots, q_K$ for an index X such that $F(X - 1) <
   U \leq F(X)$.
4. Return X and stop.

If the number of integers from p_K to q_K is small, implement step 3 with linear
search; otherwise, implement it with binary search or narrower buckets.

PROBLEM 5.2.4. Analyze the worst-case and average behavior of this algorithm and its
variants.

Almost surely, such an "indexed" search scheme is ancient folklore. The
earliest reference seems to be Chen and Asau (1974). Probably the narrow-
bucket option mentioned above is appropriate in the tails of F, especially if
these are long and thin.

Ahrens and Kohrt (1981) describe an algorithm based on the same
principles but incorporating clever computational refinements. One such
refinement is to store only p_k in bucket k if $p_k = q_k$ and to store only $-p_k$ there
otherwise. The method requires two tables, one of which is the tabulated
cdf. The second table L has length m where m is an arbitrary design parameter
set by the user and $L(k)$ denotes the contents of bucket k. Increasing m
makes the algorithm run faster. Setting m between N and $3N$ often seems to
work well. The method is easy to implement:

1. Generate U.
2. $I \leftarrow \lfloor Um \rfloor + 1$ [Comment: I is uniformly distributed on the integers
   $1, 2, \ldots, m$.]
3. If $L(I) > 0$, then return $L(I)$ and stop.
4. (Otherwise begin linear search): $X \leftarrow - L(I)$.
5. While $F(X) < U$, set $X \leftarrow X + 1$.
6. Return X and stop.

Our empirical experience, agreeing with theirs, is that this is at least as fast
as any other method. Use it if there is enough memory available.

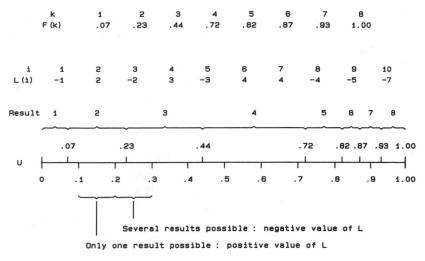

Figure 5.2.1. Example of Ahrens–Kohrt method.

Figure 5.2.1 illustrates the table set-up for a case where $N = 8$ and $m = 10$. Appendix L gives VDSETP, the subroutine for setting up L, and IVDISC, the routine for doing the actual inversion. Ahrens and Kohrt also include an efficient check for a tail region; they use a narrow-bucket scheme there. Consult their paper for a number of further refinements geared to assembly language programming.

5.2.2. Tabular Approximation of the Inverse Transform

Tabulate a sequence of pairs $(F(x_i), x_i)$ where $x_i < x_{i+1}$. Often this requires significant set-up time. The following algorithm inverts the piecewise linear approximation to F that interpolates these pairs.

1. Find X_i such that $F(X_i) \leq U \leq F(X_{i+1})$. [Comment: See §5.2.1 for ways to do this.]
2. Return

$$X = \frac{[F(X_{i+1}) - U]X_i + [U - F(X_i)]X_{i+1}}{F(X_{i+1}) - F(X_i)}.$$

PROBLEM 5.2.5. What do you suggest doing if the distribution has one or two infinite tails?

PROBLEM 5.2.6. Suppose we want to use quadratic instead of linear interpolation. Give details of how this can be done, paying particular attention to the values near the ends of the table. Repeat for spline interpolation. (Hint: See Conte and de Boor [(1980), §6.7] for example.)

Ahrens and Kohrt (1981) propose a more accurate but more complicated interpolation scheme to replace step 2. They use a sophisticated and computationally expensive set-up routine to:

(a) numerically integrate the pdf to obtain F;
(b) decide how to partition $(0, 1)$, the domain of F^{-1}, into unequal length intervals based on the behavior of F;
(c) determine the form and the coefficients of the interpolating function for each such interval.

Their paper gives further, but not complete, details. Based on their computational results, they conclude that both the speed and accuracy of their method, not counting set-up time, are about as good as the best available methods tailored to generate variates from standard distributions. Thus, if many variates are needed from a given distribution with fixed parameters, Ahrens and Kohrt provide a method that is close to ideal.

5.2.3. Empirical cdf's

Suppose we have n observations $X_1, X_2, \ldots, X_n$ from the true but unknown distribution, ordered so that $X_1 \leq X_2 \leq \cdots \leq X_n$. If we assume the approximation $F(X_i) = i/n$, then we say we are using the empirical cdf. We can use either constant, linear, or higher-order interpolation to define $F(x)$ for $X_{i-1} < x < X_i$. By constant interpolation we mean $F(x) = i/n$ if $X_{i-1} < x \leq X_i$. Using constant interpolation on the empirical cdf when generating $X = F^{-1}(U)$ corresponds to setting $X = X_i$ with probability $1/n$.

5.2.4. A Mixed Empirical and Exponential Distribution

One misgiving about using the unadorned empirical cdf is that it is discrete and that, even when interpolated, it probably gives a poor fit to the true underlying distribution in the right tail. Our quasi-empirical distribution, motivated in Section 4.6 and described below, alleviates these drawbacks.

First order the observations so that $X_1 \leq X_2 \leq \cdots \leq X_n$. Then fit a piecewise linear cdf to the first $n - k$ observations and a shifted exponential to the k largest observations. Assuming that $F(0) = 0$ and defining $X_0 = 0$, the cdf is

$$F(t) = \begin{cases} i/n + (t - X_i)/[n(X_{i+1} - X_i)] & \text{for } X_i \leq t \leq X_{i+1}, \\ & i = 0, 1, \ldots, n - k - 1 \\ 1 - (k/n)\exp(-(t - X_{n-k})/\theta) & \text{for } t > X_{n-k}, \end{cases}$$

where

$$\theta = \left(X_{n-k}/2 + \sum_{i=n-k+1}^{n} (X_i - X_{n-k})\right)\bigg/k.$$

The mean of the mixed distribution is $(X_1 + \cdots + X_n)/n$ for $1 \leq k \leq n$. Another routine calculation shows that its variance is

$$\frac{1}{3n}\left[2\sum_{i=1}^{n-k-1} X_i^2 + \sum_{i=1}^{n-k-1} X_i X_{i+1} + X_{n-k}^2 \right]$$

$$+ \frac{k}{n}[(\theta + X_{n-k})^2 + \theta^2] - \left[\frac{1}{n}\sum_{i=1}^{n} X_i\right]^2.$$

To generate a variate from this mixed distribution by inversion:

(1) generate a random U, uniform on $(0, 1)$;
(2) if $U > 1 - k/n$, then return

$$X = X_{n-k} - \theta \log(n(1 - U)/k);$$

(3) otherwise set $V \leftarrow nU$, $I \leftarrow \lfloor V \rfloor$, and return

$$X = (V - I)(X_{I+1} - X_I) + X_I.$$

In Appendix L we call the program implementing this algorithm REMPIR. In step 2 we use $1 - U$ rather than U to preserve monotonicity.

5.2.5. Rejection

Inversion requires knowledge of the cdf. In some instances one has the probability density function (pdf) explicitly but not the cdf. The normal distribution is a case in point: no closed form expression for its cdf is known. The rejection method, first suggested by von Neumann (1951), can be used when only the pdf is known.

In its simplest form the rejection method requires that the pdf f be bounded and nonzero only on some finite interval $[a, b]$. Define

$$c = \max\{f(x) | a \leq x \leq b\}.$$

Now

(1) generate X uniform on (a, b);
(2) generate Y uniform on $(0, c)$:
(3) if $Y \leq f(X)$, then output X; otherwise go to 1.

The method can be explained with the aid of Figure 5.2.2. Steps 1 and 2 generate a point (X, Y) uniformly distributed over the rectangle with dimensions c and $b - a$. If the point falls below $f(X)$, then step 3 accepts X; otherwise, step 3 rejects X and the loop begins afresh. Shortly, we generalize this method and prove that it works.

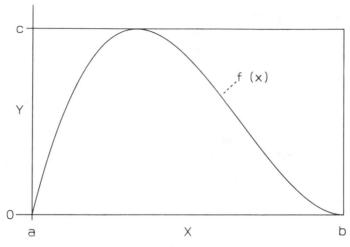

Figure 5.2.2. Simple rejection

PROBLEM 5.2.7. Let U and V be strings of uniform random numbers. We use U to generate nonuniform random numbers on run 1, and V on run 2. Let the nth nonuniform random number accepted on run 1 be generated from $\{U_i : i \in A\}$, and let the nth nonuniform random number accepted on run 2 be generated from $\{V_j : j \in B\}$. The sets A and B are all the uniform random numbers generated at steps (1) and (2) after the $(n - 1)$st acceptance up to and including the nth acceptance at step (3). If $n > 1$ and the generator uses rejection, show that A and B can be disjoint. Can this misalignment persist? Why does it vitiate the effect of common random numbers, antithetic variates, and control variates?

Remark. Franta's (1975) experiments confirm this. Schmeiser and Kachitvi-chyankul (1986) alleviate the misalignment above, but the resulting induced correlation is still weaker than with inversion. Except when nonuniform variates from one run are linearly transformed for use in another run, inversion induces the strongest correlations as theorem 2.1.2 shows. For *certain* distributions rejection techniques increase variate-generation speed but seldom enough to offset the resulting difficulties in applying variance reduction techniques. One may question (see Chapter 4) whether such distributions should be postulated. The variate-generation literature, concentrating solely on speed, gives the impression that rejection is generally superior to "old-fashioned" inversion; all things considered, the opposite seems true.

5.2.6. Generalized Rejection

The simple rejection method is inefficient when X is rejected in step 3 with significant probability, wasting two uniform variates. Generalized rejection

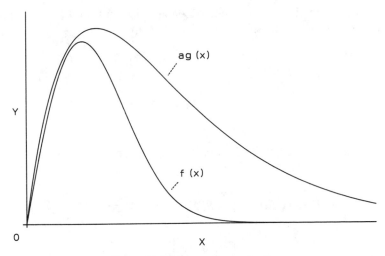

Figure 5.2.3. Generalized rejection.

simultaneously improves efficiency and relaxes the requirement that X fall in a finite interval. It requires a *majorizing density* g such that there is a constant a with $f(x) \le ag(x)$ for all x. Figure 5.2.3 illustrates this. The method is:

(1) Generate X with the pdf g.
(2) Generate Y uniform on $[0, ag(X)]$.
(3) If $Y \le f(X)$, then output X; otherwise go to 1.

Steps 1 and 2 generate points which are uniformly distributed under the curve ag. The closer that g approximates f, the higher is the probability that steps 1 and 2 will not have to be repeated. The test in step 3 is equivalent to checking whether $f(X)/ag(X) \ge U$, where U is uniform on $(0, 1)$. To expedite this test on the average, look for bounds for the ratio on the left which are more easily calculated than the ratio itself. Such bounds *squeeze* the ratio.

PROBLEM 5.2.8. Show how to take advantage of a squeeze.

A common way of getting bounds truncates series expansions of f or g, checking the sign of the remainder term. We usually try to select g by astutely patching together two or more densities, say one for the tails and the other for the midsection, to make $\sup[f(x)/g(x)]$ close to one, to make hard-to-compute factors of f and g cancel, and to make variates with pdf g faster to generate than variates with pdf f.

Now we show that any accepted X has distribution F, proving that generalized rejection works. This will follow from

$$P\{X \le t \mid Y \le f(X)\} = \frac{P\{X \le t \text{ and } Y \le f(X)\}}{P\{Y \le f(X)\}}$$

$$= \frac{F(t)/a}{1/a} = F(t)$$

as soon as we justify the second equality. To do so, condition on X and then "uncondition":

$$P[X \le t \text{ and } Y \le f(X)] = \int_{-\infty}^{\infty} P[X \le t \text{ and } Y \le f(X) \mid X = z] \, dG(z)$$

$$= \int_{-\infty}^{t} \left[\frac{f(z)}{ag(z)}\right] g(z) \, dz = F(t)/a,$$

where the correctness of the bracketed term follows from the fact that Y is (conditionally) uniform on $[0, ag(z)]$; likewise,

$$P[Y \le f(X)] = 1/a$$

completing the proof.

PROBLEM 5.2.9. What is the expected number of times that step 3 is executed?

EXAMPLE 5.2.1. Marsaglia (1964) presents an elegant use of generalized rejection for generating random variables from the tail of the standard normal distribution. Let d be positive, and suppose we want a draw X from the standard normal distribution conditional on $X \ge d$. The pdf for X then has the form $f(x) = c \cdot \exp(-x^2/2)$, where c is a constant which can be determined. An appropriate choice for g is $g(x) = x \cdot \exp(-[x^2 - d^2]/2)$. This happens to be the pdf of the square root of an exponential random variable with mean 2 displaced d^2 to the right of the origin. To see this, let Y be an exponential random variable with mean 2. What is the distribution of $(Y + d^2)^{1/2}$? The cdf is

$$G(x) = P\{(Y + d^2)^{1/2} \le x\}$$
$$= P\{Y \le x^2 - d^2\}$$
$$= 1 - \exp(-[x^2 - d^2]/2) \quad \text{for } x \ge d.$$

The pdf is simply the derivative; so

$$g(x) = x \cdot \exp(-[x^2 - d^2]/2) \quad \text{for } x \ge d.$$

Now we need a constant a such that $ag(x)/f(x) \geq 1$ for $x \geq d$. Hence we require

$$\frac{ax \cdot \exp(-[x^2 - d^2]/2)}{c \cdot \exp(-x^2/2)} \geq 1,$$

which reduces to

$$x \geq \frac{c}{a} \exp(-d^2/2).$$

To satisfy this inequality, choose

$$\frac{c}{a} \exp(-d^2/2) = d. \tag{$*$}$$

Generate a variate with pdf g by inverting G to get

$$X = \sqrt{(d^2 - 2\log(1 - U_1))},$$

which has the same distribution as

$$X = \sqrt{(d^2 - 2\log(U_1))}.$$

Accept X if

$$U_2 ag(X) \leq f(X);$$

that is, if

$$U_2 aX \cdot \exp(-[X^2 - d^2]/2) \leq c \cdot \exp(-X^2/2),$$

which reduces using $(*)$ to

$$U_2 X \leq d.$$

The generating procedure can now be summarized:

(1) Generate U_1, U_2 uniform on $(0, 1)$.
(2) Set $X = \sqrt{(d^2 - 2\log(U_1))}$.
(3) If $U_2 X \leq d$, then output X; otherwise go to 1.

If $d = 1$, then the procedure is about 66% efficient; that is, X is accepted with probability 0.66. If $d = 3$, the efficiency increases to 88%. Exercise: Show that as $d \to \infty$ the efficiency $\to 100\%$. Schmeiser (1980) gives a more general algorithm for generating variates from distribution tails based on generalized rejection.

Devroye (1984a) gives a flexible family of algorithms to generate variates from unimodal or monotone densities. They use rejection but differ from the generalized rejection algorithm given earlier. They are compact, not burdened by *ad hoc* devices for specific applications. Devroye (1984b) complements this approach with a simple algorithm for generating variates with a log-concave density.

5.2.7. Composition

Suppose that

$$f(x) = \int g_y(x)\, dH(y).$$

To generate a random number with density f by composition:

(1) Generate a random number Y from H.
(2) Generate a random number X with density g_Y.

PROBLEM 5.2.10. Show that this procedure is in general not monotone, even when inversion implements each step, and hence (in general) incompatible with common random numbers, etc. Under what conditions is it monotone?

The composition method goes back at least to Marsaglia (1961). He indicated its generality with his description of an algorithm for generating normal variates. One can generate the tails using the method of Example 5.2.1.

EXAMPLE 5.2.2. Ahrens and Dieter (1972a) and Kinderman and Ramage (1976) present similar composition algorithms for generating standard normal variates which strike a good compromise between efficiency and ease of programming. The latter method may be marginally faster on some computers. Corresponding to a discrete H above, Ahrens and Dieter partition the area under the standard normal density function as shown in Figure 5.2.4.

The five different areas under the curve comprise the following approximate proportions of the total area:

(a) 0.9195;
(b) 0.0345;
(c) 0.0155;
(d) 0.0241;
(e) 0.0064.

Ahrens and Dieter use much more accurate approximations to these probabilities.

The procedure chooses a random variate with a trapezoidal distribution like that in section (a) with approximate probability 0.9195. Such a random variate can be generated quickly using two uniform random variates: see Problem 5.5.5. Random variates with distributions like the odd-shaped areas (c), (d), and (e) are generated with approximate probabilities 0.0155, 0.0241, and 0.0064 by rejection using either a rectangular or a triangular bounding density. With probability about 0.0345 a random variate with the tail distribution (b) is generated by the generalized rejection method described earlier. The complete algorithm is presented in Appendix L as

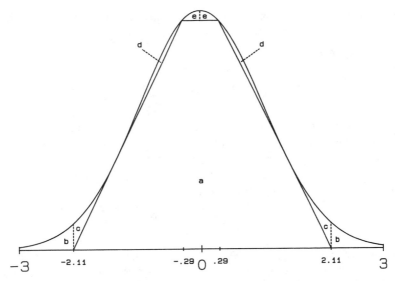

Figure 5.2.4. Ahrens–Dieter composition method applied to the standard normal.

subroutine TRPNRM. The key characteristic of this algorithm is that with probability about 0.9195 the random variate is generated very quickly. Because rejection is sometimes used, synchronization is impossible.

Fox [(1978b), p. 704] presents a composition method for discrete random variables. It uses a design parameter m which determines a table size. As m increases so does the speed and the similarity to inversion. Two tables R (remainder) and B (bucket) are required. They are set up as follows:

set $R_0 \leftarrow 0, Q_0 \leftarrow 0$;
for $i = 1, 2, \ldots, N$ set

$$q_i \leftarrow \lfloor mp_i \rfloor,$$

$$r_i \leftarrow p_i - q_i/m,$$

$$R_i \leftarrow R_{i-1} + r_i,$$

$$Q_i \leftarrow Q_{i-1} + q_i;$$

set $t \leftarrow 1 - R_N$;
for $k = 1, 2, \ldots, Q_N$ set

$$B(k) \leftarrow i \quad \text{where } Q_{i-1} < k \leq Q_i.$$

Choose m so that q_i is p_i truncated, with its decimal point or binary point then shifted all the way to the right. For example, if $m = 1000$ and $p_i = 0.7356$, then $q_i = 735$, $r_i = 0.0006$, and 735 buckets contain i. Step 2 below accesses these buckets by address calculation. It is executed with probability t, the sum of the truncated probabilities. Loosely speaking, it inverts their "distribution." The generation algorithm is:

(1) Generate U uniform on $(0, 1)$.
(2) If $U < t$, then output $X = B(\lfloor mU \rfloor)$.
(3) Otherwise output $X = k$ where $R_{k-1} \le U - t < R_k$.

This algorithm is fast if and only if m is large enough so that t is nearly one. For sufficiently large m, it is faster on average than the alias method (§5.2.8).

Step 3 inverts the "remainder distribution." It can be performed in time proportional to $\log N$ by using binary search or (faster) by the Ahrens–Kohrt method (§5.2.1) or the alias method (§5.2.8). A total of $N + Q_N + 1$ locations are required to store R and B. Note that $Q_N \le m$.

In Table 5.6.1 and Appendix L this method is called LUKBIN. It uses binary search to implement step 3.

PROBLEM 5.2.11. Show that Fox's method works. Justify calling it a composition method. On a binary machine, show that by choosing m to be a power of 2 the computations of $\lfloor mp_i \rfloor$ and $\lfloor mU \rfloor$ can be done using only shifts. What good would it do to set $R_k \leftarrow R_k + t$ for $k = 0, 1, 2, \ldots, N$?

5.2.8. The Alias Method for Discrete Distributions

Suppose the random variable X is distributed over the integers $1, 2, \ldots, n$ with $p(i) = P\{X = i\}$. The alias method for generating X is easy to program and is fast for arbitrary $p(i)$. The method requires only one uniform variate, one comparison, and at most two memory references per sample. However, it requires two tables of length n, and although it may be approximately monotonic in its transformation of the uniform variable U into the output variable X, it is not in general exactly so.

First transform U into an integer I uniformly distributed over $1, 2, \ldots, n$. Now I is a tentative value for X, but with a certain "aliasing" probability $1 - R(I)$ replace it by its "alias" $A(I)$. If we choose the aliases and the aliasing probabilities properly, then X has the desired distribution. We defer discussion of these choices.

The alias method is a streamlined composition method. We decompose the original distribution into a uniform mixture of two-point distributions. One of the two points is an alias. Having found a particular two-point distribution from which to generate, the rest is easy. Observe the simplicity of the generator:

(1) Generate V uniform on the continuous interval $(0, n)$, for example by $V = nU$.
(2) Set $I \leftarrow \lceil V \rceil$ (thus I is uniform on the integers $[1, n]$).
(3) Set $W \leftarrow I - V$. (Now W is one minus the fractional part of V and is therefore uniform on $(0, 1)$ and independent of I.)

(4) If $W \le R(I)$, then output $X = I$; otherwise, output $X = A(I)$, where $R(I)$ and $A(I)$ are tabulated values.

The remaining problem is the proper selection of $R(i)$ and $A(i)$. By the comments at steps 2 and 3 above, $P[W \le R(I), I = i] = R(i)/n$ and $P[W > R(I), I = j] = [1 - R(j)]/n$. Summing the probabilities of the mutually exclusive ways to get $X = i$, the generator gives

$$P[X = i] = R(i)/n + \sum_{\{j:A(j)=i\}} [1 - R(j)]/n.$$

To make this equal $p(i)$, select $R(i)$ and $A(i)$ using the following set-up algorithm:

0. [Initialize a "high" set H and a "low" set L; ϕ denotes the empty set]: $H \leftarrow \phi, L \leftarrow \phi$.
1. For $i = 1$ to n:
   (a) [Initialize]: set $R(i) \leftarrow np(i)$;
   (b) if $R(i) > 1$, then add i to H;
   (c) if $R(i) < 1$, then add i to L.
2. (a) If $H = \phi$, stop;
   (b) [Claim: $H \ne \phi \Rightarrow L \ne \phi$]: otherwise select an index j from L and an index k from H.
3. (a) Set $A(j) \leftarrow k$; [$A(j)$ and $R(j)$ are now finalized].
   (b) $R(k) \leftarrow R(k) + R(j) - 1$; [steps 3(a) and 3(b) together transfer probability from H to L].
   (c) if $R(k) \le 1$, remove k from H;
   (d) if $R(k) < 1$, add k to L;
   (e) remove j from L [and from further consideration].
4. Go to step 2.

This algorithm runs in $O(n)$ time because at each iteration the number of indices in the union of H and L goes down by at least one. Throughout, the total "excess" probability in H equals the total "shortage" in L; i.e.,

$$\sum_{i \in H} [R(i) - 1] = \sum_{i \in L} [1 - R(i)].$$

This justifies the claim in step 2(b). At the final iteration, there is just one element in each of H and L. Hence $R(j) - 1 = 1 - R(k)$ at step 2(b) the last time through, and at the finish $R(k) = 1$ leaving both H and L empty. To show that the final assignments of $R(i)$ and $A(i)$ are correct it suffices to verify two claims:

1. If $np(i) < 1$, then $\{j: A(j) = i\} = \phi$.
2. If $np(i) > 1$, then

$$R(i) + \sum_{\{j:A(j)=i\}} [1 - R(j)] = np(i).$$

PROBLEM 5.2.12. Do this.

The algorithms for set-up and generation as given here are discussed in Kronmal and Peterson (1979); see also Walker (1977) and references cited therein. Ahrens and Dieter (1982a) and, independently, Kronmal and Peterson (1981) invented a continuous version of the alias method called the acceptance/complement method.

Implementation Considerations. Step 3 in the generation algorithm can be deleted if step 4 of this algorithm is changed to

"If $V \geq R(I)$, then output $X = I$; otherwise output $X = A(I)$."

and an additional final step of the set-up algorithm resets $R(i) \leftarrow i - R(i)$ for $i = 1, 2, \ldots, n$.

The desirable property of monotonicity is more closely approximated if j and k in step 2 of the set-up algorithm are chosen so that $|j - k|$ is never allowed to be large. Matching the largest element in H with the largest element in L probably keeps $|j - k|$ small in most cases. If either L or H contains only one element, then minimize $|j - k|$. The alias method is compatible with synchronization when it is approximately monotone. Even though step 4 of the generator contains a rejection-like test, steps 1–4 are always executed exactly once.

In Table 5.6.1 and in Appendix L this method is called IALIAS.

5.2.9. Functional Approximations of the Inverse Transform

Even though there may be no closed form expression for F^{-1} or even for F, it may nevertheless be possible to approximate F^{-1} with a function which is easy to evaluate without using auxiliary tables. Deriving such an approximation is an exercise in numerical analysis which we will not investigate. Abramowitz and Stegun (1964) and Kennedy and Gentle [(1980), Chapter 5] list approximations for many functions appearing in probability and statistics. Most approximations for a function g are of one of the following types:

(1) Continued fraction

$$g(t) \simeq Q_0(t),$$

where

$$Q_i(t) = a_i + b_i t/(1 + Q_{i+1}(t))$$

for $i = 0, 1, 2, \ldots, n - 1$ and

$$Q_n(t) = a_n t.$$

(2) Rational approximation

$$g(t) \simeq \frac{a_0 + a_1 t + a_2 t^2 + \cdots + a_n t^n}{1 + b_1 t + b_2 t^2 + \cdots + b_m t^m}.$$

For example, Odeh and Evans (1974) give the following rational approximation of $X = F^{-1}(U)$ for the standard normal:

$$X = Y + \frac{p_0 + p_1 Y + p_2 Y^2 + p_3 Y^3 + p_4 Y^4}{q_0 + q_1 Y + q_2 Y^2 + q_3 Y^3 + q_4 Y^4},$$

where

$$Y = \sqrt{-\log([1 - U]^2)}$$

and

$p_0 = -0.322232431088,$	$q_0 = 0.099348462606,$
$p_1 = -1,$	$q_1 = 0.588581570495,$
$p_2 = -0.342242088547,$	$q_2 = 0.531103462366,$
$p_3 = -0.0204231210245,$	$q_3 = 0.10353775285,$
$p_4 = -0.0000453642210148,$	$q_4 = 0.0038560700634.$

This approximation has a relative accuracy of about six decimal digits and is valid for $0.5 < U < 1$. The symmetry of the normal allows us to extend it to $0 < U < 1$ by the transformations $U = 1 - U$ and $X = -X$. In Appendix L this algorithm is called VNORM.

5.2.10. Other Ingenious Methods

There are a number of other devices for generating random variables of nonuniform distribution which do not fit under any of the previous headings.

The Box–Muller Method for the Normal. Box and Muller (1958) show that one can generate two independent standard normal variates X and Y by setting

$$X = \cos(2\pi U_1)\sqrt{-2\log(U_2)},$$
$$Y = \sin(2\pi U_1)\sqrt{-2\log(U_2)},$$

which we justify later. This method is superficially akin to inversion. We know that there is no exact method using a single uniform random variate and inversion to generate a single normal variate. Being suitably bold we decide to try to generate two simultaneously. This, surprisingly, is possible. The method is not one-to-one, but rather two-to-two. It is not monotone in U_1.

PROBLEM 5.2.13. Can the method be modified to make it monotone?

The joint pdf of two independent standard normals is simply the product of the individual pdf's:

$$f(x, y) = \frac{1}{2\pi} \exp(-[x^2 + y^2]/2).$$

The key to the proof that the Box–Muller method works is to switch to polar coordinates. Let a point in two-space be represented by r, its distance from the origin, and θ, the angle with the x-axis of the line containing the point and the origin. We want to derive the joint pdf $f_p(r, \theta)$. Changing variables in the usual way, we find

$$f_p(r, \theta) = \frac{r}{2\pi} \exp[-r^2/2].$$

The absence of θ from the joint density function implies that θ is uniformly distributed over $[0, 2\pi]$, generated simply with $2\pi U_1$. Integrating over all possible values of θ gives the marginal pdf for R, namely,

$$f_p(r) = r \cdot \exp(-r^2/2).$$

Thus the cdf is

$$F_p(r) = \int_0^r t \cdot \exp[-t^2/2] \, dt = 1 - \exp[-r^2/2].$$

This is easy to invert. Setting

$$U_2 = 1 - \exp[-R^2/2],$$

we find

$$R = \sqrt{-2 \log(1 - U_2)},$$

which has the same distribution as

$$R = \sqrt{-2 \log(U_2)}.$$

Projecting the vector (R, θ) onto the axes now give the formulas for X and Y. Before using the Box–Muller method, however, note the warning in Section 6.7.3 about using it in conjunction with a linear congruential uniform random number generator. The sine and cosine computations can be eliminated using a rejection technique; see, for example, Knuth [(1981), p. 117]. By combining relations among the normal, chi-square, and exponential distributions, one can motivate the Box–Muller method heuristically; we leave this as an exercise for the reader.

Von Neumann's Method for the Exponential. The following algorithm, due to von Neumann (1951) and generalized by Forsythe (1972), is useful for generating exponential random variates (only) on machines with slow floating point arithmetic. It requires no complicated function such as log, square root, or sin. We look for a sequence $U_1 \geq U_2 \geq \cdots \geq U_{N-1} < U_N$, with N even.

(1) Set $k \leftarrow 0$.
(2) Generate U, uniform on $(0, 1)$, and set $U_1 \leftarrow U$.
(3) Generate U^*, uniform on $(0, 1)$.
(4) If $U < U^*$, then output $X = k + U_1$ and stop.
(5) Generate U, uniform on $(0, 1)$.
(6) If $U \leq U^*$, then go to (3).
(7) Otherwise set $k \leftarrow k + 1$ and go to (2).

The output X is an exponentially distributed random variate with mean 1 as the following argument shows. Consider the instances in which the output X is less than or equal to 1. This can occur only if step 7 is never executed. Denote the successive uniform random numbers used by the algorithm as $U_1, U_2, U_3, \ldots$. We first consider $P\{X \leq x\}$ for $x \leq 1$. By following the algorithm we can tabulate the instances for which $X \leq x$ along with their associated probabilities:

Outcome	Probability
$x \geq U_1 < U_2$	$x - x^2/2$
$x \geq U_1 \geq U_2 \geq U_3 < U_4$	$(x^3/3!) - x^4/4!$
$x \geq U_1 \geq U_2 \geq U_3 \geq U_4 \geq U_5 < U_6$	$(x^5/5!) - x^6/6!$
$\vdots$	$\vdots$

Thus

$$P\{X \leq x\} = x - (x^2/2) + (x^3/3!) - (x^4/4!)$$
$$+ (x^5/5!) - (x^6/6!) \cdots = 1 - e^{-x}$$

for $0 \leq x \leq 1$.

So the algorithm is correct for $0 \leq x \leq 1$. But now it follows that

$$P\{X > 1\} = e^{-1} = P\{\text{step 7 is executed at least once}\}.$$

If step 7 is executed, the procedure starts all over again. By induction and by the memoryless property of the exponential, namely $P\{X \leq k + x | X > k\} = 1 - e^{-x}$, the correctness of the algorithm for $X > 1$ follows.

PROBLEM 5.2.14. Show that the probabilities displayed above are correct. Show that k has the geometric distribution with parameter e^{-1}. What is the average value of k? What is the average number of uniform variates needed to generate one exponential variate? Generalize the method to generate variates with density proportional to $e^{-h(x)}$ for some h and answer the analogs of the preceding questions.

5.3. Continuous Distributions

In the following sections we summarize procedures for generating samples from most continuous distributions one is likely to use in practical simulations.

5.3.1. The Nonstandard Normal Distribution

A normal variate X with mean μ and standard deviation σ can be generated from a standard normal variate Z using the transformation $X = \mu + \sigma Z$. The pdf of X is

$$f(x) = (1/[\sigma\sqrt{2\pi}]) \exp(-[x - \mu]^2/[2\sigma^2]).$$

Methods have already been given for generating standard normal variates in Sections 5.2.7, 5.2.9, and 5.2.10; see also Knuth [(1981), pp. 117–127]. A good assembly language implementation of the rectangle–wedge–tail composition method of MacLaren et al. (1964) is probably the fastest—about three times faster than a good Fortran version of the trapezoidal method. An approximate method occasionally (wrongly) recommended is

$$Z = \left(\sum_{i=1}^{12} U_i\right) - 6,$$

which has mean zero and variance one and by the central limit theorem is approximately normal. In Appendix L it is called SUMNRM. On most computers it is slower than the Box–Muller method (§5.2.10).

5.3.2. The Multivariate (Dependent) Normal Distribution

Let $X = (X_1, X_2, \ldots, X_m)$ be a vector of dependent normal random variates, with the vector of means $\mu = (\mu_1, \mu_2, \ldots, \mu_m)$. Denote the covariance matrix by

$$V = E[(X - \mu)(X - \mu)'] = \begin{bmatrix} v_{11} & & \cdots & v_{1m} \\ & v_{22} & & \\ \vdots & & \ddots & \\ v_{m1} & & & v_{mm} \end{bmatrix}.$$

To initialize the generator, first compute a lower triangular matrix C such that $V = CC'$ (see below). Then

(1) Generate a vector Z with m independent standard normal components.
(2) Output $X = \mu + CZ$.

Barr and Slezak (1972) find this method the best of a number of procedures suggested for generating multivariate normals. Because V is symmetric, use Cholesky factorization to calculate C:

For $i = 1$ to m

for $j = 1$ to $i - 1$

$$c_{ij} = \frac{(v_{ij} - \sum_{k=1}^{j-1} c_{ik} c_{jk})}{c_{jj}};$$

$$c_{ji} = 0;$$

$$c_{ii} = \left(v_{ii} - \sum_{k=1}^{i-1} c_{ik}^2 \right)^{1/2}.$$

The easy calculation

$$V = E[(CZ)(CZ)']$$
$$= E[CZZ'C'] = CE(ZZ')C' = CC'$$

shows that the generator works. If V is such that it implies there is perfect correlation among certain of the variables (i.e., V is singular), then the above procedure must be modified slightly. In particular, if the m by m covariance matrix is based on less than m observations, then V is singular.

5.3.3. Symmetric Stable Variates

The normal distribution has the following elegant property: if two normal random variables X_1 and X_2 have means μ_1 and μ_2 and variances σ_1^2 and σ_2^2, then $X_1 + X_2$ also has a normal distribution with mean $\mu_1 + \mu_2$ and variance $\sigma_1^2 + \sigma_2^2$. This property of remaining in a class under addition is called the stable property. A distribution F is a stable distribution of class α if its characteristic function, defined as

$$\Psi(t) = \int_{-\infty}^{\infty} e^{itx} \, dF(x) = \int_{-\infty}^{\infty} e^{itx} f(x) \, dx$$

can be written

$$\Psi(t) = e^{i\mu t - |ct|^\alpha},$$

where $i = \sqrt{-1}$. In general, closed form expressions are not known for either the cdf or the pdf of stable symmetric distributions.

We must have $0 < \alpha \leq 2, c \geq 0$. These distributions are symmetric about the median μ. Sometimes a fourth parameter, allowing skewness, is also included; see Chambers *et al.* (1976). The Cauchy distribution is stable with $\alpha = 1$. When $\alpha = 2$ and $c = 1$, we have a normal distribution with mean μ and variance 2. When $\alpha < 2$, then moments $> \alpha$ do not exist. Thus, stable variates have "fat-tailed" distributions when $\alpha < 2$.

The stable class of distributions is claimed to describe percentage changes of prices in financial markets. Fama and Roll (1968) study some of its properties.

If X is a symmetric stable random variate with $c = 1$, $\mu = 0$, and some given α, then the transformation

$$Y = Xm + d$$

produces a stable random variable with $c = m$, $\mu = d$, and α unchanged. Generating stable variates with $\mu = 0$ and $c = 1$ thus suffices. Closed form expressions for the pdf or cdf of a symmetric stable distribution are known only for special cases such as the normal and the Cauchy. Bergström (1952) shows that the cdf can be written as either of two infinite series:

(1) $$F_\alpha(x) = \frac{1}{2} + \frac{1}{\pi\alpha} \sum_{k=1}^{\infty} (-1)^{k-1} \frac{\Gamma((2k-1)/\alpha)}{(2k-1)!} x^{2k-1},$$

or

(2) $$F_\alpha(x) = 1 + \left[\frac{1}{\pi} \sum_{k=1}^{\infty} (-1)^k \frac{\Gamma(\alpha k)}{k! x^{\alpha k}} \sin\left(\frac{k\pi\alpha}{2}\right)\right],$$

where Γ is the gamma function.

For $|x|$ small the sum in (1) converges rapidly. For $|x|$ large the sum in (2) converges rapidly. Thus, truncating the sum in either (1) or (2) could possibly be used to approximate $F_\alpha(x)$. There may be numerical problems because of the large values taken on by the gamma and factorial functions. Binary search can be used with (1) or (2) to approximate F^{-1}.

Chambers et al. (1976) give an elegant way of generating stable variates based on the observation that if V is uniform on $(-\pi/2, \pi/2)$ and W is exponential with mean 1, then the following expression is a stable variate with $\mu = 0, c = 1$ and $0 < \alpha \leq 2$:

$$\frac{\sin(\alpha V)}{\cos(V)^{1/\alpha}} \cdot \left(\frac{\cos(V - \alpha V)}{W}\right)^{(1-\alpha)/\alpha}.$$

As $\alpha \downarrow 1$ it reduces to $\tan(V)$ which is the inversion method for generating Cauchy variates.

When $\alpha = 2$ it reduces to half the Box–Muller method for generating normal variates. Note, however, the $\sqrt{2}$ difference in scaling between the standard normal and the standard stable. Unfortunately, for $\alpha \neq 1$ the method is not inversion.

A subroutine based on Chambers et al. for generating stable variates, both skewed and unskewed, appears in Appendix L as FUNCTION RSTAB.

Fitting the distribution to sample data is probably best done by percentile matching rather than moment matching because moments exist only for $\alpha = 2$. The location parameter μ is probably best estimated by the 50th percentile of the sample data. Fama and Roll (1971) suggest essentially the following. Call the inverse of the stable cdf $F^{-1}(u; \mu, c, \alpha)$. The difference $\Delta(c, \alpha, r, s) = F^{-1}(r; \mu, c, \alpha) - F^{-1}(s; \mu, c, \alpha)$ is independent of the location parameter μ and virtually independent of α when $r = 0.72$ and $s = 0.28$. A suitable estimation procedure is therefore

(i) choose $\hat{c}$ to satisfy

$$\Delta(\hat{c}, 1, 0.72, 0.28) = x_{0.72} - x_{0.28},$$

where x_p is the sample pth percentile;
(ii) choose $\hat{\alpha}$ to satisfy

$$\Delta(\hat{c}, \hat{\alpha}, 0.95, 0.05) = x_{0.95} - x_{0.05}.$$

Paulson *et al.* (1975) give a more complex but possibly more accurate estimating procedure.

5.3.4. The Cauchy Distribution

The Cauchy distribution, a symmetric stable distribution with $\alpha = 1$, is the t-distribution with 1 degree of freedom. It has no moments. It is occasionally recommended as a good distribution for shaking out bugs during the development of a simulation model. Use it if you need lots of extreme values. Figure 5.3.1 shows the standard normal and Cauchy distributions.

The pdf of a Cauchy distribution with median μ and scale σ is

$$f(x) = \sigma/[\pi(\sigma^2 + [x - \mu]^2)], \qquad -\infty \le x \le \infty.$$

The cdf is

$$F(x) = \frac{1}{\pi} \arctan((x - \mu)/\sigma) + \tfrac{1}{2}.$$

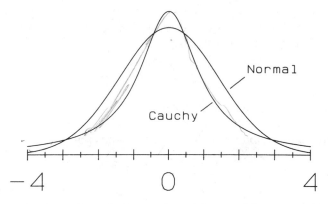

Figure 5.3.1. Cauchy and standard normal special cases of the symmetric stable class.

Generate by inversion:

$$X = \sigma \tan(\pi(U - \tfrac{1}{2})) + \mu.$$

Physically, think of spinning a pointer centered at (μ, σ) with X being the intercept on the x-axis. The ratio of two independent standard normal variables happens to be a Cauchy variable with $\mu = 0$ and $\sigma = 1$. This is, however, not an efficient method of generation. Problem 5.5.7 at the end of the chapter outlines an alternative fast method.

5.3.5. The Lognormal Distribution

A variable X is lognormal if $\log X$ is normal. The pdf for the lognormal distribution with parameters μ and σ is

$$f(x) = (1/[x\sqrt{2\pi\sigma}]) \exp[-(\log x - \mu)^2/(2\sigma^2)].$$

The mean and variance of X are

$$E[X] = \exp[\mu + \sigma^2/2],$$

$$\mathrm{Var}[X] = [\exp(\sigma^2) - 1] \exp[2\mu + \sigma^2].$$

The obvious procedure for generating a lognormal random variable is

(1) Generate Y normally distributed with mean μ and variance σ^2.
(2) Output $X = \exp Y$.

We noted in Section 4.2 that we expect a random variable to be approximately lognormal if it is the product of "enough" positive random variables. The "law of proportionate effects" holds if when I_0 is the level of an index in period 0 and i_k is the growth rate in period k, then the level I_n in period n is $I_0(1 + i_1)(1 + i_2) \cdots (1 + i_n)$, a product of random variables. This does not imply that successive values of I_n are iid, though I_n may be approximately lognormal for large n. One regularly finds, however, that the sequence of differenced logs, $\log(I_k) - \log(I_{k-1}) = \log(1 + i_k)$, does behave as a collection of iid normal random variables.

5.3.6. The Exponential Distribution

The exponential distribution with parameter λ has pdf

$$f(x) = \lambda e^{-\lambda x} \quad \text{for } 0 \le x$$

and cdf

$$F(x) = 1 - e^{-\lambda x} \quad \text{for } 0 \le x.$$

The mean, variance, and mode are respectively, $1/\lambda$, $1/\lambda^2$, and 0. On a computer inversion is fast; set

$$X = -\log(1 - U)/\lambda.$$

Because $1 - U$ has the same distribution as U, some recommend using the simpler transformation

$$X = -\log(U)/\lambda.$$

Strictly speaking, this is no longer inversion and correlation of the wrong sign for variance reduction techniques may be induced.

Ahrens and Dieter (1972a) and Knuth [(1981), p. 128] describe other generators.

5.3.7. The Hyperexponential Distribution

The hyperexponential is a high-variance generalization of the exponential distribution. Generating a hyperexponential variate corresponds to randomly choosing a variate from one of n different exponential distributions. The pdf of the hyperexponential with n branches is

$$f(x) = \sum_{i=1}^{n} p_i \lambda_i e^{-\lambda_i x} \quad \text{for } 0 \le x.$$

Its cdf is

$$F(x) = \sum_{i=1}^{n} p_i(1 - e^{-\lambda_i x}).$$

The mean, variance and mode are $\sum_{i=1}^{n} p_i/\lambda_i$, $\sum_{i=1}^{n} p_i/\lambda_i^2$ and 0, respectively. The variance always equals or exceeds the square of the mean. A comparison of the exponential pdf and a hyperexponential pdf appears in Figure 5.3.2. The parameters p_i, λ_i satisfy

$$p_i, \lambda_i \ge 0 \quad \text{and} \quad \sum_{i=1}^{n} p_i = 1.$$

Using composition to generate hyperexponential variates, we

(1) Generate U_1 and find k such that

$$\sum_{i=1}^{k-1} p_i < U_1 \le \sum_{i=1}^{k} p_i.$$

(2) Output $X = -\log(U_2)/\lambda_k$.

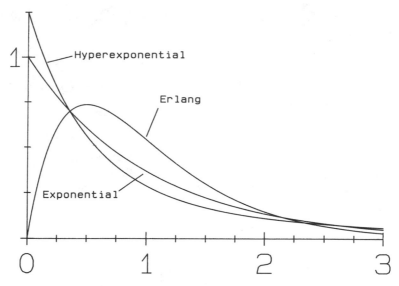

Figure 5.3.2. Hyperexponential, exponential, and Erlang-2 pdf's with mean 1.

5.3.8. The Laplace and Exponential Power Distributions

The exponential power distribution class is convenient for examining the effect of the shape of a symmetric distribution. Figure 5.3.3 shows the pdf,

$$f(x) = \mu \cdot \exp(-|\mu x|^\alpha)/[2\Gamma(1 + 1/\alpha)]$$

for

$$-\infty < x < \infty, \alpha \geq 1 \text{ and } \mu > 0,$$

for several values of α. This class includes the uniform on $(-1/\mu, 1/\mu)$ and the normal as special cases when respectively $\alpha \to \infty$ and $\alpha = 2$. When $\alpha = 1$, a variate from this distribution has the same distribution as the difference of two iid exponential variates, each with expectation $1/\mu$. In this case it is called the Laplace distribution.

The mean, variance, and kurtosis $(E[X^4]/E^2[X^2])$ of the exponential power distribution are 0, $\Gamma(3/\alpha)/[\mu^2\Gamma(1/\alpha)]$, and $\Gamma(5/\alpha)\Gamma(1/\alpha)/[\Gamma(3/\alpha)^2]$.

Fitting the distribution by moment matching is straightforward. The parameter α is determined solely by the kurtosis, and so it can be found by a simple search procedure. The variance then directly determines μ.

Tadikamalla (1980) generates exponential power variates by rejection. For the majorizing density he uses the normal if $\alpha \geq 2$ and the Laplace if $1 \leq \alpha < 2$. For $\alpha = 1$ or $\alpha = 2$, invert; e.g., for $\alpha = 1$:

(1) Generate U.
(2) If $U \leq \frac{1}{2}$, return $X = \log(2U)/\mu$; otherwise return $X = -\log(2 - 2U)/\mu$.

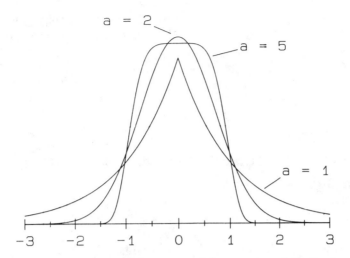

Figure 5.3.3. Examples of the exponential power class of distribution.

PROBLEM 5.3.1. Verify this.

PROBLEM 5.3.2. Y_1 and Y_2 are independent and exponentially distributed, not necessarily with the same mean, and $X = Y_1 - Y_2$. Given that $X \geq 0$, show that X has the same (conditional) distribution as Y_1.

Replacing x by $x - d$ in the expression defining $f(x)$ shifts the mean from 0 to d. Sometimes $1/\mu$ is called the scale.

5.3.9. Erlang and Gamma Distributions

The Erlang distribution has two parameters, λ and k. An Erlang random variate is the sum of k iid exponential variates, each with expectation $1/\lambda$. Figure 5.3.2 shows an Erlang distribution with $k = 2$. The obvious generating procedure is

$$X = - \sum_{i=1}^{k} \log(U_i)/\lambda,$$

which simplifies slightly to

$$X = -\log\left(\prod_{i=1}^{k} U_i\right) \Big/ \lambda.$$

The pdf is

$$f(x) = \lambda e^{-\lambda x}(\lambda x)^{k-1}/(k - 1)! \quad \text{for } 0 \leq x.$$

The cdf is

$$F(x) = 1 - \sum_{j=0}^{k-1} e^{-\lambda x}(\lambda x)^j/j!.$$

The gamma distribution is obtained by allowing k to be any nonnegative real number rather than just an integer. The pdf for the gamma is

$$f(x) = \lambda e^{-\lambda x}(\lambda x)^{k-1}/\Gamma(k) \quad \text{for } 0 \le x.$$

The mean and variance are k/λ and k/λ^2, respectively. For $k \ge 1$, the mode is $(k-1)/\lambda$ and the variance is less than or equal to the square of the mean. For $k < 1$, the mode is at 0 (where $f(x)$ is infinite) and the variance is greater than the square of the mean.

To generate a gamma variate with shape parameter k, a simple but slow algorithm described by Berman (1970) and due to Jöhnk (1964) is

(1) Set $m = \lfloor k \rfloor$ and $q = k - m$.
(2) Set $Z = -\log(\prod_{i=1}^{m} U_i)$.
(3) Generate a random variate W which is beta distributed with parameters q, $1 - q$, using an algorithm given in Section 5.3.10.
(4) Generate a uniform random number U and set $Y = \log U$.
(5) Output $X = (Z + WY)/\lambda$.

When k is large, step 2 requires a large number of uniform random variates.

Tadikamalla (1978) proposes an algorithm based on rejection which is faster at least for $k > 2$, say. He bounds the beta distribution using the Laplace distribution (§5.3.8) with mean $k - 1$ and scale $\theta = (1 + \sqrt{4k - 3})/2$. His algorithm follows:

(1) Generate a Laplace variate X with mean $k - 1$ and scale θ.
(2) If $X < 0$, go to 1.
(3) Generate U.
(4) If $U \le T(X)$ where

$$T(x) = \left| \frac{(\theta - 1)x}{\theta(k - 1)} \right|^{k-1} \exp[-x + (|x - (k - 1)| + (k - 1)(\theta + 1))/\theta],$$

output X; otherwise go to 1.

Appendix L lists faster but more complicated rejection-based generators RGS and RGKM3.

The above methods are incompatible with variance reduction techniques that require synchronization; likewise, for the faster but more complicated methods of Marsaglia (1977), Schmeiser and Lal (1980), Tadikamalla and Johnson (1981), and Ahrens and Dieter (1982a). Inversion is compatible with variance reduction techniques. Two approximations to the inverse of the gamma cdf called VCHISQ and VCHI2 are given in Appendix L.

5.3.10. The Beta Distribution

The beta distribution achieved a certain prominence when early treatments of the PERT project management technique gave the impression that activity times are approximately beta distributed. At best this is a convenient assumption, not the literal truth. The beta distribution happens to be convenient in Bayesian statistics for the "prior" distribution of the parameter of a binomial distribution.

The beta distribution has two parameters m and n, known as degrees of freedom. Define $a = m/2$, $b = n/2$. Now if Y and Z are gamma distributed with parameters λ, a and λ, b, respectively (λ can be chosen arbitrarily), then

$$X = Y/(Y + Z)$$

is a beta variate. The above is a simple though not necessarily efficient means for generating beta variates. The beta pdf is

$$f(x) = \frac{x^{a-1}(1 - x)^{b-1}}{\beta(a, b)},$$

where

$$\beta(a, b) = \int_0^1 y^{a-1}(1 - y)^{b-1}\, dy$$

is the beta function and $0 \leq x \leq 1$.

The mean and variance are $a/(a + b)$ and $ab/[(a + b + 1)(a + b)^2]$, respectively. When $a > 1$ and $b > 1$ the mode is $(a - 1)/(a + b - 2)$. Fox's (1963) algorithm works when a and b are both integers and runs fast when both are small.

(1) Generate $a + b - 1$ uniform random numbers in the interval $(0, 1)$.
(2) The value of the ath smallest of these is beta distributed with parameters a, b.

In Appendix L this algorithm is called BETAFX.

The beta distribution is convenient if one wishes to examine the effect of right or left skewness. Figure 5.3.4 indicates how the shape of the distribution is affected by the parameters a and b. It can be rescaled and translated to cover an arbitrary finite interval. As the figure shows, it does not take large values of a or b to produce a significantly skewed distribution.

When either a or b is nonintegral and neither is greater than one, then the following method described by Berman (1970) and due to Jöhnk (1964) is fairly fast:

(1) Repeat the following until $X + Y \leq 1$:
    (a) Generate a uniform random number U in $(0, 1)$, and set $X = U^{1/a}$.
    (b) Generate a uniform random number V in $(0, 1)$, and set $Y = V^{1/b}$.
(2) $X/(X + Y)$ is beta distributed with parameters a and b.

This method is inefficient if $a + b$ is large because the loop may be executed many times. Cheng (1978) presents a simple rejection procedure which is faster when a or b is greater than one. The steps are:

(0) (Initialization.) Set $\alpha = a + b$. If $\min(a, b) \leq 1$ set $\beta = 1/\min(a, b)$; otherwise set $\beta = \{(\alpha - 2)/(2ab - \alpha)\}^{1/2}$. Set $\gamma = a + 1/\beta$.

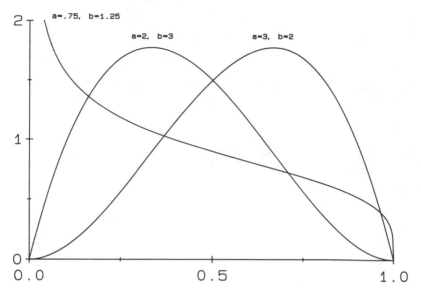

Figure 5.3.4. Beta pdf's.

(1) Generate U_1 and U_2, and set $V = \beta \log\{U_1/(1 - U_1)\}$, $W = a \cdot \exp(V)$.
(2) If $\alpha \log\{\alpha/(b + W)\} + \gamma V - \log(4) < \log(U_1 U_2 U_3)$, go to (1).
(3) Return $X = W/(b + W)$.

Step 0 need be performed only at the first call for a given pair a, b. This method is called BETACH in Appendix L. Schmeiser and Babu (1980) give more complicated methods. Not counting set-up time, these are faster according to their empirical results. An approximation to the inverse beta cdf appears in Appendix L as VBETA.

5.3.11. The Chi-square Distribution

The chi-square distribution has a single positive integer parameter n, the number of degrees of freedom. If $\{Z_i\}$ is a sequence of independent standard normal variates, then

$$X = \sum_{i=1}^{n} Z_i^2$$

has a chi-square distribution with n degrees of freedom. The mean and variance are n and $2n$, respectively, and the mode is $\max(0, n - 2)$. The pdf for the chi-square distribution is

$$f(x) = [x^{(n/2)-1} \exp(-x/2)]/[\Gamma(n/2)2^{n/2}], \qquad x \geq 0.$$

This is a gamma distribution with $\lambda = 1/2$ and $k = n/2$. If n is large, generate chi-square variates using one of the methods cited near the end of Section 5.3.9.

If n is even and $\{Y_i\}$ is a sequence of independent exponential variates, each with expectation 1, then

$$X = 2 \sum_{i=1}^{n/2} Y_i$$

also has a chi-square distribution with n degrees of freedom. If n is even and not large, this is an efficient means of generating X. If n is odd and not large, an efficient procedure is

$$X = \sum_{i=1}^{\lfloor n/2 \rfloor} Y_i + Z^2,$$

where Z is a standard normal variate. In Table 5.6.1 and in Appendix L, VCHISQ denotes a quick and dirty approximation to the inverse of the chi-square cdf and VCHI2 denotes an approximation accurate to about five decimal places but requiring more computation.

5.3.12. The F-Distribution

The F-distribution is a two-parameter distribution important in statistical hypothesis testing, e.g., in analysis of variance. Let Y_1 and Y_2 be chi-square with w and v degrees of freedom, respectively. Then

$$X = (Y_1/w)/(Y_2/v)$$

is F-distributed with parameters w and v. The pdf is

$$f(x) = \frac{(v/w)^{v/2}}{\beta(w/2, v/2)} x^{(w/2)-1}(x + v/w)^{-(w+v)/2}, \qquad x \geq 0,$$

where $\beta(a, b)$ is the beta function. The mean, variance, and mode are $v/(v - 2)$ (for $v > 2$), $2v^2(v + w - 2)/[w(v - 4)(v - 2)^2]$ (for $v > 4$), and $v(w - 2)/[w(v + 2)]$ (for $v > 2$), respectively. To generate an F-variate, transform a beta variate:

(1) Generate Z from a beta distribution with parameters $a/2$ and $b/2$.
(2) Return $F(a, b) = (b/a)Z/(1 - Z)$, using an obvious notation.

Alternatively, use (approximate) inversion—called VF in Appendix L.

5.3.13. The t-Distribution

Let Y_1 be standard normal, and let Y_2 be independent of Y_1 and be chi-square with v degrees of freedom; then $X = Y_1/\sqrt{Y_2/v}$ is t-distributed with v degrees of freedom. The pdf of the t-distribution is

$$f(x) = \frac{\Gamma((v + 1)/2)}{\Gamma(v/2)\sqrt{v\pi}} (1 + x^2/v)^{-(v+1)/2}.$$

The mean and mode are both 0, while the variance is $v/(v - 2)$ (for $v > 2$). As $v \to \infty$, the t-distribution approaches the normal distribution; when $v = 1$ the t-distribution is Cauchy. A t variate squared is an F variate with parameters 1 and v. In Appendix L VSTUD is a functional approximation to the inverse of the cdf of the t-distribution.

5.3.14. The Weibull Distribution

The Weibull generalizes the exponential. Its cdf and pdf with parameters $\lambda > 0, c > 0$ are

$$F(x) = 1 - \exp[-(\lambda x)^c],$$

$$f(x) = \lambda c(\lambda x)^{c-1} \exp[-(\lambda x)^c], \qquad x \geq 0.$$

To generate Weibull variates, invert:

$$X = ([-\log(1 - U)]^{1/c})/\lambda,$$

where U is uniform on $(0, 1)$. The mean and variance are $[\Gamma(1 + 1/c)]/\lambda$ and $[\Gamma(1 + 2/c) - (\Gamma(1 + 1/c))^2]/\lambda^2$, respectively. Figure 5.3.5 illustrates the Weibull pdf for various values of c. If $c < 3.602$ the distribution has a long right tail. If $c = 3.602$ it closely approximates a normal. If $c > 3.602$ it has a long left tail. Section 4.5 sketches how the Weibull distribution arises as a limit distribution of minima.

Fitting the Weibull by matching moments is reasonably straightforward. Notice that σ/μ is a function of only c. Thus, a simple search procedure can find c given σ/μ. Subroutine ALOGAM in Appendix L can be used (after

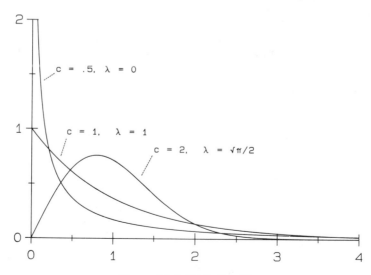

Figure 5.3.5. Weibull pdf's.

exponentiation) to evaluate $\Gamma(\ )$. Once c is determined, $1/\lambda$ is a linear function of the mean.

5.3.15. The Gumbel Distribution

When X is the maximum of a collection of random variables, then a class of distributions studied by Gumbel (1958) and Galambos (1978) plays a role analogous to that of the normal and the Weibull for addition and minimization, respectively. The so-called Gumbel or extreme value distributions are limiting distributions under less general conditions than before; however, they have the expected property that the maximum of two Gumbel random variables of the same class also has a Gumbel distribution.

Two types of Gumbel distributions are of interest. The first, called the type I extreme value distribution by Barlow and Proschan (1965), has cdf

$$F(x) = \exp[-be^{-ax}],$$

with pdf

$$f(x) = ab \cdot \exp[-(be^{-ax} + ax)].$$

The maximum of a collection of exponential-like random variables tends to a type I distribution.

A second type of Gumbel distribution may arise when taking the maximum of random variables with "fat-tailed" distributions. The type II distribution of class k has cdf

$$F(x) = \exp[-bx^{-k}],$$

with pdf

$$f(x) = bkx^{-(k+1)} \cdot \exp(-bx^{-k}).$$

Let $X = \max(Y_1, Y_2)$, where both Y_1 and Y_2 are Gumbel type II of class k with parameters b_1 and b_2. Then

$$P\{X \le t\} = P\{Y_1 \le t\}P\{Y_2 \le t\}$$
$$= \exp(-b_1 t^{-k}) \cdot \exp(-b_2 t^{-k}) = \exp(-[b_1 + b_2]t^{-k}).$$

Thus X is Gumbel of class k with parameter $b_1 + b_2$. Moments higher than the kth do not exist for a type II distribution.

One might expect such things as completion times of projects to be approximately Gumbel distributed if each project consists of a large number of independent activities in parallel. Rothkopf (1969) uses the Gumbel distribution to approximate the distribution of winning bids in a model of highest-bid-wins auctions.

To generate type I variates, invert:

$$X = [\log(b) - \log(-\log U)]/a;$$

likewise, for type II variates,

$$X = [-b/\log U]^{1/k}.$$

Comparing methods for generating variates, we see that a type II Gumbel variate is the reciprocal of a Weibull variate.

5.3.16. The Logistic Distribution

The logistic distribution has cdf and pdf

$$F(x) = 1/[1 + \exp(-[x - a]/b)],$$

$$f(x) = \exp(-[x - a]/b)/[b(1 + \exp(-[x - a]/b))^2].$$

The mean, variance, and mode are a, $(b\pi)^2/3$ and a. Variates are easily generated by inversion:

$$X = a - b \log[(1/U) - 1].$$

The logistic distribution is an asymptotic distribution for the midrange of a set, defined to be the average of the smallest and the largest of a set of iid observations.

5.3.17. The Generalized Lambda Distribution

Ramberg and Schmeiser (1974) introduce a four-parameter "generalized lambda" distribution to approximate a wide class of other distributions or data by matching moments. A generalized lambda variate X is obtained by

$$X = \lambda_1 + [U^{\lambda_3} - (1 - U)^{\lambda_4}]/\lambda_2.$$

The mean can be shifted to any value by changing λ_1. The variance can be shifted to any nonnegative value by changing λ_2. The skewness and kurtosis (or peakedness) are controlled by λ_3 and λ_4. If $\lambda_3 = \lambda_4$, then the distribution is symmetric with mean λ_1 and variance $2(1/(2\lambda_3 + 1) - \beta(\lambda_3 + 1, \lambda_3 + 1))/\lambda_2^2$, where $\beta(a, b)$ is the beta function. Ramberg and Schmeiser give tables of $\lambda_1, \lambda_2, \lambda_3$, and λ_4 for approximating such distributions as the normal, gamma, t, and Cauchy. In view of recent developments in variate-generation algorithms, the generalized lambda distribution seems appropriate only for matching data.

5.3.18. Nonhomogeneous Poisson Processes

In many systems the arrival rate of jobs, events, etc., varies with the time of day. If the times between successive arrivals are nevertheless exponential and independent, the arrival process is a nonhomogeneous Poisson process.

Suppose the instantaneous arrival rate at time t is $\lambda(t)$. Then the expected number of arrivals in any interval (b, c) is $\int_b^c \lambda(t)\, dt$. The integrated arrival rate function is $a(x) = \int_0^x \lambda(t)\, dt$.

There are at least two methods for generating nonhomogeneous Poisson events.

Method I: [Thinning by rejection, cf. Lewis and Shedler (1979a).] Assume an interarrival time is to be generated at time s.

(1) Set $\bar\lambda \leftarrow \sup\{\lambda(t)|t \geq s\}$.
(2) Generate an exponential random variable X with mean $1/\bar\lambda$.
(3) Generate U uniform on $(0, 1)$.
(4) If $U \leq \lambda(s + X)/\bar\lambda$, then make the next arrival occur at time $s + X$; otherwise set $s \leftarrow s + X$ and go to (1).

PROBLEM 5.3.3. Show that thinning works. Hint: Show (almost trivially) that the arrivals generated by the thinning algorithm form a nonhomogeneous Poisson process. Next show that on an arbitrary interval (b, c) the expected number of arrivals from this process is $a(c) - a(b)$. To do this, condition on k arrivals in that interval from the homogeneous process generated at step 2; then remove the condition. To condition, use Problem 4.9.14 to show that the conditional joint density of the k values of $\lambda(s + X)/\bar\lambda$ at step 4 such that $b \leq s + X \leq c$ is

$$k!\left[\int_b^c \frac{\lambda(t)}{\bar\lambda(c - b)}\, dt\right]^k = k!\left[\frac{a(c) - a(b)}{\bar\lambda(c - b)}\right]^k.$$

This says that the ith arrival from the homogeneous process is selected, independently of the others, with probability

$$p = [a(c) - a(b)]/[\bar\lambda(c - b)].$$

Therefore, given k such arrivals, the expected number selected is kp. Finally, $E[kp] = a(c) - a(b)$ as was to be shown.

Method II relies on Proposition 4.9.1. The next event occurs at $a^{-1}(s + X)$, where X is an exponential variate with mean one and s is the current time.

In a typical application λ, the derivative of a, is approximated by a piecewise constant function which may cycle. This is illustrated in Figure 5.3.6 for a cycle with three segments. A finite cycle length often seems reasonable; otherwise, we are extrapolating a trend infinitely far into the future—a speculation which cannot be validated. We give below an algorithm for generating an interevent time from a nonhomogeneous Poisson process with a piecewise-constant arrival rate.

Suppose that at time s we want to generate the time for the next event. We have to generate an exponential variate X with mean one and then evaluate $a^{-1}(s + X)$. Now a^{-1} is a continuous, piecewise-linear function with breakpoints $r_0, r_1, \ldots, r_N, \ldots$ where N is the number of segments in a cycle and $r_0 = 0$. In Figure 5.3.6, $N = 3$ and $r_1 = \lambda_1 t_1, r_2 = r_1 + \lambda_2(t_2 - t_1)$,

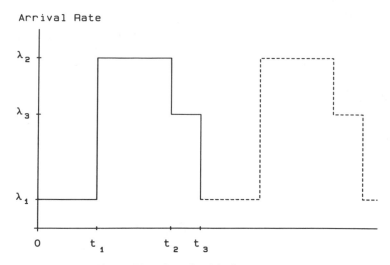

Figure 5.3.6. A cycle with three segments.

and $r_3 = r_2 + \lambda_3(t_3 - t_2)$. So we have first to find the piece of a^{-1} where $s + X$ falls, and then to do inverse linear interpolation. If the average inter-event time is much shorter than the typical segment length, the myopic search shown below for the right piece of a^{-1} seems reasonable. Otherwise, a myopic search wastes time examining segments in which no event occurs; in this case, use binary search or a hybrid bucket system (see §5.2.1).

The breakpoints of a^{-1} and λ are related by $r_i = a(t_i)$. Writing $s + X$ in the form $jr_N + y$ where $0 \leq y < r_N$ gives $a^{-1}(s + X) = jt_N + a^{-1}(y)$. To compute $a^{-1}(y)$, for a given $y < r_N$, proceed as follows:

1. (Put $z = r_1$): Set $z \leftarrow \lambda_1 t_1$.
2. (Check first segment): If $y \leq z$, then output y/λ_1 and stop.
3. (Set index): Set $i \leftarrow 2$.
4. (Put $w = r_i$): Set $w \leftarrow z + \lambda_i t_i$.
5. (Check ith segment): If $y \leq w$, then output

$$t_{i-1} + (y - z)/\lambda_i$$

and stop. [Comment: If y falls in this segment,

$$a^{-1}(y) = a^{-1}(z + (y - z))$$
$$= t_{i-1} + (y - z)/\lambda_i.]$$

6. (Try next segment): Set $i \leftarrow i + 1$, $z \leftarrow w$, and go to step 4.

PROBLEM 5.3.4. Suppose we want to generate a series of interevent times. Modify the algorithm to take advantage of saving the most recent values of i and z between calls.

Thinning may be faster than inversion if λ is hard to integrate, a is hard to invert, or the expected number of arrivals per cycle of λ is less than, say, one. The first two difficulties may be caused by postulating (probably unnecessarily) complex forms for λ or a.

5.4. Discrete Distributions

In many situations a random variable of interest takes only integer values. We now consider the generation of random variables with discrete distributions.

5.4.1. The Binomial Distribution

The binomial distribution has two parameters, n and p. It describes the number of successes in n independent trials where p is the probability of success at any given trial. A simple algorithm requiring no set-up but $O(n)$ marginal time follows immediately:

(1) Set $X \leftarrow 0$.
(2) Do the following n times:

$$\text{Generate } U \text{ uniform on } (0, 1).$$
$$\text{If } U \le p, \text{ set } X \leftarrow X + 1.$$

(3) Output X.

For large n, this is intolerably slow when more than a few variates are needed.

The probability mass function (pmf) is

$$f(x) = P\{X = x\}$$

$$= \frac{n!}{(n - x)!\, x!}\, p^x (1 - p)^{n-x}, \qquad x = 0, 1, \ldots, n.$$

The mean and variance are np and $np(1 - p)$, respectively. To tabulate the cdf, use the routine TBINOM (Appendix L); to generate variates by inversion, then use the routine IVDISC (§5.2.1 and Appendix L). This requires $O(n)$ time to set up a table of size n and then $O(1)$ marginal time.

For large n, Ahrens and Dieter (1980) and Devroye (1980) give methods with average time $O(1)$ instead of $O(n)$—including set-up. These methods, however, are not inversion. Their marginal speeds are slower than inversion.

5.4.2. The Poisson Distribution

The Poisson distribution has a single parameter θ. The pmf is

$$f(x) = e^{-\theta}\theta^x/x!, \qquad x = 0, 1, \dots.$$

The mean and variance are both θ. Because X has an infinite range, inversion is potentially slow—even when the cdf is calculated efficiently as needed. The binomial distribution approaches the Poisson when $n \to \infty$ and $p \to 0$ with $np = \theta$ remaining constant.

The Poisson is also related to the exponential. If interarrival times are independent and exponential with mean 1, then the number of arrivals occurring in the interval $(0, \theta)$ is Poisson with parameter θ. We exploit this to derive a generating procedure. If $\{Y_i\}$ is a sequence of independent exponentials, each with expectation 1, then we wish to find X such that the Xth arrival occurs before θ but the $X + 1$st occurs after θ; that is,

$$\sum_{i=1}^{X} Y_i \le \theta < \sum_{i=1}^{X+1} Y_i.$$

Recalling how we generate exponentials, this is

$$-\sum_{i=1}^{X} \log U_i \le \theta < -\sum_{i=1}^{X+1} \log U_i,$$

or equivalently

$$\prod_{i=1}^{X} U_i \ge e^{-\theta} > \prod_{i=1}^{X+1} U_i.$$

This gives the generator:

(1) Set $X \leftarrow -1$, $m \leftarrow \exp(\theta)$.
(2) Repeat the following until $m < 1$:

        Generate U uniform on $(0, 1)$.
        Set $X \leftarrow X + 1$, $m \leftarrow mU$.

(3) Output X.

When θ is large this procedure is slow. In this case, however, Poisson variates are approximately normally distributed with mean and variance θ: consider seriously using the normal distribution instead. Ahrens and Dieter (1982b), Devroye (1981a), and Schmeiser and Kachitvichyanukul (1981) give $O(1)$ average time exact methods; they are fast but complicated and not inversion.

Do you really want to use a Poisson variate equal to 100 times its mean? We recommend clipping tails to avoid statistical anomalies. If a clipped

Poisson distribution is acceptable, then a cdf table can be set up using algorithm TTPOSN in Appendix L. Generate a variate using the inversion algorithm IVDISC. The marginal time is $O(1)$. To obtain setup time and space requirement both $O(\sqrt{\theta})$, use the method of Fox and Glynn (1986c) to choose left and right truncation points as input to TTPOSN. Fox and Glynn show how to do this so that the truncation error is negligible in a precise sense.

5.4.3. Compound Poisson Distributions

In many inventory systems we can assume that the times between demands are independent and exponentially distributed. Thus, the number of demand points in a given interval is Poisson distributed. However, at each demand point the number of units ordered follows some arbitrary distribution. The number of units ordered in an interval then has a compound Poisson distribution.

If the arrival rate of demand points is λ and the quantities ordered at each point are independent and identically distributed with mean μ and variance σ^2, then the mean and variance of the number ordered in an interval of length t are $\lambda t\mu$ and $\lambda t[\sigma^2 + \mu^2]$.

The obvious generating procedure for the compound Poisson is to generate K from the Poisson distribution with parameter $\theta = \lambda t$ and then sum K draws from the arbitrary distribution compounding the Poisson. If the compounding distribution is discrete, it is sometimes faster to calculate the pmf for the compound Poisson and then to invert. Define

$$b(j) = P[\text{batch size at a demand point} = j] \qquad j = 0, 1, 2, \ldots,$$

$$r(j, t) = P[\text{total demand in an interval of length } t \text{ is equal to } j].$$

Adelson (1966) shows that $r(j, t)$ can be calculated by the recursion

$$r(0, t) = \exp(-\lambda t[1 - b(0)]),$$

$$r(j, t) = (\lambda t/j) \sum_{k=1}^{j} kb(k)r(j - k, t), \qquad j \geq 1.$$

5.4.4. The Hypergeometric Distribution

The binomial distribution can also be derived in terms of sampling from the famous probabilist's urn. The urn contains m balls, a fraction p of which are gamboge with the remainder chartreuse. Make n draws from the urn, returning the ball just drawn and remixing after each draw. The number of gamboge balls drawn is binomially distributed.

If, however, the sampling is without replacement, then the total m makes a difference, and the number of gamboge balls drawn has the hypergeometric distribution with parameters m, n, and p. Writing $q = 1 - p$, the pmf is

$$f(x) = \frac{\binom{mp}{x}\binom{mq}{n-x}}{\binom{m}{n}}, \qquad x = 0, 1, \ldots, n.$$

The mean and variance are np and $npq(m - n)/(m - 1)$, respectively. The obvious generating procedure follows:

(1) Set $X \leftarrow 0$, $g \leftarrow pm$, $c \leftarrow m - g$
    (X is the number of gamboge balls drawn, and g and c are respectively the number of gamboge and chartreuse balls remaining in the urn).
(2) Do the following n times:
        Generate U uniform on $(0, 1)$.
        If $U \leq g/m$, then set $X \leftarrow X + 1$, $g \leftarrow g - 1$; otherwise set $c \leftarrow c - 1$.
        Set $m \leftarrow m - 1$.
(3) Output X.

Tabulating the cdf and then inverting via IVDISC has higher marginal speed. Kachitvichyanukul and Schmeiser (1985) give a rejection algorithm uniformly fast in the parameters m, n, and p.

5.4.5. The Geometric Distribution

The geometric distribution is the discrete analog of the exponential, the two being memoryless. It has a single parameter p, the probability of success at each trial. Writing $q = 1 - p$, the pmf and cdf are:

$$f(x) = q^x p,$$
$$F(x) = 1 - q^{x+1}, \qquad x = 0, 1, 2, \ldots.$$

Thus, X is the number of trials before the first success. The mean, variance, and mode are respectively q/p, q/p^2, and 0. To generate geometric variates

(1) Set $X \leftarrow -1$.
(2) Repeat the following until $U \leq p$:
        Set $X \leftarrow X + 1$.
        Generate U uniform on $(0, 1)$.
(3) Output X.

Alternatively, invert: simply set $X = \lfloor \log(1 - U)/\log q \rfloor$. For small values of p this is faster. It always simplifies synchronization.

PROBLEM 5.4.1. Show that this works. (Hint: Show that

$$X = n \Leftrightarrow n + 1 > \log(1 - U)/\log q \geq n$$
$$\Leftrightarrow q^{n+1} < 1 - U \leq q^n$$
$$\Leftrightarrow 1 - q^n \leq U < 1 - q^{n+1},$$

and hence that X has the required distribution.)

PROBLEM 5.4.2. Show that $\log(1 - U)$ can be replaced by the negative of an exponential variate.

PROBLEM 5.4.3. For the case $p = \frac{1}{2}$, show how to exploit a binary computer having an instruction that (effectively) counts the number of leading zero bits in a word.

PROBLEM 5.4.4. Generate the outcomes of a series of Bernoulli trials via a shorter series of geometric variates.

PROBLEM 5.4.5. In a uniformized, continuous-time Markov chain generate each sequence of (null) jumps from a state to itself using just one uniform variate. (Hint: See Problem 5.4.4.)

5.4.6. The Negative Binomial and Pascal Distributions

The negative binomial distribution is the discrete analog of the Erlang. It has two parameters, p and k. If k is an integer, it is called the Pascal distribution. A Pascal variate is the sum of k independent geometric variates each with parameter p. Stated another way, it is the number of failures until k successes.

Writing $q = 1 - p$, the pmf is

$$f(x) = \frac{\Gamma(k + x)}{\Gamma(x + 1)\Gamma(k)} p^k q^k, \qquad x = 0, 1, 2, \ldots .$$

The mean and variance are kq/p and kq/p^2, respectively.

For arbitrary positive k, composition works:

(1) Generate θ from a gamma distribution with parameters k and $\lambda = p/q$.
(2) Generate X from a Poisson distribution with parameter θ. Now X is negative binomial with parameters p and k.

To verify this, observe that

$$P\{X = x\} = \int_0^\infty P\{X = x|\theta\}[\lambda^k \theta^{k-1} e^{-\lambda\theta}/\Gamma(k)] \, d\theta$$

$$= \int_0^\infty \frac{e^{-\theta}\theta^x}{x!} [\lambda^k \theta^{k-1} e^{-\lambda\theta}/\Gamma(k)] \, d\theta$$

$$= \frac{\Gamma(k + x)}{\Gamma(x + 1)\Gamma(k)} \left(\frac{\lambda}{1 + \lambda}\right)^k \left(\frac{1}{1 + \lambda}\right)^x$$

$$= \frac{\Gamma(k + x)}{\Gamma(x + 1)\Gamma(k)} p^k q^x.$$

Figure 5.4.1. Two-dimensional Poisson variates.

Fitting the negative binomial to data is simplified if we note that σ^2/μ is a function of p only. Thus, a simple search procedure can find p given σ^2/μ. Once p is determined, k is linear in μ.

5.4.7. Multidimensional Poisson Distributions

A generalization of the Poisson distribution to two dimensions follows. If θ is the expected number of variates in a rectangle with sides a and b, then to get these variates:

(1) Generate a random variable N from the Poisson distribution with mean θ.
(2) For $i = 1, 2, \ldots, N$, the coordinates of the ith point are two random variables, X_i uniform on $(0, b)$ and Y_i uniform on $(0, a)$.

Figure 5.4.1 shows a typically haphazard scattering of two-dimensional Poisson variates. Lewis and Shedler (1979a) give a thinning algorithm for the nonhomogeneous Poisson generalization.

5.5. Problems

PROBLEM 5.5.1. The width, height, and depth of a box are independent random variables. What distribution might approximate the volume of the box? (Hint: Perhaps stretching a point, suppose that 3 is a large number.)

PROBLEM 5.5.2. As described in Section 5.3.7, the composition method for hyper-exponential random variables requires two uniform random variables. Reformulate the method to require only one.

PROBLEM 5.5.3. Given that a standard Cauchy variate is a ratio of two independent standard normal variates, show that $X = \tan(U\pi)$ generates Cauchy variates but not monotonely.

PROBLEM 5.5.4. Use inversion to show the same thing. (Hint: $\tan \theta$ has period π.)

PROBLEM 5.5.5. Show that if $X = aU_1 + bU_2, 0 \leq a \leq b$, then the density function of X is trapezoidal; specifically:

$$
f(x) = \begin{cases}
x/(ab), & 0 \leq x \leq a, \\
1/b, & a \leq x \leq b, \\
(a + b - x)/(ab), & b \leq x \leq a + b.
\end{cases}
$$

PROBLEM 5.5.6. If $\{U_i\}$ is a sequence of independent random variables uniformly distributed in $(0, 1)$, $X_1 = U_1 + U_2 - 1$ is triangularly distributed over $(-1, 1)$ with mean 0. Show that $X_2 = U_1 - U_2$ has the same distribution as X_1 and that X_1 and X_2 are uncorrelated. Are they also independent?

PROBLEM 5.5.7. Let (V_1, V_2) be uniformly distributed in a circle centered at the origin. Show that V_1/V_2 has a Cauchy distribution. Use this fact to construct an algorithm to generate Cauchy variates.

PROBLEM 5.5.8 [Sunder (1978)]. Show that to generate two dependent standard normal variates with correlation ρ, the following variation of the Box–Muller technique works:

$$
Z_1 = \sin(2\pi U_1)\sqrt{-2 \log U_2},
$$
$$
Z_2 = \sin(2\pi U_1 - \phi)\sqrt{-2 \log U_2},
$$

where $\cos \phi = \rho$, i.e. verify that the expected value of $Z_1 Z_2$ is ρ and that Z_1 and Z_2 are standard normal variates.

PROBLEM 5.5.9. Prove that $\sqrt{U_1}$ and $\max(U_1, U_2)$ have the same distribution if U_1 and U_2 are independent random variables uniformly distributed on $(0, 1)$.

PROBLEM 5.5.10. Exactly N people arrive at a building every morning. The arrival times are uniformly distributed over the interval 8 A.M. to 9 A.M. Give three ways of generating

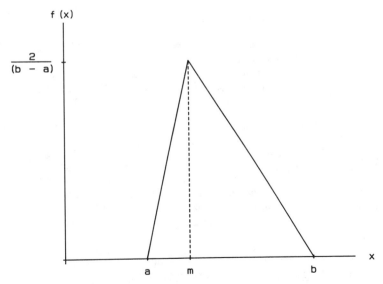

Figure 5.5.1. Triangular distribution.

these arrival times. To be useful, the arrival times must be arranged in increasing order. How much computation is required for each method as a function of N? (See Problem 1.9.12.)

PROBLEM 5.5.11. In PERT project management models an activity's probability distribution is described by three parameters, a, m, and b, which are respectively the smallest possible, most likely, and largest possible time. The triangular distribution (Figure 5.5.1) is often assumed:

(a) Give several methods for generating random variables with this distribution. For each method, how could antithetic variates be generated, and how effective would they be?
(b) What if the triangle is symmetric?

PROBLEM 5.5.12. The duration of a certain activity is a random variable with the following distribution:

Time in weeks	Probability
2	0.1
3	0.3
4	0.3
5	0.2
6	0.1

Six variates from the uniform distribution over the integers 0 to 99 follow: 97, 55, 24, 13, 86, 91. Generate six simulated durations for the activity above.

PROBLEM 5.5.13. A random variable X has cdf

$$F(x) = 1 - e^{-3\sqrt{x}}.$$

Give a method for generating X.

PROBLEM 5.5.14. In a disk unit the time for the read/write head to move from one track to another, known as the seek time, is equal to h times the distance between the two tracks. The tracks are equally spaced. At the completion of one operation the head does not move until the next. Track requests are uniformly distributed over all the tracks. Derive the density function of the seek times, and the expected seek time.

PROBLEM 5.5.15. Some distributions appear in this chapter for the benefit of those who do not necessarily accept the recommendations of Chapter 4. Barring *exceptional* circumstances, identify them.

PROBLEM 5.5.16. Suppose that your random number generator is an unbiased coin; thus it has only two possible outcomes: H and T. What is an efficient way of using the coin to simulate the roll of a die which has six equally likely outcomes: 1, 2, 3, 4, 5, 6? What is an efficient way of using the coin to simulate the roll of two dice where only the sum of the outcomes is of interest?

PROBLEM 5.5.17. A ball goes in urn i with probability p_i. With n urns and k iid balls, let N_i be the number of balls in urn i. Generate the multinomial variates $(N_1, \ldots, N_n)$.

(Hint: Set $R_i = N_1 + \cdots + N_i$, generate binomial variates $B_1, \ldots, B_{n-1}$ where B_i has parameters $(k - R_{i-1}, \; p_i/(p_i + \cdots + p_n))$, and put $N_i = B_i$, $i = 1, \ldots, n - 1$ and $N_n = k - R_{n-1}$. Here $R_0 = 0$.) How fast can this be done as a function of k and n? (Give the computational complexity.)

PROBLEM 5.5.18. Given the transition probabilities in a row of a Markov chain and the corresponding alias table, how fast can you generate a transition from the corresponding state? Give some standard queueing examples where the transition probabilities and alias table for the row can both be generated in $O(1)$ time, independently of the number of states.

PROBLEM 5.5.19 (Continuation). Repeat Problem 5.5.18 for Example 1.4.1. (Hint: Define the states properly—"Markovianize". See Section 4.3.) Does this seem more efficient than using a conventional clock? Try it.

PROBLEM 5.5.20 (Continuation). Specialize Problem 3.6.20 to Example 1.4.1.

PROBLEM 5.5.21. For *truncated* quasi-empirical distributions, give the counterpart to the development in Section 5.2.4.

PROBLEM 5.5.22 (Computational complexity). Suppose we simulate a continuous-time Markov chain with n states using the regenerative method to analyze output and the alias method to generate transitions. The expected number of transitions per cycle is L. We want the root mean square error to be less than δ. Show that, for unstructured matrices, the expected work is $O(n^2 + L/\delta^2)$. Show that, for a deterministic numerical approach that computes the vector π of steady-state probabilities, the work is $O(n^3)$. What about sparse matrices? row generators?

5.6. Timings

Table 5.6.1 lists the timings of generators discussed in this chapter. The self-explanatory names are keyed to section numbers. Figure 6.5.2 shows the uniform random number generator used, and Figure 6.7.1 the random integer generator. The other codes are in Appendix L. All runs were done using the same compiler and operating environment. The main interest in these timings is their relative, not their absolute, values. These timings are benchmarks for the generators tested and for others that may be tested. Extrapolation to other computers or to other distribution parameters and, for LUKBIN and IVDISC, to other table sizes is left to the reader.

Table 5.6.1. Time in Seconds to Generate and Sum 100,000 Variates on a
DEC 2060. (Summation done in integer arithmetic if the random variable is
integer.)

Distribution/Generator	Using assembly-coded uniform generator	Using portable Fortran uniform generator
1. —/Null [call dummy subroutine and sum]	0.52	0.52
2. Uniform on (0, 1) [6.5.2]	1.7	3.2
3. Uniform on integers from 1 to 11/IVUNIP [6.7.1]	—	13.3
4. Normal/VNORM [5.2.9]	16.6	18.4
5. Normal/SUMNRM [12 uniforms, 5.3.1]	19.7	37.4
6. Normal/BOXNRM [Box–Muller, 5.2.10]	13.0	14.0
7. Normal/TRPNRM [Ahrens–Dieter, 5.2.7]	6.2	9.2
8. Binomial/IALIAS [$n = 10, p = 0.5$; 5.2.8, 5.4.1]	7.1	8.5
9. Binomial/LUKBIN [$n = 10, p = 0.5$; 5.2.7, $m = 11$]	7.0	8.4
10. Binomial/LUKBIN [$n = 10, p = 0.5$; 5.2.7, $m = 21$]	6.9	8.3
11. Binomial/IVDISC [$n = 10, p = 0.5$; 5.2.1, $m = 11$]	7.3	8.7
12. Binomial/IVDISC [$n = 10, p = 0.5$; 5.2.1, $m = 21$]	7.2	8.6
13. Binomial/BINSRH [$n = 10, p = 0.5$; 5.2.1]	7.4	8.8
14. Beta/BETACH [$a = 2, b = 3$; 5.3. 10]	29.2	32.2
15. Beta/BETACH [$a = 2, b = 10$; 5.3.10]	31.4	34.6
16. Beta/BETAFX [$a = 2, b = 3$; 5.3.10]	13.2	18.8
17. Beta/BETAFX [$a = 2, b = 10$; 5.3.10]	25.8	41.2
18. Beta/VBETA [$a = 2, b = 3$; approximate inversion]	32.6	34.1
19. Beta/VBETA [$a = 2, b = 10$; approximate inversion]	32.6	34.1
20. Chi-square/VCHISQ [$DF = 1$; 5.3.11, based on squared normal]	17.5	18.9

Distribution/Generator	Using assembly-coded uniform generator	Using portable Fortran uniform generator
21. Chi-square/VCHISQ [$DF = 2$; 5.3.9, exact-based based on exponential]	7.7	9.1
22. Chi-square/VCHISQ [$DF = 3$; 5.3.11, approximate inversion]	35.4	36.8
23. Chi-square/VCHISQ [$DF = 30$; 5.3.11, approximate inversion]	35.4	36.8
24. Chi-square/VCHI2 [$DF = 3$; 5.3.11, approximate inversion]	177.8	181.0
25. Gamma/RGS [$k = 0.5$]	28.8	33.3
26. Gamma/RGKM3 [$k = 1.5$]	16.0	20.1
27. Gamma/RGKM3 [$k = 15$]	18.6	23.4
28. F/VF [$DF\ 1 = 4, DF\ 2 = 6$; approximate inversion]	36.6	38.1
29. F/VF [$DF1 = 1, DF\ 2 = 3$; approximate inversion]	47.4	49.9
30. Stable/RSTAB [$d = 2$]	48.3	51.2
31. t/VSTUD [$DF = 1$; 5.3.4, 5.3.13—exact, Cauchy cdf]	14.6	16.6
32. t/VSTUD [$DF = 2$; 5.3.13—exact cdf except for use of VNORM]	8.0	9.4
33. t/VSTUD [$DF = 3$; 5.3.13, approximate inversion]	46.7	47.2
34. t/VSTUD [$DF = 30$; 5.3.13, approximate inversion]	47.0	48.9
35. Uniform on (0, 1)/UNIFL [Appendix L, composite generator]		5.2

Note: the timer accuracy is about $\pm 4\%$.

CHAPTER 6

Uniform Random Numbers

6.1. Random Introductory Remarks

Generating random numbers uniformly distributed in a specified interval is fundamental to simulation: every procedure discussed in Chapter 5 for generating a nonuniform random number transforms one or more uniform random numbers. In this chapter a random number is a variate uniformly distributed over (0, 1).

Most computers have functions in the program library for producing supposedly random numbers. If your faith in them is firm, read only Sections 6.7 and 6.8. Sadly, your faith may well be misplaced. If your faith is shaky, read the whole chapter. We shall see that there are still some fundamental issues regarding the proper generation of random numbers. In many respects, Knuth (1981) is the bible of random number generation. We draw on that excellent book in a number of places, yet several of our conclusions differ.

6.2. What Constitutes Randomness

In any introduction to random number generation, it is customary to expatiate on what constitutes a random number. For example, suppose that a particular generator is advertised as producing integers uniformly distributed over the interval 0 to 999999. We ask it to produce a random number and it produces the number 999999. Most people feel that 999999 is not a very random number; they are happier with a number like 951413. Nevertheless, with any uniform random number generator worthy of the name, 999999 has the same likelihood as 951413. We are driven to conclude that we cannot speak of a number produced arithmetically or by table look-up as being random but rather we can only speak of a generator as producing a random sequence. Almost any definition of randomness involves examining longer and longer sequences, looking for patterns or the lack of them. All practical "random number" generators produce only a finite sequence which is then repeated. These periodic sequences are clearly not random. Though commonly disparaged, a *pseudorandom* number generator that produces an apparently genuine random number sequence is in many ways better than a truly random number generator as Section 1.5 points out.

How do we assess the goodness of a pseudorandom number sequence? We use the following qualitative approach:

Condition 1. The numbers when treated as points on the line segment from 0 to 1 are approximately uniformly distributed;
Condition 2. Successive nonoverlapping pairs when treated as points in the unit square are approximately uniformly distributed;

$$\vdots$$

Condition n. Successive nonoverlapping n-tuples of numbers when treated as points in the n-dimensional hypercube are approximately uniformly distributed.

A generator is accepted as sufficiently random for practical purposes if its output satisfies these n conditions for an n large enough for the application at hand. Unfortunately, empirical tests of these conditions are not statistically powerful and for $n > 2$ the only theoretical results yet established for practical generators are somewhat negative. Moreover, the work to apply any particular test generally becomes prohibitive for n greater than 10 at most.

An interesting alternative definition of randomness has been studied by several authors; see Bennett (1979) for a good introduction and references. In this approach, a series is "random" if, informally, it cannot be specified by an algorithm requiring less bits than the series itself. Thus the series

$$0101000000101001011011 \ldots$$

is random if the quickest way to describe it is simply to write it out, whereas the series

$$01010101010101010101 \ldots$$

can be specified by a short algorithm. Though interesting mathematically, this approach does not suggest how to write random number generators for a real computer.

PROBLEM 6.2.1. The decimal expansion of an irrational number is called *normal* if each digit 0 through 9, and indeed each block of digits of any length, occurs with equal asymptotic frequency. It has been conjectured that π, e, $\sqrt{2}$, and so on, are normal.

One provably normal number is Champernoune's C:

$$C = 0.1234567891011121314151617181920212223\ldots$$

Can you use this fact to build a practical random number generator? If so, how; if not, why not?

For numerical integration with a fixed-dimensional integrand or for global optimization, neither of these definitions of randomness quite hits the mark. Consider all boxes contained in the cube with sides parallel to the axes and one corner at the origin. The *discrepancy* of n points in the s-dimensional unit cube is the supremum over these boxes of the absolute difference between the number of these points inside the box and n times its volume. Sequences that have low discrepancy themselves and whose projections onto each lower-dimensional unit cube also have low discrepancy are called *quasirandom*, a misnomer since they do not necessarily purport to emulate real randomness. Niederreiter (1978) and Sobol' (1982) show that quasirandom sequences are suitable. Bratley and Fox (1986) and Fox (1986) implement appropriate generators with discrepancy $O(\log^s n)$ whose projections onto d-dimensional unit cubes have discrepancy $O(\log^d n)$.

PROBLEM 6.2.2. Show that a regular s-dimensional grid with n points has discrepancy $O(n^{1-1/s})$ and, even worse (why?), that its projections onto lower-dimensional unit cubes all have discrepancy $O(n^{1-1/s})$. Why are grids reasonable when $s = 1$?

PROBLEM 6.2.3. Why are typical queueing simulations not fixed dimensional? What does this imply about the relevance of quasirandom sequences to them? What about Problems 1.9.6–1.9.9? Markov-chain simulations?

6.3. Classes of Generators

There are various ways of generating apparently random numbers. We shall discuss the following generators:

(a) output of some apparently random device, e.g., the result of a coin flip, the output from a cosmic ray counter, etc.;
(b) tables of random numbers;
(c) midsquare (only historical interest);
(d) Fibonacci (only historical interest);
(e) linear congruential (the most commonly used today);
(f) linear recursion mod 2 (Tausworthe);
(g) hybrid or combination.

Except for type (a) generators, the output is computer generated. Because a computer has only a finite number of states, it must eventually return to a previous state and at this point it begins to cycle. One criterion of goodness for a generator is its cycle length, a surrogate for Condition 1 in Section 6.2.

Some people have suggested using the infinite number of digits in the decimal expression for such irrational numbers as π and e. Partly because of computational considerations, these methods have not been used to achieve a potentially infinite sequence of nonrepeating random digits.

6.3.1. Random Devices

In the 1950s there was considerable interest in apparently random devices to produce random numbers. To do this, some early computers contained (intentionally) random components such as cosmic ray counters. Others used the least significant bits of the digital clock. These methods were never popular because one could not duplicate the sequence of random numbers used from one run to the next unless they were stored. This capability to repeat a random sequence is important for at least two reasons: (1) it is easier to debug a simulation program if one can exactly repeat a previous run, except for some additional output statements; and (2) statistically, the comparison of two alternative policies can be done more precisely if each policy is simulated under the same "random" external events. For lotteries, however, using random physical devices reduces the chance of rigging. For encryption via bitwise exclusive-or with the *cleartext*, most pseudorandom sequences are not secure.

6.3.2. Tables

Before the widespread use of computers, most books discussing probability theory, statistics, or Monte Carlo contained an appendix listing several pages of random numbers. The practitioner simply proceeded sequentially through it. The most widely used table is RAND (1955). With computers it is typically easier to generate random numbers by an arithmetic process as needed rather than to read the numbers from a stored table.

6.3.3. The Midsquare Method

The midsquare method of generating pseudorandom numbers is apparently the first method proposed for use on digital computers. A reason for the method's fame is that it was proposed by John von Neumann.

Suppose we wish to generate four digit integers and the last number generated was 8234. To obtain the next number in the sequence we square the last one and use the middle four digits of the product. In this case the product is 67798756 so the next pseudorandom number is 7987. The next few numbers in the sequence are 7921, 7422, 0860.

The method has poor statistical qualities and the initial number (the *seed*) must be chosen with care. In the above case, for example, if a two-digit number ever results, then all future numbers also have at most two digits: consider the sequence 99, 98, 96, 92, 84, 70, 49, 24, 5, 0, 0, 0. Notice its apparent positive serial correlation. Even worse, once a zero arises, all future numbers are also zero.

6.3.4. Fibonacci and Additive Congruential Generators

Another poor method adds two or more previous numbers together and then takes the remainder when this sum is divided by a number called the modulus. This general form is called additive congruential. Adding the two preceding numbers is called the Fibonacci method because of its similarity to the recursion defining the Fibonacci series.

If X_i is the ith number generated and m is the modulus, then the Fibonacci method is

$$X_i = (X_{i-1} + X_{i-2}) \bmod m.$$

Dieter (1982) pointed out to us a damning drawback of Fibonacci generators: the permutations $X_{i-1} < X_{i+1} < X_i$ and $X_i < X_{i+1} < X_{i-1}$ never appear, though each should have probability 1/6. Another major weakness is conspicuous serial correlation. For example, suppose that $m = 1000$, $X_0 = 1$, and $X_1 = 1$; then the next sequence of numbers is 2, 3, 5, 8, 13, 21, 34, 55, 89, 144, 233, 377, 610, 987, 597, 584, 181, etc. Small numbers tend to follow small numbers.

6.3.5. Linear Congruential Generators

The most commonly used generators are linear congruential generators. They compute the ith integer X_i in the pseudorandom sequence from X_{i-1} by the recursion

$$X_i = (aX_{i-1} + c) \bmod m.$$

The parameters a, c, and m determine the statistical quality of the generator.

If c is zero, then the resulting generator is called a (pure) "multiplicative congruential" generator. The multiplicative generators with $m = 2^{31} - 1 = 2147483647$ and either $a = 16807$ or $a = 630360016$ are widely used. For example, the generator with $a = 16807$ is used in the APL system from IBM [Katzan (1971)], the scientific library from IMSL (1980), and in the SIMPL/I system [IBM (1972a, b)]. The generator with $a = 630360016$ is used in the Simscript II.5 system and in the DEC-20 Fortran system. See Section 6.8.2 for a list of apparently better multipliers.

In the late 1960s linear congruential generators for which m was a power of 2 were popular. The most common values were $m = 2^{31}$ and $m = 2^{35}$. For example, Simula on the UNIVAC 1108 uses $m = 2^{35}$ and $a = 5^{13}$. This generator has a cycle length of 2^{33}.

When m is a power of 2, one can perform the mod m operation by retaining only the last $\log_2 m$ bits of $aX_{i-1} + c$. Even today most large computers have a word length of either 32 or 36 bits. One bit represents the sign of the number, leaving either 31 or 35 bits. Thus, mod 2^{31} or mod 2^{35} is particularly easy to perform. A particular version (RANDU) widely used was

$$X_i = 65539X_{i-1} \bmod 2^{31}.$$

Unfortunately, it produces observably nonrandom output.

Marsaglia (1968) points out a major weakness of linear congruential generators. He shows that successive overlapping sequences of n numbers from a multiplicative generator all fall on at most $(n!\,m)^{1/n}$ parallel hyperplanes. For example, if you plot sequences of three numbers as points in an otherwise transparent cube and view the cube from the proper angle, then the points plotted appear as parallel lines.

The approximate value of $(n!\,m)^{1/n}$ for $m = 2^{31}$ is tabulated below for various values of n.

n	No. of distinct hyperplanes $(n!\,m)^{1/n}$
1	2^{31}
2	2^{16}
3	~ 2344
4	~ 476
5	~ 192
6	~ 108

For certain applications, a small number of covering hyperplanes can give grossly wrong results. This occurs in simulating the location of points in space and in Monte Carlo integration, for example. Also, recall that in Chapters 2 and 3 we advocated generating all the attributes of a transaction as it enters the system; to get these attributes, we transform successive random numbers. As we saw in Chapter 5, some methods for generating a nonuniform random number use a (possibly random) number of successive uniform random numbers. Even when inversion is used to transform n successive uniform random numbers, it certainly seems possible that the resulting n nonuniform random numbers can have bad n-dimensional structure.

EXAMPLE 6.3.1 (GPSS random number generator). The random number generator supplied in the GPSS system (§7.3) from IBM is unusual. It is similar to a multiplicative congruential generator but it is complicated by a shuffling feature which appears to be an attempt to make the output "more random." The major effect, however, is to make the generator difficult to evaluate analytically. Thus, it is difficult to state beforehand whether the generator has good statistical behavior. Empirical tests suggest that the generator is nonrandom in an undesirable way.

The GPSS literature speaks of eight distinct generators within GPSS, confusing some users. These eight generators are identical and produce the same sequence unless the user explicitly initializes each generator uniquely. The same generator is used to produce both integers in the range [0, 999] and six-digit fractions in the range [0.000000, 0.999999].

The components of the GPSS/360 generator [see IBM (1971)] are eight 32-bit words called base numbers, a 32-bit word called the multiplier, and a 3-bit number called the index. The eight base numbers are addressed 0 through 7. When a new random number is needed, the following steps are performed:

(1) The base number pointed to by the index is multiplied by the multiplier, giving a 64-bit product. Number the bits from least to most significant as $1, 2, \ldots, 64$.
(2) If bit 32 is 1, then a 2's complement is performed on 32 least significant bits, i.e., each bit is complemented and then a 1 is added to this when interpreted as a binary number. As a result, bit 32 will always be zero. The low-order 31 bits are stored back in the multiplier to be used for the next random number.
(3) Bits 49–51 are stored in the index to be used for the next random number.
(4) If a random integer in the range [0, 999] is desired, the *middle* 32 bits, treated as an integer, are divided by 1000 and the remainder becomes the three-digit random integer.
(5) If a random fraction in the range [0.000000, 0.999999] is desired, the *middle* 32 bits of the product, treated as an integer, are divided by 10^6 and the remainder gives the six digits of the random fraction.

The GPSS user can change the initial multiplier if he wants a different stream. The base numbers are fixed, common to all eight generators. It is not clear how their values were chosen. Babad (1975) points out that step 2 causes the middle of the middle 32 bits to be nonrandom. Thus, step 5 in particular might be expected to produce nonrandom results. He further shows that the GPSS/360 generator fails some reasonable empirical tests.

6.3.6. Linear Recursion mod 2 Generators

A linear congruential generator calculates X_i solely from X_{i-1}. Additive congruential generators use several previous values of X. Both these methods are special cases of the general formula

$$X_i = (a_1 X_{i-1} + a_2 X_{i-2} + \cdots + a_n X_{i-n} + c) \bmod m.$$

In this section we consider the generators obtained when the modulus m equals 2. Because X can then equal only 0 or 1, such generators produce a bit stream $\{b_i\}$. Furthermore, the only values that need be considered for the a_j are also 0 and 1. Thus b_i is obtained by adding modulo 2 several of the preceding bits in the stream. Modulo 2 addition is of course the exclusive-OR operation provided by most computers; we denote it by XOR.

We digress for a moment to consider the implementation of such generators using a shift register with feedback and a *primitive polynomial h* of degree k, defined below. As an example we take

$$h(x) = x^4 + x + 1,$$

which has degree 4. This polynomial specifies a feedback shift register as shown in Figure 6.3.1. Each box is a one-bit memory holding 0 or 1. At each iteration the register is shifted one place right, the boxes corresponding to the terms in h are added modulo 2, and the sum is fed back into the left-hand box. The output satisfies the recursion

$$b_i = (b_{i-3} + b_{i-4}) \bmod 2.$$

As a second example, we take

$$h(x) = x^5 + x^4 + x^3 + x^2 + 1$$

of degree 5. This specifies the feedback shift register shown in Figure 6.3.2. In this case the bits generated at the output satisfy

$$b_i = (b_{i-1} + b_{i-2} + b_{i-3} + b_{i-5}) \bmod 2.$$

PROBLEM 6.3.1. Justify.

Since each of the k boxes can hold 0 or 1, there are 2^k possible states for the shift register. Thus, the sequence $\{b_i\}$ must be periodic. Since the all-zero state generates only zeros, the maximum possible period is $2^k - 1$. We can now supply the definition promised above: h is a primitive polynomial if the shift register corresponding to h generates a sequence with period $2^k - 1$.

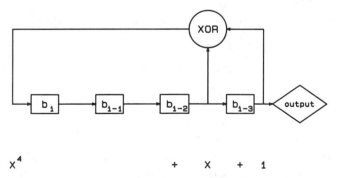

Figure 6.3.1. Feedback shift register corresponding to $x^4 + x + 1$.

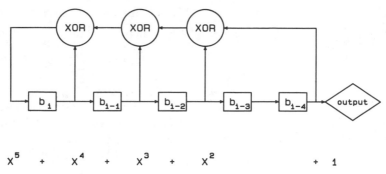

$$x^5 \quad + \quad x^4 \quad + \quad x^3 \quad + \quad x^2 \qquad\qquad + \quad 1$$

Figure 6.3.2. Feedback shift register corresponding to $x^5 + x^4 + x^3 + x^2 + 1$.

EXAMPLE 6.3.2. If the shift register of Figure 6.3.1 is initialized to the state 1000, its successive states and outputs are:

State Output	State Output	State Output
1. 1000	6. 0110 0	11. 1110 1
2. 0100 0	7. 1011 0	12. 1111 0
3. 0010 0	8. 0101 1	13. 0111 1
4. 1001 0	9. 1010 1	14. 0011 1
5. 1100 1	10. 1101 0	15. 0001 1
		16. 1000 1
		etc.

All 15 possible nonzero states are achieved, so $x^4 + x + 1$ is a primitive polynomial.

PROBLEM 6.3.2. Verify that $x^5 + x^4 + x^3 + x^2 + 1$ is a primitive polynomial.

PROBLEM 6.3.3. Verify that $x^4 + x^2 + 1$ is not a primitive polynomial.

There exist primitive polynomials of degree k for every k, and they are widely tabulated. Stahnke (1973) gives a primitive polynomial for every $k \le 168$, and Zierler (1969) gives primitive polynomials for some special values of k.

The sequence of bits $\{b_i\}$ produced at the output is called a pseudorandom sequence, a pseudo-noise sequence, or a maximal-length shift register sequence. MacWilliams and Sloane (1976) summarize their many useful properties; in particular:

(a) A complete period contains 2^{k-1} ones and $2^{k-1} - 1$ zeros.
(b) If a window of length k slides along a pseudorandom sequence of period $2^k - 1$, each of the $2^k - 1$ nonzero binary k-tuples appears exactly once in a complete period.

(c) Half the runs of identical symbols have length 1, one quarter have length 2, and so on (as long as these fractions give an integral number of runs). In each case the number of runs of 0's equals the number of runs of 1's.
(d) Let A be the number of places where $b_0 \cdots b_{n-1}$ and the cyclic shift $b_i b_{i+1} \cdots b_{n+i-1}$ agree, and D the number of places where they disagree. Define the autocorrelation function $\rho(i)$ by

$$\rho(i) = \frac{A - D}{n}.$$

Then

$$\rho(0) = 1$$
$$\rho(i) = -1/(2^k - 1), \qquad 1 \le i \le 2^k - 2,$$

and it can be shown that this is the best possible autocorrelation function of any binary sequence of length 2^{k-1} in the sense of minimizing

$$\max\{\rho(i): 0 < i < 2^k - 1\}.$$

These properties justify calling $\{b_i\}$ pseudorandom, with no pejorative connotation.

EXAMPLE 6.3.3. The output of the shift register shown in Figure 6.3.1, when initialized to the state 1000, is

$$000100110101111 \qquad 000100\ldots$$

(cf. Example 6.3.2). The four properties (a)–(d) above are easily verified.

PROBLEM 6.3.4. Verify them for the output of the shift register shown in Figure 6.3.2.

Tausworthe (1965) shows that if sequences of n bits generated in this way are considered as n-bit integers, then these integers are approximately uniformly distributed, they have approximately zero one-step serial correlation, and they do not have certain multidimensional nonuniformities associated with linear congruential generators.

One elegant way to generate such integers takes k equal to the number of bits in the computer word (not counting the sign bit) and chooses a primitive polynomial with only three terms, say

$$h(x) = x^k + x^q + 1$$

such that $k \ge 2q$. These exist for most k, but not all. Now if the computer can do full-word logical operations, components of the sequence $\{b_i\}$ can be generated k bits at a time as required using only two shifts and two exclusive-OR operations as follows:

(1) $Y \leftarrow X$. (X is formed by bits $b_{i+k-1} b_{i+k-2} \cdots b_i$.)
(2) Right shift Y by q bits, filling with zeros.

(3) $Y \leftarrow X \leftarrow Y$ XOR X. (The low-order bits of X have now been updated.)

(4) Left shift Y by $k - q$ bits, filling with zeros.

(5) $X \leftarrow Y$ XOR X. (X is now formed by bits $b_{i+2k-1}b_{i+2k-2} \cdots b_{i+k}$, i.e., X is the next required integer.)

If the word holding X has a sign bit, set it positive before X is used. Payne (1970) gives a Fortran implementation of this algorithm; see also Section 6.5.3.

PROBLEM 6.3.5. Show that this algorithm works.

For a 32-bit word including a sign bit, $k = 31$, and we may take $q = 3, 6, 7$ or 13. The generator in this case is full cycle, that is, all the integers in $[1, 2^{31} - 1]$ are produced.

The Tausworthe generator has not been thoroughly tested. Toothill et al. (1971) report some negative results using the runs up and down test; we describe the latter in Section 6.6.3. There is some indication that it has bad high-order serial correlation. According to Knuth [(1981), p. 30], this generator when used alone is a "poor source of random [whole-word] fractions, even though the bits are individually quite random." When combinations of generators are used (see next section) it is often recommended that one, but no more, of the generators be of the Tausworthe type.

Lewis and Payne (1973) suggest a refinement of this method. Begin by choosing a suitable primitive polynomial. For ease of implementation this is usually a trinomial; for statistical reasons k is usually chosen to be a Mersenne exponent, i.e., $2^k - 1$ is prime. For example,

$$x^{89} + x^{38} + 1$$

is a primitive polynomial [Zierler (1969)] and $2^{89} - 1$ is prime. To generate random integers, use the recursion

$$X_i = X_{i-k+q} \text{ XOR } X_{i-k},$$

where the X_i are now computer words, and XOR is a full-word exclusive-OR operation. Thus, each individual bit of X cycles with period $2^k - 1$ as though it were being generated by a shift register. To implement this recursion efficiently, use an array $X[1:k]$ and the following algorithm:

(1) Initialize $X[1]$ to $X[k]$: see below.

(2) Initialize $j \leftarrow k - q, i \leftarrow k$.

(3) Set $X[i] \leftarrow X[j]$ XOR $X[i]$, and output $X[i]$.

(4) Decrease j and i by 1. If $j = 0$, set $j \leftarrow k$; if $i = 0$, set $i \leftarrow k$.

(5) If enough random integers have been generated, stop; otherwise, go to step 3.

Lewis and Payne initialize the array X using the underlying generator to generate the individual bits of each word; thus, $X[1]$ gets the bits $b_1 b_{1+d} b_{1+2d} \ldots$, where d is a suitable "delay," $X[2]$ gets $b_2 b_{2+d} b_{2+2d} \ldots$, and so on.

Such generators are fast, easy to implement, and offer astronomical cycle lengths regardless of the word-size of the computer being used. Using k words of memory, a cycle length of $2^k - 1$ can be achieved; values of k and q are available at least up to $k = 9689$. They are claimed to have good statistical properties if k, q, and d are chosen properly. For further consideration of these generators, see Bright and Enison (1979).

6.3.7. Combinations of Generators

Combining two pseudorandom sequences $\{X_i\}$ and $\{Y_i\}$ to produce a third pseudorandom sequence $\{Z_i\}$ aims to reduce nonrandomness. If X_i and Y_i are distributed over the integers from 0 to $m - 1$, typical suggestions are:

(a) set $Z_i = (X_i + Y_i) \bmod m$;
(b) use $\{Y_i\}$ to randomly shuffle $\{X_i\}$ and then set $\{Z_i\}$ equal to the shuffled sequence;
(c) set $Z_i = X_i \operatorname{XOR} Y_i$.

With respect to (a) or (c), if either $\{X_i\}$ or $\{Y_i\}$ is in fact perfectly randomly distributed and these two sequences are statistically independent, then $(X_i + Y_i) \bmod m$ and $X_i \operatorname{XOR} Y_i$ are also perfectly randomly distributed. Thus, neither (a) nor (c) will hurt a good generator provided that the other generator uses a completely different mechanism. Unfortunately, one cannot define precisely the notion of "completely different" in this context. Ideally, we would like the two generators to produce statistically and logically independent sequences. Because in practice the two sequences are deterministic, statistical independence occurs vacuously but logical independence is not well defined. Optimists believe that either (a) or (c) improves two faulty generators. This is supported by the analysis of Brown and Solomon (1979), though they study an idealized model of random number generation. Combining a linear congruential generator and a Tausworthe generator may attenuate the multidimensional pathologies of either used alone.

Wichmann and Hill (1982) report good results using essentially method (a) and the result in Problem 1.9.19 to combine three generators:

$$X_{i+1} = 171 X_i \bmod 30269,$$
$$Y_{i+1} = 172 Y_i \bmod 30307,$$
$$Z_{i+1} = 170 Z_i \bmod 30323.$$

Using the methods of Section 6.5.2, these computations never require more than 16 bits at any step. This suggests that combining the three generators above produces a good, portable composite generator for computers with 16-bit words. Based on the spectral test (§6.4.4) and on extensive empirical checks, L'Ecuyer (1986) recommends combining the two generators

$$X_{i+1} = 40014 * X_i \bmod 2147483563,$$

$$Y_{i+1} = 40692 * Y_i \bmod 2147483399,$$

as

$$Z_{i+1} = (X_{i+1} + Y_{i+1}) \bmod 2147483563.$$

For computers with words at least 32 bits, the short multipliers above allow a fast, portable implementation using principles explained in Section 6.5.2. See the program UNIFL in Appendix L for details.

PROBLEM 6.3.6 (L'Ecuyer (1986)). Show that this generator has the (huge) period $2147483562 * 2147483398/2$. Show that its output does not have the pathology (pointed out in Section 6.3.5) for pure linear congruential generators of falling on relatively few hyperplanes. Show that it does not combine pathologically with the Box–Muller method (cf. Section 6.7.3).

With respect to (b), Marsaglia and Bray (1968) suggest using $\{X_i\}$ to initialize a table, say, of size 100. When a new random number is needed, a new Y_i is generated and appropriately scaled to provide an index into the table. The number at this location is output and another X_i is generated and stored at the location just used. Taking the cycle lengths of $\{X_i\}$ and $\{Y_i\}$ relatively prime produces an output sequence with cycle length equal to their product. Knuth [(1981), p. 32] describes another shuffling method due to Bays and Durham. Unlike the Marsaglia–Bray method, it uses no auxiliary second sequence. To shuffle just one input sequence, Bays and Durham

   (i) generate a random number U, not outputting it;
  (ii) use a few of U's bits to provide a table address A;
 (iii) output the random number V in A;
 (iv) store (the whole word) U in A.

Empirical results [Nance and Overstreet (1978)] suggest that shuffling does lead to an improvement, at least for mediocre generators. It is generally difficult to analyze combinations of generators to predict their statistical behavior. The statistical quality of a generator is not directly related to the generator's complexity.

Knuth [(1981), p. 4] describes his early attempt to develop a "fantastically good" random number generator program. Part of the justification for its goodness was to be that it was "so complicated that a man reading a listing of it without explanatory comments wouldn't know what the program was doing." He discovered empirically that the generator has a rather short cycle, sometimes of length 1, depending upon the seed. His advice is that a generator should not be chosen at random; some theory should be used. Nevertheless, he says [Knuth (1981), p. 32] that "on intuitive grounds it appears safe to predict that [shuffling the output of a generator backed up by theory] will satisfy virtually *anyone's* requirements for randomness in a computer-generated sequence . . . ." Rosenblatt (1975), using idealized models of multiplicative generators and shuffling, indicates that shuffling may make the output distributions of consecutive pairs and triples more uniform but says nothing about higher dimensions. Apart from this, there is little to support Knuth's claim other than that shuffling can give a long cycle length; shuffling is not well understood. Shuffling genuinely random numbers achieves nothing. Shuffling a sequence of pseudorandom numbers aims to get a better multidimensional structure but could conceivably produce a worse one.

6.4. Choosing a Good Generator Based on Theoretical Considerations

There is a respectable body of knowledge available to help choose a uniform random number generator. Even though multidimensional uniformity of sequences of random numbers is the ultimate measure of goodness, it is usually difficult to predict this behavior analytically. Thus, we look at related measures of goodness such as one-step serial correlation and cycle length. The former is a surrogate for Condition 2 of Section 6.2. It is calculated under the idealistic assumption that the seed is *genuinely* random.

6.4.1. Serial Correlation of Linear Congruential Generators

Greenberger (1961) loosely bounds the one-step serial correlation of a linear congruential generator from above by

$$\frac{1}{a}\left[1 - \frac{6c}{m} + 6\left(\frac{c}{m}\right)^2\right] + \frac{a + 6}{m}.$$

Obviously, if m is large, it is easy to choose a and c so that this bound is close to zero. When $c = 0$, it achieves its minimum when $a = \sqrt{m}$. Thus, a frequent (dubious) recommendation was that a should be of the order of $\sqrt{m}$. Coveyou and MacPherson (1967) suggest that a should *not* be extremely close to $\sqrt{m}$ or some simple rational multiple of m, e.g., $a = m/2$ or $2m/3$. Even though the one-step serial correlation may be low, other kinds of non-randomness may creep in, e.g., serial correlation of numbers two or more steps apart.

Moreover, Knuth [(1981), pp. 84 *et seq.*] points out that the above bound is rather poor; nearly all multipliers give a serial correlation substantially less than $1/\sqrt{m}$. Likewise the idea of choosing

$$\frac{c}{m} = \frac{1}{2} \pm \frac{1}{6}\sqrt{3}$$

on the grounds that these are the roots of $1 - 6x + 6x^2 = 0$ does little good: the value of c has hardly any influence on the serial correlation when a is a good multiplier. Knuth gives a much better bound, depending only on a and m, for the serial correlation in terms of Dedekind sums. See also Dieter (1971) and Dieter and Ahrens (1971). According to Dieter (1971, 1982), a bound on the "discrepancy" (a measure closely related to one-step serial correlation) leads to a simple rule for choosing good multipliers a when $c = 0$:

> Select a so that the quotients in the Euclidean algorithm for a and $m/4$ (if m is a power of 2) or a and m (if m is prime) are small.

This gives small discrepancy, assuring very nearly uniform distribution of pairs whenever the singletons appear uniformly distributed (see §6.4.2 below). Dieter (1982) says that in his experiments with J. Ahrens it always also produced good distributions of triplets, quadruplets, quintuples, and sextuples.

6.4.2. Cycle Length of Linear Congruential Generators

First, consider the generator

$$X_{i+1} = aX_i \bmod m,$$

where m is prime. The generator is full cycle, generating every integer in $[1, m - 1]$ before repeating, if a is a primitive element modulo m; that is, if $a^i - 1$ is a multiple of m for $i = m - 1$ but for no smaller i. For a proof, see Knuth [(1981), p. 19] for example. A full-cycle generator with m large has a long period and that is clearly good: the one-dimensional (marginal) distribution of the output appears uniform. Not generating 0 and m is good. For example, when generating exponential variates by inversion, the smallest and largest variates generated are $-\log((m - 1)/m)$ and $-\log(1/m)$ respectively rather than $-\log(0)$ and $-\log(1)$. The former causes an overflow error on most computers. The latter can jam clock mechanisms. A less then full-cycle generator might have large or irregularly-spaced gaps in the sequence of possible values $1, \ldots, m - 1$ and that would be bad.

PROBLEM 6.4.1. Verify that 5 is a primitive element modulo 7, and thus that the generator

$$X_{i+1} = 5X_i \bmod 7$$

is full cycle.

Knuth [(1981), p. 20], gives conditions for testing whether a is a primitive element modulo m; the conditions are suitable for computer implementation.
    Second, consider the case when m is not prime. In particular, the generator

$$X_{i+1} = aX_i \bmod 2^n,$$

with $n > 3$ can have a cycle length of at most 2^{n-2}. This is achieved if X_0 is odd and a has the form $8k + 3$ or $8k + 5$ for some integer k, as follows from Knuth [(1981), p. 19] for example.

PROBLEM 6.4.2. Verify that the generator

$$X_{i+1} = 11X_i \bmod 32$$

has a cycle length of 8 if X_0 is odd. What happens: (i) if $a = 9$; (ii) if X_0 is even?

Next, the generator

$$X_{i+1} = (aX_i + c) \bmod m,$$

with $c > 0$ is full cycle, i.e., generates every integer in $[0, m - 1]$ before repeating, if (Knuth [(1981), p. 16], for example)

(i) c and m are relatively prime;
(ii) $a - 1$ is a multiple of every prime p which divides m;
(iii) $a - 1$ is a multiple of 4 if 4 divides m.

In particular, a generator of the form

$$X_{i+1} = (aX_i + c) \bmod 2^n,$$

with $c > 0$, $n > 1$ is full cycle if c is odd and a has the form $4k + 1$.

PROBLEM 6.4.3. Show that the generator

$$X_{i+1} = (121X_i + 567) \bmod 1000$$

is full cycle. Calculate the first 20 or 30 values of X_i starting from $X_0 = 0$. Writing X_i as a three-digit number (e.g., 007), observe: (i) the first digits; and (ii) the last digits of the X_i-sequence. What do you conclude?

PROBLEM 6.4.4. Let $X_{i+1} = (aX_i + c) \bmod 2^n$ and $R_{i+1} = X_{i+1} \bmod 2^k$, $k < n$. Thus R_{i+1} is the k low-order bits of X_{i+1}. Show that the sequence $\{R_i\}$ has period at most 2^k. Under what conditions does its period equal 2^k?

PROBLEM 6.4.5. Let $X_{i+1} = aX_i \bmod 2^n$ and $R_{i+1} = X_{i+1} \bmod 2^k$, $k < n$. Show that the sequence $\{R_i\}$ is a string of all ones or all zeros when $k = 1$ and has period at most 2^{k-2} when $k > 1$. Under what conditions does its period equal 2^{k-2}?

PROBLEM 6.4.6. Let $X_{i+1} = aX_i \bmod(2^n - 1)$ and $R_{i+1} = X_{i+1} \bmod 2^k$, $k < n$. What is the maximum period of the sequence $\{R_i\}$? Under what conditions is the maximum achieved?

The moral of these problems is that the low-order bits for generators using mod 2^n have much shorter cycles than the full words generated. If only part of a generated random word is to be used, take it from the most significant bits.

6.4.3. Cycle Length for Tausworthe Generators

We saw above that cycles of length $2^m - 1$ can be generated easily for any m. If successive nonoverlapping sets of n bits are to be used, the cycle length will still be $2^m - 1$ if n is relatively prime to this number. To ensure this readily, take m as a Mersenne exponent, so that $2^m - 1$ is itself prime. If the fast algorithm for generating m bits at a time is used (see above, §6.3.6), a full cycle is obtained when m is relatively prime to $2^m - 1$; the latter holds for most word-lengths of interest.

6.4.4. The Spectral Test

The spectral test has become the most respected theoretical test of a linear congruential random number generator. The test and its name were first proposed by Coveyou and MacPherson (1967). The test was originally motivated by the consideration of possible nonrandom wave structure in the output of a random number generator. Coveyou and MacPherson were concerned about the possible departures from a flat frequency spectrum which should not be but might be found in the output of a generator. Knuth [(1981), p. 110] cites several papers that point out that the spectral test can be understood on much more straightforward geometrical grounds. Essentially, the spectral test is another way of measuring the k-dimensional uniformity of a complete cycle of the output of a generator. Marsaglia (1968) bounds the number of parallel hyperplanes that cover all the (possibly overlapping) k-tuples of random numbers generated; see Section 6.3.5. The spectral test considers a different but related problem. It determines the maximum distance between adjacent hyperplanes, the maximum being taken over all sets of covering parallel hyperplanes. The larger this maximum, the worse the generator. Dieter's (1975) insightful paper treats the covering and maximum-distance problems well. Niederreiter (1978) discusses other aspects of the distribution of random numbers in higher dimensions.

To motivate the test, consider generators of the form $X_{i+1} = aX_i \bmod m$. Complete cycles of the output for two specifications of this generator are listed below:

Case $a = 7, m = 11$:

$$X_i = 1, 7, 5, 2, 3, 10, 4, 6, 9, 8.$$

Case $a = 6, m = 11$:

$$X_i = 1, 6, 3, 7, 9, 10, 5, 8, 4, 2.$$

Both generators satisfy one-dimensional uniformity as well as possible, i.e., every integer in the interval [1, 10] is generated exactly once in a cycle.

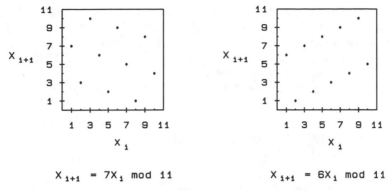

$$X_{i+1} = 7X_i \bmod 11 \qquad\qquad X_{i+1} = 6X_i \bmod 11$$

Figure 6.4.1. Plot of output pairs of two generators.

In two dimensions, uniformity collapses. Figure 6.4.1 plots the pairs of consecutive numbers generated by the two generators.

In each case the points in 2-space can be covered by a family of parallel lines. For the generator $X_{i+1} = 6X_i \bmod 11$, all the points can be covered by either two parallel lines of positive slope or five parallel lines of negative slope. For the generator $X_{i+1} = 7X_i \bmod 11$, all the points can be covered by either four lines of positive slope or three lines of negative slope.

The spectral test considers all sets of parallel lines which cover the points, and chooses the set for which the distance between adjacent lines is maximum. For our two examples, the output from $X_{i+1} = 7X_i \bmod 11$ appears more random in 2-space than does the output from $X_{i+1} = 6X_i \bmod 11$. Either with a ruler or by means of a little algebra and arithmetic, you may verify that the maximum distance between parallel lines is about 3.479 for the former and 4.919 for the latter. Thus, the spectral test agrees with our eyesight for this little example.

Application of the Spectral Method to Linear Congruential Generators. To test the k-dimensional uniformity of a linear congruential generator of the form $X_{i+1} = aX_i \bmod m$ using the spectral method, we must solve the optimization problem

$$S_k: V_k = \min\sqrt{X_1^2 + X_2^2 + \cdots + X_k^2},$$

subject to

$$X_1 + aX_2 + \cdots + a^{k-1}X_k = mq,$$

with $X_1, X_2, \ldots, X_k$, and q integer (possibly negative) and $q \neq 0$.

As we show shortly, m/V_k is the maximum distance between adjacent parallel hyperplanes together covering all the points in k-dimensional space generated by $X_{i+1} = aX_i \bmod m$.

Solving this optimization problem for $k = 2$ for the generator $X_{i+1} = 7X_i \bmod 11$ gives $X_1 = 1$, $X_2 = 3$. Thus $V_2 \approx 3.1623$ and the spacing between lines is $m/V_2 = 3.479$, as we saw earlier. For the generator $X_{i+1} = 6X_i \bmod 11$, the solution is $X_1 = -1$, $X_2 = 2$; now $V_2 \approx 2.236$ and the maximum spacing between lines is 4.919, again as we saw earlier.

For our two small examples it is easy to solve the optimization problem by enumeration. For "real" generators with m of the magnitude 2^{31}, complete enumeration is out of the question. Knuth [(1981), §3.3.4] gives tricks to reduce the computation. Even with these tricks, it is difficult to calculate the spectral measure for dimension $k > 10$.

Derivation of the Spectral Calculation. Any member of a family of parallel hyperplanes in k-space has the form

$$\alpha_1 X_1 + \alpha_2 X_2 + \cdots + \alpha_k X_k = mq \quad \text{for } q = 0, 1, 2, \ldots.$$

The value of q determines which hyperplane of the family is chosen. In each family of equally-spaced parallel hyperplanes that covers the output from a generator $X_{i+1} = aX_i \bmod m$, one hyperplane passes through the origin. Thus, finding the spacing between adjacent hyperplanes in a family is equivalent to finding the length of the line segment from the origin to the next hyperplane in the family: we solve

$$\min\sqrt{X_1^2 + X_2^2 + \cdots + X_k^2},$$

$$\text{subject to} \quad \alpha_1 X_1 + \alpha_2 X_2 + \cdots + \alpha_k X_k = m.$$

Using a Lagrange multiplier and elementary calculus, we find the solution $X_j = m\alpha_j/\sum_{i=1}^k \alpha_j^2$. The distance between adjacent hyperplanes is therefore

$$m/\sqrt{\alpha_1^2 + \alpha_2^2 + \cdots + \alpha_k^2}$$

for this particular family.

Among all families of parallel hyperplanes that cover the points the spectral test seeks to maximize this distance, i.e., to minimize

$$\sqrt{\alpha_1^2 + \alpha_2^2 + \cdots + \alpha_k^2},$$

subject to the α_i's corresponding to some family of parallel hyperplanes. How do we determine this? Assume that the generator generates the output 1 for some X_i; then $X_{i+1} = a$, $X_{i+2} = a^2 \bmod m$, etc. The point in k-space $(X_i, X_{i+1}, \ldots, X_{i+k-1})$ must be on one of the hyperplanes and hence

$$\alpha_1 X_i + \alpha_2 X_{i+1} + \cdots + \alpha_k X_{i+k-1} = \alpha_1 + a\alpha_2 + \cdots + a^{k-1}\alpha_k$$

must equal mq for some integer q. This condition justifies the optimization problem S_k, after replacing α_i by X_i for $i = 1, \ldots, k$.

6.5. Implementation of Uniform Random Number Generators

Implementing uniform random number generators in a high-level language not designed for doing bit operations requires cunning, as shown below.

6.5.1. Multiplicative Generators With Modulus 2^n

The easiest generators to implement are the multiplicative congruential generators for which m is a power of two. Suppose our computer uses 32 bits, including the sign bit at the left, to represent an integer and we want to use $m = 2^{31}$. Thus, after performing the multiplication $X_i = aX_{i-1}$, we want to retain the least significant 31 bits. If the product is not more than 31 bits long, we are done. Otherwise, the 31 low-order bits are still correct; however, the top (leftmost) bit, interpreted as the sign bit, is computed just as any other bit during a multiplication. Half the time this bit will be 1, causing the number to be interpreted as negative. In this case we want to reverse the leftmost bit but no others.

How we do this depends whether our machine uses two's complement arithmetic, like an IBM 370, or one's complement arithmetic, like a CDC machine. On a two's-complement computer with n data bits (not counting the sign bit) the difference between a negative number and a positive number with the same bit configuration is 2^n. On a one's-complement computer the difference is $2^n - 1$.

These observations suggest the following Fortran implementation for two's complement computers of the generator

$$X_{i+1} = 65539 X_i \bmod 2^{31}:$$

```
      FUNCTION  UNIF(IX)
C
C  INPUT: IX, A RANDOM INTEGER,
C     0 < IX < 2**31-1
C
C  OUTPUTS: IX, A NEW RANDOM INTEGER
C     UNIF, A RANDOM FRACTION,
C     0 < UNIF < 1
C
C  2**31 IS 2147483648
C  2**(-31) IS ABOUT 0.4656613E-9
C
      IX = IX*65539
      IF (IX .LT. 0), IX = 1+(IX+2147483647)
C
C  THE PARENTHESES ON THE RIGHT INSURE THAT ALL
C  QUANTITIES THERE ARE MACHINE-REPRESENTABLE
C
      UNIF = IX*.4656613E-9
      RETURN
      END
```

Despite its simplicity, we recommend *against* using this generator. The low-order bits of the output are far from random (see Problem 6.4.5). Its high-order bits are observably nonrandom too.

6.5.2. Multiplicative Generators With Prime Modulus

When m is not a power of two, one cannot perform the mod operation by simply discarding high-order bits. If the computer has a data type such as DOUBLE PRECISION with at least as many bits as the product aX_i, and if a mod operation is allowed on this data type, then implementation is simple.

Consider the prime-modulus generator

$$X_{i+1} = 16807 X_i \bmod (2^{31} - 1).$$

Figure 6.5.1 implements this generator on most computers.

```
      FUNCTION UNIF(IX)
C
C  INPUT: IX, A RANDOM INTEGER,
C     0 < IX < 2**31-1
C
C  OUTPUT: IX, A NEW RANDOM INTEGER
C     UNIF, A RANDOM FRACTION,
C     0 < UNIF < 1
C
C  2**31-1 IS 2147483647
C
      IX = DMOD(16807D0*IX,2147483647D0)
      UNIF = IX/2147483647.0
      RETURN
      END
```

Figure 6.5.1. A simple Fortran random number generator.

The fairly computer-independent IMSL subroutine library uses this method. It will not work on an IBM 370 class computer if the multiplier is changed from 16807 to the often recommended 630360016, because DOUBLE PRECISION on the IBM 370 carries only 53 bits of accuracy. The product 16807*IX has at most a 46 bit result, while the product 630360016*IX can have more than 53 bits.

Payne *et al.* (1969) describe a clever, faster method, attributing it to D.H. Lehmer. Suppose we wish to compute

$$X_{i+1} = aX_i \bmod P, \tag{i}$$

but that it is easier to compute

$$Z = aX_i \bmod E, \tag{ii}$$

which would be true, for example, if $P = 2^n - 1$ and $E = 2^n$. Let $E > P$ and define g and k as follows:

$$g = E - P > 0, \tag{iii}$$

$$k = \lfloor aX_i/E \rfloor. \tag{iv}$$

Adding kg to both sides of (ii) gives

$$Z + kg = aX_i - k(E - g) = aX_i - kP. \tag{v}$$

The general method is

$$X_{i+1} \leftarrow Z + kg.$$

$$\text{If } X_{i+1} \geq P, \quad \text{then } X_{i+1} \leftarrow X_{i+1} - P$$

as we now justify.

We must show that $Z + kg < 2P$. From (ii), $Z \leq E - 1$, so that for $Z + kg < 2P$ it suffices to have

$$E - 1 + kg < 2P. \tag{vi}$$

Because $X_i < P$, from (iv) we find that a sufficient condition for (vi) is

$$E - 1 + (aP/E)g < 2P,$$

or also

$$P + g - 1 + (aP/E)g < 2P,$$

or also

$$(aP/E) < \frac{P - g + 1}{g}.$$

We finally obtain that a sufficient condition for the general method to work is

$$a < \frac{E(P - g + 1)}{gP} = E\left(\frac{1}{g} - \frac{1}{P} + \frac{1}{gP}\right).$$

For example, if $P = 2^{31} - 1$, then $g = 1$ and the sufficient condition is trivial: $a < E + 2^{31}$.

If the mod $2^{31} - 1$ implementation can be written so that no intermediate result uses more than 31 bits, it will run on many modern computers. Schrage (1979) uses the fact that any 31-bit number can be written as $\alpha 2^{16} + \beta$ where α and β have 15 and 16 bits, respectively. Because 16807 requires only 15 bits, computations can be arranged so that intermediate results need at most 31 bits. Schrage's method will not work with a multiplier of more than 15 bits. Figure L.11 of Appendix L shows a Fortran implementation.

The following 31-bit implementation is about twice as fast. Let b and c be integers satisfying

$$0 < b < P,$$

$$0 \leq c < a,$$

$$ab + c = P.$$

From these definitions, we get

$$X_{i+1} = aX_i \bmod P$$
$$= aX_i - kP, \qquad\qquad\qquad\qquad \text{(vii)}$$

where

$$k = \lfloor aX_i/P \rfloor = \lfloor aX_i/(ab + c) \rfloor.$$

Alternatively, compute

$$X_{i+1} = Z + (k_1 - k)P, \qquad\qquad\qquad\qquad \text{(viii)}$$

where

$$k_1 = \lfloor X_i/b \rfloor$$

and

$$Z = aX_i - k_1 P$$
$$= aX_i - k_1(ab + c)$$
$$= a(X_i - k_1 b) - k_1 c$$
$$= a(X_i \bmod b) - k_1 c,$$

the last equality from the definition of k_1. If c is small relative to ab, then comparing the definitions of k and k_1 there is reason to hope that a, b, and c can be chosen so that $k_1 - k$ equals 0 or 1 for every integer X_i between 0 and P; see Problem 6.5.1 below. Because we choose c small relative to ab, we say that the method relies on approximately factoring P. This principle is implicit in Wichmann and Hill (1982).

If Z can be computed using at most 31 bits at each step and $k_1 - k = 0$ or 1 always, then the simple algorithm

$$X_{i+1} \leftarrow a(X_i \bmod b) - k_1 c.$$

If $X_{i+1} < 0$,   then set

$$X_{i+1} \leftarrow X_{i+1} + P$$

works. Figure 6.5.2 illustrates it, using $a = 16807$, $P = 2^{31} - 1$, $b = 127773$, and $c = 2836$. L'Ecuyer (1986) recommends, with justification, instead setting $a = 40692$, $P = 2147483399$, $b = 52774$, and $c = 3791$.

PROBLEM 6.5.1. Show that (vii) and (viii) always give the same answer. Show that the program in Figure 6.5.2 is valid, i.e., that

$$a(X_i \bmod b) < 2^{31},$$
$$k_1 c < 2^{31}, \qquad k_1 - k = 0 \text{ or } 1.$$

For arbitrary a, b, c satisfying $ab + c = P$ but with $a > \sqrt{P}$, why would you expect the algorithm above not to work?

PROBLEM 6.5.2. Write an assembly language implementation for a machine that has a register with at least 47 bits, not counting the sign. Write an assembly language implementation for a machine that has a register with just 31 bits, not counting the sign. [See also Knuth (1981), pp. 11, 12, 15 (Problem 8), 522.]

```
      FUNCTION UNIF( IX)
C
C  PORTABLE RANDOM NUMBER GENERATOR IMPLEMENTING THE RECURSION:
C     IX = 16807 * IX MOD (2**(31) - 1)
C  USING ONLY 32 BITS, INCLUDING SIGN.
C
C  SOME COMPILERS REQUIRE THE DECLARATION:
C     INTEGER*4 IX, K1
C
C  INPUT:
C     IX = INTEGER GREATER THAN 0 AND LESS THAN 2147483647
C
C  OUTPUTS:
C     IX = NEW PSEUDORANDOM VALUE,
C     UNIF = A UNIFORM FRACTION BETWEEN 0 AND 1.
C
      K1 = IX/127773
      IX = 16807*( IX - K1*127773) - K1 * 2836
      IF ( IX .LT. 0) IX = IX + 2147483647
      UNIF = IX*4.656612875E-10
      RETURN
      END
```

Figure 6.5.2. A portable Fortran random number generator.

Table 5.6.1 compares speeds of (i) a Fortran and (ii) an assembly language implementation of the generator $X_{i+1} = 16807X_i \bmod(2^{31} - 1)$. It also compares speeds of Fortran implementations of nonuniform random number generators that use (i) and (ii), respectively. Figure 6.5.2 shows the Fortran code for the uniform random number generator used.

Knuth [(1981), pp. 171–172] recommends a quite different "portable" random number generator despite admitting that "very little has yet been proved about [its] randomness properties." Actually, Knuth did not ignore his maxim that "some theory should be used" to select a generator. The theoretical basis for the method rests on a plausible, but unproved, conjecture of J. F. Reiser discussed in Knuth [(1981), p. 36, Problem 26].

PROBLEM 6.5.3. Write an assembly language implementation of the generator

$$X_{i+1} = (69069X_i + 1) \bmod 2^{31}$$

for a machine that has a register with at least 47 bits, not counting the sign. Repeat for a machine that has a register that has just 31 bits, not counting the sign. Give a Fortran code that is at least as portable as that in Figure 6.5.2. (According to Knuth [(1981), p. 104], Marsaglia calls 69069 "a candidate for the best of all multipliers." It has nice symmetry, but we do not vouch for its scientific merit. Niederreiter [(1978), pp. 1026–1027] pans it.)

6.5.3. Implementing the Tausworthe Generator

Figure 6.5.3 gives a modified version of the Tausworthe generator in Payne (1970). A recommended initial value for IX is 524287. The correctness of this coding is compiler and machine dependent. The coding works with a

```
      FUNCTION UNIF( IX)
C
      LOGICAL A, B, ACOMP, BCOMP
      EQUIVALENCE (I,A), (J,B), (ACOMP,ICOMP), (BCOMP,JCOMP)
      DATA K/2147483647/, FK/2147483647./, N/262144/, M/8192/
C
      I = IX
C
C  DO THE RIGHT SHIFT
C
      J = I/M
C
C  GET READY TO DO AN EXCLUSIVE OR
C
      ICOMP = K-I
      JCOMP = K-J
C
C  DO THE XOR USING INCLUSIVE OR,
C  COMPLEMENT AND .AND.
C
      B = A .AND. BCOMP .OR. ACOMP .AND. B
C
C  LOW ORDER BITS ARE DONE
C  NOW DO THE REST
C
      I = J
C
C  DO THE LEFT SHIFT
C
      J = J*N
      ICOMP = K-I
      JCOMP = K-J
      A = A .AND. BCOMP .OR. ACOMP .AND. B
      IX = I
      UNIF = I/FK
      RETURN
      END
```

Figure 6.5.3. A Tausworthe generator for a computer with full-word logical operations, 32-bit words, and two's complement arithmetic.

computer which uses full-word logical operations, two's complement arithmetic and carries exactly 31 data bits in a word.

PROBLEM 6.5.4. Under these conditions, show in detail that the code works correctly. When any of these conditions does not hold, give counterexamples.

6.6. Empirical Testing of Uniform Random Number Generators

The analytical results regarding the quality of random numbers deal only with a few facets of randomness. Therefore, generators must pass a battery of empirical tests before they become widely accepted.

Random bit sequences have iid bits and $P[$a given bit $= 1] = P[$a given bit $= 0] = \frac{1}{2}$. When in a linear congruential generator m is a power of 2, the high-order bits supposedly have these properties. When m has the form $2^n - 1$ and n is large, the high-order bits supposedly differ negligibly from a sequence with these properties. These facts can be used to test pseudorandom number sequences by breaking the numbers into bits, but little testing has been done using this approach; see Gavish and Merchant (1978) for an exception. Perhaps contrary to intuition, not all patterns in random bit sequences are spurious. A lot is known about them. The recent results of Guibas and Odlyzko (1980) regarding patterns may be the basis of a powerful test of randomness in bit strings. Requiring that pseudorandom sequences have these patterns and other esoteric qualities of randomness is probably asking too much. No sequence produced arithmetically can have *every* such property. We can test only those relatively few properties that seem relevant to practical computation.

A linear congruential generator with full cycle of length N generates integers on $1, \ldots, N$ say without duplicates on N successive draws. A truly random sequence of k draws, however, would have some duplicates with probability about $1 - \exp(-k^2/(2N))$ as k approaches N. Heath and Sanchez (1986) analyze this so-called "birthday" problem and conclude that k^2/N should be small to avoid easily-detected statistical anomalies. If in your application you think that such anomalies are intolerable and k^2/N would not be small, try a composite generator as discussed in Section 6.3.7—keeping in mind the caveats stated there.

PROBLEM 6.6.1. If m has the form $2^n + 1$, show that there is a bias in the top bit.

6.6.1. Chi-square Tests

Some tests of the randomness of uniform random number generators are based on the chi-square test. This test has low power: it rejects only a small proportion of the cases which violate the hypothesis being tested. The test can justify rejecting a hypothesis, but provides little justification for accepting one. Nevertheless, it is generally applicable, and the statistic calculated as part of the test can provide a useful ranking of the plausibility of alternative hypotheses, even though it may not reject any.

The chi-square test applies in the following situation. Assume the event space (e.g., the possible values of the random numbers drawn) can be partitioned into n subsets (e.g., $U \leq 0.1$, $0.1 < U \leq 0.2$, etc.). From a sample of M independent observations, let f_i be the number of outcomes falling into subset i. Let $\bar{f_i}$ be the expected number in the ith subset under the hypothesized distribution. If the hypothesis is true, then as M increases the statistic

$$\chi^2 = \sum_{i=1}^{n} \frac{(f_i - \bar{f_i})^2}{\bar{f_i}}$$

has asymptotically the chi-square distribution with $n - k - 1$ degrees of freedom, where k is the number of parameters estimated from the data; for testing a uniform distribution over a known range, $k = 0$. The approximation tends to be reasonable if $\min\{f_i\} \geq 5$. If χ^2 is large, reject.

For example, to test a uniform generator we might define cells by dividing the line (0, 1) into several intervals. Not all cells need have the same probability associated with them, or equivalently, the interval lengths need not be the same. We might wish to make the intervals shorter near 0 and near 1 because it is usually the extreme values which are important (especially when transforming to nonuniform random numbers by inversion).

PROBLEM 6.6.2. If $\chi^2 = 0$, what might you suspect? Would you ever reject a fit if χ^2 is merely very small?

Choosing the Number of Cells for the Chi-square Test. Mann and Wald (1942) point out that the power of the chi-square test to reject a false hypothesis depends upon the number of cells used. The greater this number, the greater the "resolving" power. The test, however, is based on an approximation which is bad if the expected number of outcomes in a cell is small.

With equal cell sizes, a quick and dirty approximation to the Mann–Wald rule uses $4M^{2/5}$ cells. For example, with 50,000 observations, this suggests using about 300 cells. Gorenstein (1967) gives a more accurate expression.

6.6.2. Kolmogorov–Smirnov Tests

Like the chi-square test, this test allows one to make a statement about the probability of an observed sample being drawn from a specified distribution. It works only for continuous distributions. Unlike the chi-square test, not even the asymptotic distribution of the KS statistic is known analytically if any parameters are estimated from the data; but see Lilliefors (1967, 1969).

With n observations $X_1, X_2, \ldots, X_n$, order them so that $X_{i+1} \geq X_i$ for $i = 1, \ldots, n - 1$. Compute

$$K_n^+ = \sqrt{n} \max_j [(j/n) - F(X_j)], \qquad j = 1, 2, \ldots, n,$$

$$K_n^- = \sqrt{n} \max_j [F(X_j) - (j - 1)/n], \qquad j = 1, 2, \ldots, n.$$

These are respectively the maximum positive and negative differences between the empirical cdf and the hypothesized cdf. Compare K_n^+ and K_n^- with values tabulated for example in Knuth [(1981), p. 48]. If K_n^+ or K_n^- is too large or too small at the desired confidence level, then reject the hypothesis that the observations are drawn from the distribution with cdf F.

If F is continuous and has no parameters calculated from the data, then the KS test is probably better than the chi-square test. The KS test is exact for any value of n whereas the chi-square test is based on an approximation which is good only when n is large. Anderson and Darling (1954) present another test similar in spirit to the KS test but based on the squared differences between the hypothesized and the empirical cdf rather than the maximum difference. There seems to be no strong evidence that it is more powerful than the KS test. We know of no specific studies of relative power of tests of (putatively) random number sequences.

6.6.3. Tests Specifically for Uniform Random Number Sequences

k-Dimensional Uniformity. The most common test partitions a number sequence into nonoverlapping subsequences of length k and then interprets each subsequence as a point in a k-dimensional cube. Each dimension of the cube is partitioned into s subintervals, thus partitioning the cube into s^k cells. A chi-square test is applied to the number of points falling in each cell. The number of degrees of freedom for the test is $s^k - 1$. For generators shown analytically to have long, full cycles and low one-step serial correlation, testing k-dimensional uniformity is probably superfluous for $k \leq 2$ but may reveal unexpected nonuniformity for $k > 2$.

The difficulty is the amount of storage required when k is large. For example, it seems reasonable to check the first 3 bits of random numbers for uniformity. This corresponds to partitioning each dimension into 8 subintervals. Similarly, it seems reasonable to base the analysis on subsequences of length 6. The number of cells required is then $8^6 = 262,144$. Thus we consider additional tests which examine other facets of randomness.

Runs Up and Down. Consider the sequence of 11 numbers:

$$1 \quad 4 \quad 3 \quad 2 \quad 9 \quad 8 \quad 7 \quad 8 \quad 6 \quad 5 \quad 3.$$
$$+ \quad - \quad - \quad + \quad - \quad - \quad + \quad - \quad - \quad -$$

Below each pair we indicate the direction of change with a $+$ or a $-$. A sequence of N numbers has $N - 1$ such changes. This particular sequence has three runs up of length 1, two runs down of length 2, and one run down of length 3.

Assuming that ties between adjacent numbers have probability zero and that the numbers are iid, then the number of runs up and down r in a sequence of N numbers is asymptotically normally distributed with mean and variance $(2N - 1)/3$ and $(16N - 29)/90$, respectively. A test of the hypothesis that the numbers are iid follows immediately, assuming that the statistic is in fact normally distributed.

Runs Above and Below the Median. Another way of defining a run is the number of successive outcomes falling on the same side of the median. The distribution of the length of this kind of run is geometric with $P\{\text{run length} = n\} = 1/2^n$.

The number of runs of either kind is asymptotically normally distributed with mean $1 + N/2$ and variance $N/2$.

Permutation Test. Generate N sets of random numbers, each set containing k numbers. Sort each set. This produces a permutation of the indices in each set. There are $k!$ possible permutations for each set. If the random numbers are independent, then all permutations are equally likely. Count the number of each permutation occurring for these N sets and apply a chi-square test.

Kolmogorov–Smirnov Test Applied to Extreme Values. Another good test is the Kolmogorov–Smirnov test applied to the maxima of sets of numbers. If $X_1, X_2, \ldots, X_n$ are iid with cdf F, then the maximum of $(X_1, X_2, \ldots, X_n)$ has cdf

$$F_n(x) = F(x)^n.$$

For example, one might generate 50 uniform random numbers and obtain 10 transformed random numbers by finding the maximum of each set of 5. These maxima should have the cdf $F_5(x) = x^5$ if the original numbers are independent and uniform on $(0, 1)$.

6.7. Proper Use of a Uniform Random Number Generator

Even though a uniform random number generator may have desirable features, things can go wrong when applying it to a simulation. Some subtle pitfalls are discussed below.

6.7.1. Generating Random Integers Uniform Over an Arbitrary Interval

Suppose we want integers uniformly distributed over the interval $[0, K - 1]$. A seemingly sensible way of doing this is to perform $J = \text{K*UNIF(IX)}$, where UNIF(IX) is a function returning a floating point number uniformly distributed over $(0, 1)$. For simplicity, assume that UNIF(IX) is a linear congruential generator with IX iterating over the successive integers in the sequence. Using floating point arithmetic for what is fundamentally an integer operation can cause problems:

(i) Different computers may use different floating point formats; thus even though UNIF may generate the same sequence IX on two different computers, the J sequence may differ.

(ii) If K is large, the roundoff inherent in floating point operations may cause the J's to be nonuniformly distributed. In particular, the values for J near K may be systematically over or under represented.

Both weaknesses can be virtually avoided by using the mod operation:

$$\text{JUNK} = \text{UNIF(IX)}$$

$$J = \text{MOD(IX,K)}$$

in Fortran. But see Section 6.7.2.

This method is not a monotone transformation of the original sequence. As shown in Chapter 2, this can be a significant impediment to effective use of variance reduction techniques. Figure 6.7.1 shows a machine-independent method which is monotone. Essentially, it is an infinite precision implementation of the method $J = K*\text{UNIF(IX)}$ which avoids the problems due to finite precision floating point arithmetic.

PROBLEM 6.7.1. Show in detail that it works correctly. (Hint: P15*P16 = P + 1, P15 requires 15 bits, P16 requires 16 bits, IX = U*P15 + V, $0 \le V <$ P15, L = X*P16 + Y, $0 \le Y <$ P16.) Show that its running time is $O(1)$ independent of L, provided that your computer's multiplication time is independent of the representation of its operands (such as the number of 1-bits) and L $< \lfloor M/(M - P) \rfloor$ with M = 2**31.

6.7.2. Nonrandomness in the Low-order Bits of Multiplicative Generators

A drawback of generators of the form $X_{i+1} = aX_i \bmod 2^n$ is that the least significant bit of $X_i, X_{i+1}, \ldots$ is always 1. Recall that X_0 should be odd and then all subsequent X_i will be odd. See also Problems 6.4.4 et seq. Do not use transformations such as MOD(IX,4) (or algebraically IX mod 4) which depend solely on the low-order bits. Outputs of generators with prime modulus do not have obvious problems with their low-order bits, but the behavior of these bits has not been carefully studied.

```
      FUNCTION IVUNIP( IX, L, P)
      IMPLICIT INTEGER(A-Z)
C
C  SOME COMPILERS REQUIRE THE ABOVE TO BE INTEGER*4
C
C  PORTABLE CODE TO GENERATE AN INTEGER UNIFORM OVER [ 1, L],
C  USING INVERSION.
C
C  INPUTS:
C    IX = A RANDOM INTEGER, 0 =< IX < P
C    L = AN INTEGER, 0 < L < P,
C    P = AN INTEGER, 1 < P < 2**31 , E.G., MODULUS OF RANDOM NUMBER
C          GENERATOR.
C
C  OUTPUTS:
C    IVUNIP = INTEGER PART OF( IX * L / P) + 1,
C           EFFECTIVELY, A RANDOM INTEGER UNIFORM ON [ 1, L].
C    PROCEDURE IS VERY SLOW FOR P << 2**31.
C
C  REFERENCE: PROF. A. PERKO, UNIVERSITY OF JYVASKYLA, FINLAND
      DATA M/2147483647/, P15/32768/, P16/65536/
C
C  DATE 30 SEPT 1986
C
C  CALCULATE IX* L IN DOUBLE PRECISION, I.E., FIND A, B  SO:
C    IX* L  =  A*(2**31) + B
C
      X = L/ P16
      Y = L - P16* X
      U = IX/ P15
      V = IX - P15* U
C
C  NOW CALCULATE:
C    A*(2**31) + B  =  Y * V  +  V * X * P16  +  U * P15 * R * 2
C    +  U * P15 * S  +  U * P15 * X* P16.
C
      YV = Y * V
      YV1 = YV/ P15
      YV2 = YV - P15* YV1
      R = Y/ 2
      S = Y - 2* R
      UR = U * R
      UR1 = UR/ P15
      UR2 = UR - P15* UR1
C
C  EXPLOIT THE POSSIBILITY THAT X = 0:
C
      IF ( X .GT. 0) GO TO 100
      VX1 = 0
      VX2 = 0
      GO TO 200
  100 VX = V* X
      VX1 =  VX/ P15
      VX2 = VX - P15* VX1
      VX1 = VX1 + U * X
  200 AB = YV1 + 2* ( UR2 + VX2) + S* U
      A2 = AB/ P16
      B = P15 *( AB - P16* A2) + YV2
      A = A2 + UR1 + VX1
C
C  NOW THAT WE HAVE A AND B
C    NOTE THAT ( A * 2**31 + B)/ P =
C    ( A * ( P + IE) + B)/ P = A + ( A * IE + B ) / P, SO
C    COMPUTE K = INTEGER PART OF ( ( A * IE + B )/ P)
C    NOTE THAT K = SMALLEST INTEGER SATISFYING:
C    A * IE + B - ( K + 1) * P < 0
```

```
C
           IE = M - P + 1
           MOE = M / IE
           INA = A
           K = 0
           KUM = B - P
     300  IF ( KUM .LT. 0) GO TO  400
           KUM = KUM - P
           K = K + 1
           GO TO  300
C
C   ADD IN SOME MORE IE'S UNTIL A OF THEM HAVE BEEN ADDED
C
     400  IF ( INA .EQ. 0) GO TO  500
           IEFIT = MIN( MOE, INA)
           INA = INA - IEFIT
           KUM = KUM + IE * IEFIT
           GO TO  300
C
     500  IVUNIP = A + K + 1
           RETURN
           END
C
C


C
```

Figure 6.7.1. Portable Fortran random integer generator.

6.7.3. Linear Congruential Generators and the Box–Muller Method

The Box–Muller method (§5.2.10) generates a pair of variables (Y_i, Z_i) which are supposedly independently normally distributed by the transformations

$$Y_i = \cos(2\pi U_{i+1})\sqrt{-2 \log U_i},$$

$$Z_i = \sin(2\pi U_{i+1})\sqrt{-2 \log U_i},$$

where U_{i+1} and U_i are supposed uniform on $(0, 1)$ and i is odd. Thus, (Y_i, Z_i) is defined only for odd i. Now suppose in fact

$$U_i = X_i/m,$$

$$U_{i+1} = X_{i+1}/m,$$

and

$$X_{i+1} = (aX_i + c) \bmod m.$$

Because $\cos \theta$ and $\sin \theta$ are periodic, simple algebra shows that

$$Y_i = \cos(2\pi(aU_i + c/m))\sqrt{-2 \log U_i},$$

$$Z_i = \sin(2\pi(aU_i + c/m))\sqrt{-2 \log U_i}.$$

All possible values of (Y_i, Z_i) fall on a spiral as Figure 6.7.2 illustrates. As an approximation to a pair of *independent* variates, (Y_i, Z_i) is terrible. As an

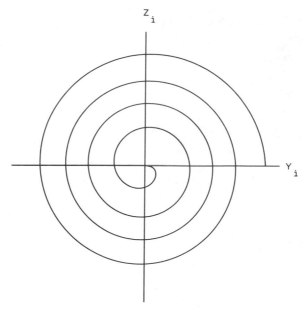

Figure 6.7.2. Output from a Box–Muller linear congruential generator.

approximation to *normal* variates, Y_i and Z_i are poor. One can salvage the combination of the Box–Muller method and linear congruential generators by modifying the latter. But the main point is that linear congruential generators are not universally reliable. This is a case where a pathology of linear congruential generators can be demonstrated analytically. In other cases, pathologies may be present but harder to detect. Seek more protection by using a hybrid generator not known to have pathologies but having theoretical support.

6.8. Exploiting Special Features of Uniform Generators

Because the generators discussed thus far are deterministic, we may alter them to meet special needs.

6.8.1. Generating Antithetic Variates With a Multiplicative Congruential Generator

Suppose we use the recursion

$$X_{i+1} = aX_i \bmod m,$$
$$U_{i+1} = X_{i+1}/m.$$

Further, we would like to generate an antithetic sequence $\{U'_i\}$, satisfying $U'_i = 1 - U_i$ for $i = 1, 2, \ldots$ This additional subtraction can be inserted into a simulation; however, Kleijnen (1974) attributes to M. Euwe the following easier method. Compute the integer sequence X'_i by the same recursion, i.e.,

$$X'_{i+1} = aX'_i \bmod m,$$

$$U'_{i+1} = X'_{i+1}/m,$$

but initialize $X'_0 = m - X_0$. Thus, $U'_0 = (m - X_0)/m = 1 - U_0$ as desired. Now for the induction step. Suppose that $X'_i = m - X_i$ and thus $U'_i = 1 - U_i$. There exist two integers p and q such that

$$X_{i+1} = aX_i \bmod m = aX_i - pm,$$

$$X'_{i+1} = aX'_i \bmod m = aX'_i - qm.$$

Adding these two gives

$$X_{i+1} + X'_{i+1} = a(X_i + X'_i) - (p + q)m.$$

Because $X_i + X'_i = m$ the right-hand side is a multiple of m; therefore the left is also. The only possibility is $X'_{i+1} + X_{i+1} = m$ as was to be shown.

6.8.2. Generating a Random Number Stream in Reverse Order

For a multiplicative generator of the form

$$X_{i+1} = aX_i \bmod m,$$

it is easy to determine another multiplier a' which generates the same stream but in reverse order. The relationship between a' and a is

$$a' = (a^{c-1}) \bmod m,$$

where c is the cycle length. Consider the generator $X_{i+1} = 3X_i \bmod 7$. It generates the sequence 1 3 2 6 4 5 1. Now $a' = 3^{6-1} \bmod 7 = 5$, which generates the sequence 1 5 4 6 2 3 1. By almost any measure of randomness the two multipliers are equally good.

If efficiency is important, however, examine both a' and a. The multiplier with the fewer 1's in its binary representation may execute faster.

Based on number-theoretic and empirical tests of *every* possible multiplier for pure linear congruential generators with modulus $2^{31} - 1$, Fishman and Moore (1986) recommend just the following five:

$$742938285,$$
$$950706376,$$
$$1226874159,$$
$$62089911,$$
$$1343714438.$$

PROBLEM 6.8.1. In view of the above observations about reversing multipliers, can you explain how the above list contains an *odd* number of recommended multipliers? Find the reversing multipliers for the above list and count the respective number of 1-bits. What can you say about the above multipliers in view of Problem 6.5.2? (L'Ecuyer (1986) performs tests, corresponding to those of Fishman and Moore (1986), on multipliers satisfying the conditions of Problem 6.5.1. His recommendation appears just before Problem 6.5.1. He goes on to consider hybrid generators.) Consider using $m - a$ or $m - a'$ as multipliers.

It is easy to modify the Tausworthe generator so that it generates a bit stream in reverse order; however, it is in general not possible to generate whole words in reverse order.

PROBLEM 6.8.2. Modify the generator of Example 6.3.2 to produce its bit stream in reverse order.

6.8.3. Generating Disjoint Sequences

A single underlying random number generator can produce disjoint random number sequences provided that these sequences are not "too" long. To do this, choose seeds for the respective sequences "far enough" apart. Table 6.8.1 provides seeds spaced 131,072 apart in the cycle of the generator $X_{i+1} = 16807 X_i \bmod(2^{31} - 1)$. When do we want sequences that do not overlap? Examples:

1. To make independent simulation runs.
2. To drive several processes, each with a distinct sequence, to facilitate synchronization as discussed in Chapter 2. In this case we use several sequences in a *single* run.

PROBLEM 6.8.3. Give an efficient method to construct Table 6.8.1. Write a program to do so.

Table 6.8.1. Seeds spaced 131,072
apart for the generator
$X_{i+1} = 16807 X_i \bmod(2^{31} - 1)$.

Sequence	Seed
1	748932582
2	1985072130
3	1631331038
4	67377721
5	366304404
6	1094585182
7	1767585417
8	1980520317
9	392682216
10	64298628
11	250756106
12	1025663860
13	186056398
14	522237216
15	213453332
16	1651217741
17	909094944
18	2095891343
19	203905359
20	2001697019
21	431442774
22	1659181395
23	400219676
24	1904711401
25	263704907
26	350425820
27	873344587
28	1416387147
29	1881263549
30	1456845529

CHAPTER 7
Simulation Programming

There is nothing difficult about writing a simulation program. The essential components, a clock mechanism, a source of random numbers, and data structures to represent transactions, resources, and queues, can all be implemented without serious problems in any general-purpose high-level programming language. To support this contention, we begin the chapter with a brief account of how to write a simulation in Fortran.

Although it is not hard to write simulations in a general-purpose language, it may be tedious. This has its own dangers: it is easier to overlook mistakes in a long program than in a short one. In the older languages, too, the only ways available to represent such structures as ordered sets may be clumsy. Several dozen languages exist that aim to make writing simulations more concise and more transparent. Some were designed from the outset for simulation; others are specially-tailored extensions of general-purpose languages. In this chapter we look at three of the major ones: Simscript, GPSS, and Simula.

Each has a badly flawed design. Although all have passed through several versions, none is a modern language. In the last section of this chapter, therefore, we consider what a modern simulation language *ought* to look like. To some extent this is begging the question: perhaps we should really consider what a modern general-purpose programming language ought to look like, and what special features are desirable for simulation programming. However, we shall not stress this distinction.

Until this paragon of languages appears, programmers have to make do with what exists. Our aim in this chapter is partly to inform the reader of the kind of tool commonly available; it is more, however, to make him critical of these tools. As we have hinted, a consumers' guide to simulation languages would run to several hundred pages. By examining instead a few well-known examples we hope to illustrate the main features to look for.

Programming is such a matter of personal style and taste that each reader must come to his own conclusion. Sad to say, ours is negative: for any important large-scale real application we would write the programs in a standard general-purpose language, and avoid all the simulation languages we know. The reason is simple: we would not be comfortable writing simulation programs (or any others, for that matter) in a language whose behavior we are not able to understand and predict in detail. The reader who cares to take our word for this should read Section 7.1; he may then skip to Section 7.5 with an easy conscience.

For less critical applications, or for implementing approximate and ephemeral models in the early stages of a project, we would, like most people, be prepared to sacrifice a little control in return for convenience. Simula using the Demos package (see §7.4.4) is largely comprehensible. Most minor questions about its behavior can be settled quickly by looking at the source listing of Demos, and it offers a tolerable automatic report generator. There are occasions when we would be tempted to use it for simple models.

GPSS offers all the fun of hefting an ax instead of a scalpel: it can be learned quickly, it produces lots of results in no time at all, and provided you can stand quite a lot of approximation in your output, it may be just what you need. When a precision tool is needed, however, it is a nonstarter.

Against our criticisms, the reader should set the fact that Simscript, GPSS, and Simula *are* widely-used languages, so many people do not share our views. Having entered our caveat, in the rest of the chapter we let the languages speak for themselves.

7.1. Simulation With General-Purpose Languages

We begin this section with a fairly complete account of how to write a simple simulation in Fortran. Fortran has numerous faults, and one great merit: its availability. The language is well-standardized, and Fortran compilers exist for virtually every computer on the market, including some microcomputers. Our examples use the current ANSI standard Fortran, usually called Fortran 77 [ANSI (1978)].

Other, more modern general-purpose languages are more powerful and elegant. In the following paragraphs we give examples of how to take advantage of their special features when writing simulations. Section 7.1.7 in particular shows how the methods illustrated for Fortran can be adapted for the increasingly-popular Pascal. However, this is usually self-evident to a programmer accustomed to a particular language. We chose to use Fortran in the examples not only on account of its ubiquity, but also because it is easy to adapt from Fortran to a better language.

The techniques illustrated are indeed simple enough to be adapted to a less powerful language if need be, to Basic for example. The user of a severely-limited microcomputer might find this necessary; otherwise the procedure has nothing to recommend it. For a system to be used often, it may just be worthwhile to write the basic clock mechanisms in assembler, but even this is not certain.

7.1.1. The Simplest Possible Clock Mechanism

At the heart of any simulation program lies a mechanism for advancing simulated time. The earliest systems advanced the clock in constant increments. This is called synchronous simulation. Nowadays, however, it is usual to keep a schedule of events to be simulated in the future, and, when the moment comes to advance the simulated clock, to advance it immediately to the time of the next scheduled event. This is asynchronous simulation. For a comparison of the two techniques, see Section 1.4.1.

Events are added to the schedule randomly, and they are removed in order of scheduled event time. The choice of a data structure to hold this schedule can be important in large simulations where many events are pending simultaneously. Section 1.7 mentions some of the proposals which have been made for minimizing the overhead involved.

Often, although the total number of events in a simulation may be large, the number actually present simultaneously in the schedule is quite small. For instance, the usual way of scheduling arrival events, such as customers arriving at a bank, jobs arriving at a computer center, and so on, is for each arrival to cause its successor to be added to the schedule. Thus even if several hundred arrivals are to be simulated, only one of them is in the schedule at once. In this case the choice of a data structure to hold the schedule is not critical: for simplicity, we might as well use an array.

For each event in the schedule, we need to know what kind of event it is, when it is to happen, and perhaps some other parameters. The event type can be coded as an integer, and the event time is a real number. If only a few other parameters are involved, it may be preferable to store them with the schedule; if there are many other parameters, store a pointer to a complete event description elsewhere.

Figures 7.1.1 and 7.1.2 give two possible subroutines for running a simulation schedule in Fortran. A call of PUT(EVTYPE, EVTIME, AUX) adds an event of type EVTYPE to the schedule at time EVTIME. As discussed above, AUX may be a parameter, or a pointer to a record elsewhere. A call of GET(EVTYPE, AUX) retrieves the type of the next event and its parameters from the schedule, and advances the common variable TIME to the simulated time of this event.

A few remarks about these routines are in order. First, we have chosen to ignore the Fortran type conventions in favor of declaring the type of each

```
      SUBROUTINE PUT(EVTYPE, EVTIME, AUX)
      COMMON / TIMCOM / TIME
      COMMON / SCHED / WHAT(50), WHEN(50), PARAM(50), NEXT
      SAVE / SCHED /
      INTEGER EVTYPE, AUX, WHAT, PARAM, NEXT, I
      REAL EVTIME, TIME, WHEN
C
      IF (EVTIME.LT.TIME) THEN
        PRINT '('' CLOCK RUNNING BACKWARDS '')'
        STOP
      ENDIF
C
      IF (NEXT.EQ.50) THEN
        PRINT '('' SCHEDULE FULL '')'
        STOP
      ENDIF
C
C SEARCH THE EVENT LIST FROM TAIL TO HEAD
C TO FIND WHERE TO INSERT THE NEW EVENT
C
      NEXT = NEXT + 1
      DO 1  I = NEXT, 2, -1
      IF (EVTIME.LE.WHEN(I-1)) GO TO 2
      WHAT(I) = WHAT(I-1)
      WHEN(I) = WHEN(I-1)
      PARAM(I) = PARAM(I-1)
    1 CONTINUE
    2 WHAT(I) = EVTYPE
      WHEN(I) = EVTIME
      PARAM(I) = AUX
      RETURN
      END
```

Figure 7.1.1. Subroutine PUT.

```
      SUBROUTINE GET(EVTYPE, AUX)
      COMMON / TIMCOM / TIME
      COMMON / SCHED / WHAT(50), WHEN(50), PARAM(50), NEXT
      SAVE / SCHED /
      INTEGER EVTYPE, AUX, WHAT, PARAM, NEXT
      REAL TIME, WHEN
C
      IF (NEXT.EQ.0) THEN
        PRINT '('' SCHEDULE EMPTY '')'
        STOP
      ENDIF
C
C WHAT NEXT?
C
      EVTYPE = WHAT(NEXT)
      TIME = WHEN(NEXT)
      AUX = PARAM(NEXT)
      NEXT = NEXT - 1
      RETURN
      END
```

Figure 7.1.2. Subroutine GET.

variable explicitly. This is a matter of taste. The schedule is held in three parallel arrays WHAT, WHEN, and PARAM in descending order of event time: WHAT(1) is the event furthest away in time, and WHAT(NEXT) the next event to be simulated. The schedule shown accommodates a maximum of 50 events: clearly the necessary limit depends on the application. We include crude tests for schedule overflow and underflow, and to prevent scheduling events after they should have happened. Our tests simply stop the program if an error arises: often it is possible to do better than this.

The variable TIME appears in a common block since it is needed throughout any simulation program. Paradoxically, the variables representing the schedule (WHAT, WHEN, PARAM, and NEXT) are in a common block for exactly the opposite reason: by *forbidding* programmers to declare this block elsewhere, we ensure that changes in the simulation schedule are made only by the routines provided for that purpose. So what we have is "a language notation which associates a set of procedures with a shared variable and enables a compiler to check that these are the only operations carried out on that variable": exactly Brinch Hansen's definition of a monitor [Brinch Hansen (1973), p. 12]. Since /SCHED/ will therefore not appear in the main program, it must be saved explicitly by each subroutine which uses it.

PROBLEM 7.1.1. We also need a subroutine, called INIT, say, to initialize the schedule. Write such a subroutine.

PROBLEM 7.1.2. By adding an extra array of links to the schedule, we can organize the events in a linked list [Knuth (1973), p. 251] to avoid moving event descriptions from one place to another. Rewrite PUT, GET, and INIT in this way.

Some programmers may find it clearer (and there are a few less lines of code to write) if PUT, GET, and INIT are made three separate ENTRY points within the same program unit, thus underlining their close relationship.

7.1.2. Generating Random Variates

We have seen in earlier chapters that to generate random variates we need a reliable source of uniform random numbers. Many Fortran systems provide a uniform random number generator in the system library. This should never be used uncritically. Test both the raw random number sequences and the transformed random number sequences to be used in any particular application. Avoid random number generators whose underlying algorithm is unknown.

If no random number generator is available, it is easy to define one. Chapter 6 gives several methods for doing this. Implementation may be best done in assembly code, since the algorithm is usually trivial, and it is often unclear what a high-level language does when, for instance, fixed-point arithmetic operations overflow. Alternatively, a user who is not too concerned

```
      REAL FUNCTION NORMAL (MEAN, STDEV, SEED)
      REAL MEAN, STDEV, UNIF, PHI, VNORM
      INTEGER SEED, IFAULT
      EXTERNAL UNIF, VNORM
C
C  GENERATE PHI UNIFORM ON (0,1) AND THEN
C  INVERT USING VNORM.  NO NEED TO CHECK
C  IFAULT IF WE HAVE FAITH IN UNIF.
C
      PHI = UNIF(SEED)
      NORMAL = MEAN + STDEV*VNORM(PHI,IFAULT)
      RETURN
      END
```

Figure 7.1.3. Real function NORMAL.

about efficiency may use a preprogrammed generator such as our portable Fortran random number generator shown in Figure 6.5.2.

We assume from now on that the user has available a real function UNIF(SEED), where SEED is an integer, returning random numbers uniformly distributed over (0, 1). Usually SEED is updated by each call of the function, so that between calls it should not normally be altered. Now Figure 7.1.3 shows, as an example, one way of coding a generator for normal variates. Following our own precepts, we generate a variate uniform on (0, 1) and then invert, using the approximate function VNORM from Appendix L. Most other distributions encountered in practice can be taken care of equally simply using Chapter 5 and Appendix L as references.

PROBLEM 7.1.3. Write a Fortran program to generate Poisson variates using the method of Section 5.4.2. Now show how to use the routines TPOISN and IVDISC of Appendix L to achieve the same result. Compare these methods for convenience, speed, and ease of maintaining synchronization.

7.1.3. Data Structures in Fortran

One weakness of Fortran as a simulation vehicle is the absence of any adequate means of specifying data structures. To take a simple example, suppose we want to represent a queue of transactions, each having a transaction number, a priority, and an entry time into the queue. Suppose further that the operations to be performed on the queue are such that it is convenient to have both forward and backward links. Then the obvious way to represent this structure in Fortran is to use arrays: for instance

```
      INTEGER TRANUM(50), PRIORY(50), PRED(50), SUCC(50)
      REAL TIMEIN(50)
```

if we know that the queue can never contain more than 50 elements.

PROBLEM 7.1.4. Show how to construct a list that can be traversed both forwards and backwards using just one link word per list element. (Hint: Treat the list as circular, but retain pointers to both the head and the tail. Now each link word may contain, for example, the sum of the addresses of the preceding and the following items.)

In practice this is little more than an amusing trick unless space is at a premium because of the increased difficulty of inserting and deleting elements.

If this is the only kind of queue and the only kind of transaction we have to worry about, then the situation is not too bad. However, if several sorts of transactions are present in our model, and if each sort can be put into a variety of queues, then the complexity of the program can increase enormously. Among other considerations, as storage requirements increase, it may no longer bc possible to allocate permanently sufficient space for the maximum size of each queue. In this case we are forced to use some kind of dynamic allocation scheme. This can be a fruitful source of error and inefficiency if great care is not exercised.

When this situation arises, the Fortran user has two options: he may continue to write his simulations in a general-purpose language, but using a more modern language designed for handling data structures with elegance, or he may opt to use a special-purpose simulation language. As we shall see, it is not always clear which is the better choice.

7.1.4. A Complete Simulation in Fortran

EXAMPLE 7.1.1 [Adapted from Mahl and Boussard (1977).]. An anti-aircraft battery consists of three guns sited close together. Five enemy aircraft, flying low at 300 m/s, arrive at 2-s intervals. The aircraft fly along a straight line directly over the battery.

When a gun fires at a particular aircraft, the shell travels at 1000 m/s. If the shell and its target meet at distance d from the battery, the aircraft is destroyed instantaneously with probability $\max(0, 0.3 - d/10000)$. After firing a shot, reloading the gun takes time normally distributed with a mean of 5 s and standard deviation 0.5 s.

When an aircraft passes directly over the battery, it drops a clutch of bombs destroying each gun with probability 0.2 independently of what happens to its neighbors.

We intend to simulate the following strategy to determine how many aircraft are shot down. All the guns fire independently. Each gun first fires at the leading aircraft as it comes within some given range R from the battery. Thereafter, each gun fires as soon as it is ready: at the nearest oncoming aircraft, if there is one, and otherwise at the nearest departing aircraft.

If we measure distances from the battery in the direction of travel of the aircraft, then Figure 7.1.4 shows the probability of hitting a target as a function of its distance away at the time of firing. If time 0 is the moment when the first aircraft arrives 3900 m from the battery, at extreme range, then the action will be over after at most 31 s.

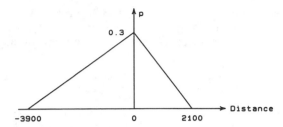

Figure 7.1.4. Hit probability as a function of firing range.

PROBLEM 7.1.5. Verify this.

Figure X.7.1 in Appendix X gives a simple program for simulating the required strategy. Five different kinds of events can happen in the simulation: the arrival of an aircraft within range R, an aircraft flying over the battery, a gun finishing loading, a shell bursting, and the end of the simulation. We have arbitrarily given integer codes to these events. The program is much clearer if we give them names as well: INRNGE, BOMB, etc. in our example.

On this occasion we have no need of elaborate data structures. The array PLANOK tells us whether a plane is in range and still flying, while GUNOK and GUNRDY tell us respectively whether a gun is still in action, and whether it is ready to fire. None of the variables used by the scheduling routines PUT and GET (except TIME, of course) appears in the main program.

We initialize the schedule by inserting an event to represent the arrival of aircraft 1 at the firing line, and another event to end the simulation. Each arriving aircraft causes its successor to be scheduled: in this way the schedule is kept short. Now the main loop of the program consists simply of extracting the next event from the schedule, using GET (which also advances TIME), and branching to appropriate code to simulate the effects of this event. In this simple example we have written the code directly in the main program; for more complicated cases it is usual to write a separate subroutine corresponding to each event.

We avoid canceling scheduled events. For instance, if a gun is destroyed while it is loading, the corresponding READY event still occurs, and we begin by testing whether the gun is in fact still in action. In more complicated cases this may not be possible, requiring an extra routine to take previously scheduled events out of the schedule.

To complete the program given we need two further real functions, both of which use the variable TIME passed in a common block: DIST(N) calculates the distance to aircraft N at a given moment, and TOFFLT(N) calculates the time of flight of a shell fired at aircraft N. Remember that the closing speed of the shell and an aircraft differs for approaching and departing planes.

PROBLEM 7.1.6. Write the functions DIST and TOFFLT, and run the complete program.

PROBLEM 7.1.7. Run the program several times with different random number streams, and observe the distribution of h, the number of aircraft hit. How many runs will be necessary to estimate $\bar{h}$, the mean number of hits, with reasonable accuracy (define this notion)?

PROBLEM 7.1.8. Choose a variance reduction technique which offers hope of improving the estimate of $\bar{h}$, program it, and observe the behavior of the modified program.

PROBLEM 7.1.9. We wish to optimize the choice of R with reasonable accuracy (again, define this). Using data from your previous experiments, plus any additional runs you need, design a simulation experiment for achieving this. Estimate how much this experiment will cost to perform.

PROBLEM 7.1.10. It is clear (why?) that choosing R negative, i.e., opening fire only when an aircraft has passed, cannot be optimal. This is fortunate, since the program will fail if $R < 0$. Why?

Even in this over-simplified example, actually writing a simulation program involves much less effort than getting useful results out of it once it is written. For simplicity we showed a program with virtually no output. In practice we usually want to provide informatory messages, at least while the program is being debugged, and there are many more statistics to collect. Nevertheless, the task of writing the program is only a small part of the whole process, especially if the basic subroutines exist already. The relative improvement to be expected from using even a good simulation language is minimal.

This is still more evident if we first have to define a model before we can use it. Our example is pure fiction, based on imagination and not at all on real data about guns, shells, and aircraft. If our task includes gathering facts about speeds, hit probabilities, loading times, and so on, then the time spent programming will be negligible in comparison.

7.1.5. Fortran Simulation Packages

Routines such as GET and PUT, or UNIF and NORMAL, are likely to be used in any simulation written in Fortran. From this observation, it is a short step to providing packages of subroutines which try to anticipate the simulation user's most likely needs. Several such packages exist.

One of the better ones, despite its age, is GASP II [Pritsker and Kiviat (1969)]. This package contains the definitions of a number of arrays used for holding the schedule, for representing queues, and so on, and also of about two dozen routines for manipulating the GASP II data structures, for generating random variates from different distributions, and for collecting and printing statistics. The package was entirely written in ANSI Fortran using the old standard. Few changes are needed to adapt it for a modern compiler. It can run on a rather small machine. Listings of all the routines are given in the cited reference.

A strong point in favor of GASP II is its succinctness and its transparency. All the routines are well documented, and the scale of the system is sufficiently small that a competent programmer can easily grasp the relationships between the different components. As a result it is easy and practical to tailor the system to a user's special requirements, or to enhance the performance of critical routines. If you intend to write simulations in Fortran or any other general-purpose language, look at GASP II before you begin. Even if you do not adopt the whole system, you may pick up some useful ideas.

It is hard to leave well alone: GASP II has had numerous offspring. GASP IV [Pritsker (1974)], for instance, can be used for discrete, continuous, and combined discrete-continuous modeling. Q-GERT [Pritsker (1977)] is designed for network simulations. SLAM [Pritsker (1986)] is the most general of this family, attracting a substantial number of users. Although we feel that for most applications the extra features do not outweigh the resulting loss in clarity, not everyone shares our opinion. Another descendant of GASP II is WIDES [Thesen (1977)]. Although WIDES may suit a particular class of users, again a rather small increase in power seems to lead to a less perspicuous system.

7.1.6. Our Standard Example—the Naive Approach

Example 1.4.1 of Chapter 1 presented a simple model of a bank. In this section we show how to program the model in Fortran in the most straightforward way. In Chapter 8 we pick up the example again, and show how to improve the simulation design in various ways to increase the accuracy of the resulting estimators or to reduce the necessary computer time. To avoid cluttering the text, the programs for this model are grouped in Appendix X. For comparison, we also programmed this standard example in Simscript, GPSS, and Simula: see Sections 7.2.4, 7.3.5, and 7.4.5, respectively. The reader should begin by reviewing Example 1.4.1.

Figure X.7.2 shows a Fortran program which implements the model using the simple techniques illustrated above. As described in Chapter 1, four event types are possible: the arrival of a customer, the end of a period of service, opening the door of the bank at 10 A.M., and closing the door at 3 P.M. As in the previous example, events are so simple that we have written the code directly in the main program instead of creating a subroutine for each one. In the program, we have chosen to measure time in minutes, and $t = 0$ corresponds to 9 A.M.

The program contains comments to clarify its structure. It also includes a number of PRINT statements controlled by a logical variable TRACE. When this variable has the value .TRUE., an ample trace of the states of the model is printed, helping us check that it is working correctly. For convenience, we chose to initialize this variable in the program; it is almost as easy to read the value of TRACE at run-time.

The routines GET and PUT, and the clock mechanism, are as described above in Section 7.1.1, except that the auxiliary parameter AUX, and the corresponding array PARAM, are not needed and have been dropped. The uniform random number generator used, called UNIF, is not shown: in fact, we used an assembly-coded version of the generator given in Figure 6.5.2. The function ERLANG is shown in Figure X.7.3. It is a direct implementation of the method suggested in Section 5.3.9. The uniform generator used has the convenient property that it does not return the value 0, so we may take logs of UNIF with confidence; one can verify *a priori* that no overflow or underflow will occur.

PROBLEM 7.1.11. As we remark in Section 5.3.9, this generating procedure for Erlang variates is obvious. However, although it does in fact transform the underlying uniform random numbers in a monotone way, it uses more than one uniform variate per Erlang variate generated. Which procedures of Appendix L could be used to generate Erlang variates by (approximate) inversion? (Hint: Consider the relationship between the Erlang and chi-square distributions.)

Arrival times are generated by the function NXTARR, shown in Figure X.7.4. It is convenient to generate absolute arrival times rather than the intervals between successive arrivals. The method used is similar to Method II of Section 5.3.18; see also Section 4.9. The integrated arrival rate function is supplied in tabular form. In the model, the arrival rate after 3 P.M., when the door of the bank closes, is irrelevant, so it is convenient to suppose that the last segment of the piecewise-linear function continues indefinitely.

PROBLEM 7.1.12. If at some time the arrival rate falls to zero and stays there, a "next arrival time" function such as NXTARR should return an infinite value (or none at all). How could you implement this cleanly and efficiently?

PROBLEM 7.1.13. The function NXTARR, or its equivalent, can also be implemented using VNHOMO from Appendix L. Do this. Comment on the advantages and disadvantages of using a general routine rather than one tailor-made for the problem in hand.

Other programming details are straightforward. Verify that the test used to decide whether a customer joins the queue or not works correctly even when the queue length is less than five. It may seem that we "waste" a random value, but we shall see that this may in fact be desirable for statistical reasons. Finally, on each run we save the number of tellers at work, the number of customers arriving, and the number of customers served. (In the naive approach, the second of these figures is not used.) As noted in Section 1.8, it is generally easier to dump observations from a simulation onto a file, to be recovered and processed later, than it is to complicate the program by trying to incorporate the required statistical calculations into the same run. These calculations transform the raw observations to the final output.

We use separate uniform random number streams to generate four streams of nonuniform random numbers: one for interarrival times, one for service times, one to determine how many tellers are at work, and one to determine which customers balk at long queues. As we shall see, using different random number streams allows us to carry out statistical experiments more easily: for this simple model, it hardly matters whether we separate the four streams or not. Ideally, for four different streams we should use four different generators, to be sure that the streams are independent. In practice, since it is hard to find four different, well-tested generators, it is common to use the same generator with four different seeds. In this case we must be sure that the sequences generated by each seed do not overlap. One way to do this is to pick a starting seed X_0 arbitrarily, and to generate beforehand the sequence X_0, X_1, X_2, etc. [For a faster way which works for multiplicative generators, see Ohlin (1977).] By choosing the other seeds sufficiently far apart in the sequence, we can be sure that the generated segments do not overlap: see Section 6.8.3.

In our example we shall be performing about 200 runs, each involving about 250 arrivals, services, and so on. Choosing the seeds 200,000 values apart leaves a comfortable margin, and this is what we did. Thus the seed $X_0 = 1234567890$ was chosen arbitrarily, the next seed 1933985544 is X_{200000}, and so on.

Before running the complete model, we ran a few simple tests of the random generators. The basic uniform generator has been extensively studied, so its reliability was taken on trust [except of course that we verified the programming using test-values given by Schrage (1979)]. However, we ran a small number of runs of a Kolmogorov–Smirnov test, using the method detailed in Section 6.6.2 and in Knuth [(1981), pp. 45 *et seq.*], of the function ERLANG with the mean and shape parameter to be used. Probably this is more useful as a check on our programming than as a check on the random generator, unless many more runs are made. We also ran a small number of tests of an exponential generator, since this is essentially what underlies the function NXTARR. In all cases, the generators appeared to be behaving correctly.

PROBLEM 7.1.14. How could you test NXTARR itself, rather than merely the underlying exponential generator?

Next, we ran the complete model a small number of times with the trace on, and examined its behavior in detail. Finally, when we were confident that it was working correctly, we made 200 runs with the trace off. (The figure 200 was chosen arbitrarily—it seemed reasonable in view of the machine time we had available.) The results are given below. We remind the reader that this is not a recommended way of carrying out simulations. Our purpose was to establish a basic performance figure for the naivest possible approach, so that the gain by using better methods can be appreciated. As we have already pointed out, such better methods form the substance of Chapter 8.

Number of tellers	Number of runs	$\hat{\theta} = \bar{x}$	Est. var. $(\hat{\theta}) = s^2/n$
1, 2 or 3	200	240.93	2.85

(a)

Number of tellers	Number of runs n_i	$\bar{x}_i$	s_i^2	True prob. p_i
1	8	151.13	72.13	0.05
2	28	229.11	110.62	0.15
3	164	247.32	212.97	0.80

$$\bar{x}_i = \sum_{j=1}^{n_i} x_{ij}/n_i,$$

$$s_i^2 = \sum_{j=1}^{n_i} (x_{ij} - \bar{x}_i)^2/(n_i - 1),$$

$$\hat{\theta} = \sum p_i \bar{x}_i = 239.78.$$

Estimated variance $(\hat{\theta}) = \sum p_i^2 s_i^2/n_i = 0.94$ (biased, probably low)

(b)

Figure 7.1.5. Results of the naive simulation.

The naivest possible approach. Figure 7.1.5(a) gives the results of our 200 runs. If x is the number of customers served on each run, we calculate simply $\hat{\theta} = \bar{x} = 240.93$ and our estimate of the variance of $\hat{\theta} = s^2/n = 2.85$.

A slightly less naive approach (see Problem 2.4.5). In our 200 runs there were 8 with one teller at work, 28 with two tellers, and 164 with three tellers. Since we might expect the number of customers served to depend heavily on the number of tellers, we may hope to get a better estimate of θ, the overall mean, by calculating separately $\bar{x}_1$, the mean when just one teller is at work, etc., and then weighting and summing these separate means. For example, the weight for $\bar{x}_1$ is the known probability that just one teller shows up. Figure 7.1.5(b) shows the results of this approach. Using the empirical variance of our estimator as a (biased) measure, method (b) seems noticeably better than method (a), the work involved being essentially the same in the two cases.

```
110-119
120-129
130-139      *
140-149      **
150-159      ****
160-169      *
170-179
180-189
190-199
200-209      **
210-219      *******
220-229      **********************
230-239      *******************************************
240-249      **********************************************
250-259      *******************************************
260-269      *************************
270-279      ******
280-289      **
290-299
```

Figure 7.1.6. Customers served.

Before leaving the naive approach, we note that, as usual, more information may be obtained from looking at a complete distribution than from
simply considering a mean value. Figure 7.1.6 is a histogram showing the
number of customers served at each run. It is evident that the mean alone
conveys no adequate feeling for what is going on.

7.1.7. Simulation Using Pascal

We adopt the common usage of printing Pascal reserved words in bold face.
Identifiers mentioned in the text are in italics.

Apart from the convenience of such so-called structured constructs as
if . . . **then** . . . **else**, **case**, **repeat** . . . **until**, and so on, Pascal [see Jensen and
Wirth (1976)] offers facilities for defining new types, and for defining data
structures. Both these facilities are useful when writing simulations.

In our standard example, for instance, instead of declaring an integer
variable for each type of event (ARRVAL, ENDSRV, etc.) and giving these
the values 1, 2, 3, . . . , essentially the same effect can be obtained by defining:

```
type event = (arrival, endservice, opendoor, shutdoor);
```

If we then declare

```
var nextevent : event;
```

the main loop of the program can have the following structure:

```
repeat get (nextevent);
case nextevent of
arrival : ...
endservice : ...
opendoor : ...
shutdoor : ...
end
until nextevent = shutdoor;
```

Although we have not saved much writing, the structure of the program is more transparent than in Fortran.

As an example of the use of data structures, consider a queue of the kind described in Section 7.1.3. This can easily be described in Pascal using a declaration such as

```
type transaction = record
                   transnumber, priority : integer;
                   pred, succ : ↑transaction;
                   timein : real
                   end;
```

Here the type ↑*transaction* indicates that *pred* and *succ* (for predecessor and successor, respectively) are pointer variables which link each transaction in the queue to its neighbors. If we also declare

```
var newtrans : ↑transaction;
```

then to acquire space for a new transaction we need only call

```
new (newtrans);
```

However many kinds of queues and transactions we have to worry about, the Pascal run-time system takes care of the dynamic allocation and recovery of memory.

PROBLEM 7.1.15. Define an event notice by

```
type event notice = record
                    whatkind : event;
                    when : real;
                    aux : integer;
                    link : ↑eventnotice
                    end
```

and an event list by

```
var eventlist : ↑eventnotice;
```

Write procedures *put* and *get* like those of Figures 7.1.1 and 7.1.2, treating the event list as a singly-linked ordered list, the next scheduled event being at the head of the list.

PROBLEM 7.1.16. Write a uniform random number generator in Pascal, either by adapting the portable Fortran generator shown in Figure 6.5.2 or otherwise. Your code should be as machine-independent as possible. Test your generator.

PROBLEM 7.1.17. Rewrite the example of Section 7.1.4 in Pascal, and test the resulting program.

The language Pascal Plus [Welsh and McKeag (1980)] is an extension of Pascal intended for writing simulations and also for writing operating systems: the basic operations of scheduling events and subsequently executing them or modeling them one by one at the right time occur in both applications. Although described by its authors as "a combination of the merits of Simula and Pascal," Pascal Plus does not seem to offer enough new features to justify wide acceptance.

Shearn (1975) gives another account of simulation using a member of the Algol family of languages.

7.2. Simscript

In the beginning Simscript had its roots in Fortran. The current version, Simscript II.5 [documented chiefly in Braun (1983), Mullarney and Johnson (1983), and Russell (1983)], still shows traces of its origin here and there. However, any structure the language once had is now completely hidden under layer after layer of later accretions. Although this may be deplorable in theory, it is understandable in the real world. Simscript and CACI Inc. (who support the language) have a large number of customers who would not tolerate reprogramming a model which works. When additions are made to the language, this means that the old constructs must stay in as well. A possible metaphor is that Simscript is the pearl in CACI's oyster: when a feature of the language proves particularly abrasive, a new, smoother feature is secreted over it, but the original irritant is still present. Unlike pearls, however, the result has little aesthetic appeal.

This is not to deny that good, readable programs can be written in Simscript. They can. The problem is that it is also easy to write obscure programs, or, more dangerously, programs which seem to do one thing but in fact do another. This can be avoided if programmers have the sense and the discipline to steer clear of ambiguous constructs. Particularly for beginners, however, or for programmers relying solely on a manual to learn about the language, the danger is always there.

To avoid laboring the point, we give just one example of Simscript's eccenticities before going on to consider the langauge's more positive aspects. The attempt to provide an "English-like" language, together with the necessity for distinguishing between a large number of constructs which have been tumbled pell-mell into the language, too often leads to confusion. Try, for instance, to guess the difference between

```
        ROUTINE EXAMPLE GIVING X
```

and

```
        ROUTINE EXAMPLE YIELDING X .
```

In the former case X is an input parameter to the routine EXAMPLE, while in the latter it is an output parameter. Although this may cause no difficulty for an experienced user, it is far from transparent.

PROBLEM 7.2.1. One striking feature of Simscript is that it has no end-of-statement indicator. While the end of a Fortran statement is signalled by the end of a line, a Cobol sentence ends with a period, and the Algol family use the semicolon, no such marker exists in Simscript.

What problems does this pose to a compiler writer? How would you give a formal representation of the syntax of such a language?

PROBLEM 7.2.2. One of the few control statements not implemented in Simscript II.5 is Clark's COME FROM [Clark (1973)]. Give examples to show how this construct might be useful in discrete digital simulation.

On the positive side Simscript contains many useful constructs. In the following paragraphs we introduce some of the most important. Indeed, while the implementation of Simscript leaves much to be desired, the underlying "world view" is coherent and clear.

7.2.1. Data Structures in Simscript

A Simscript programmer uses entities, attributes, and sets. In essence, an entity is a structured variable, and its attributes are the fields within this structure. The declaration

```
EVERY MAN HAS AN AGE, SOME DEPENDENTS, AND A NUMBER
```

indicates that the program will use records of type MAN, each containing the fields AGE, DEPENDENTS, and NUMBER. Sets are normally doubly-linked lists with a head cell. Thus

```
EVERY MAN OWNS SOME CHILDREN
```

indicates that entities of type MAN can serve as head cells for a doubly-linked list; the entities which can belong to this list might be declared by

```
EVERY CHILD MAY BELONG TO THE CHILDREN .
```

In this case records of type MAN will include pointer fields F. CHILDREN and L. CHILDREN to the first and last elements of an associated list, while a CHILD record will include pointer fields P. CHILDREN and S. CHILDREN for a predecessor and a successor, respectively.

Entities can be permanent or temporary. Permanent entities are represented by arrays, exactly in the fashion suggested above (§7.1.3). Temporary entities share dynamic storage, and are created and destroyed explicitly by the programmer. The attribute fields of an entity normally occupy one word each, but if necessary they can be packed more closely. Take care when different entity types share common attributes that the relative position of these attributes within each record is the same. For instance

```
EVERY MAN HAS AN AGE AND BELONGS TO A FAMILY

EVERY WOMAN BELONGS TO A FAMILY
```

may be incorrect, because the implicit predecessor and successor pointers for the set FAMILY will be in different places within a MAN record or a WOMAN record. More recent versions of the compiler on some computers no longer require attributes with the same name to be in the same position.

Simscript provides the usual operations for inserting entities into sets at various points, for testing membership, and for removing entities. It is clear from the mechanism used to implement sets that an entity may not belong to more than one set of a given type. For instance, with

```
EVERY CITY OWNS A CLUB

EVERY MAN MAY BELONG TO A CLUB
```

then a MAN can be in at most one CLUB. Convenient control statements are available for examining sets. Thus with the above declarations we might write

```
FOR EACH MAN IN CLUB DO ...  LOOP
```

with the obvious meaning.

PROBLEM 7.2.3. If we would like to allow a CLUB to have more than one MAN as a member and also to allow a MAN to belong to more than one CLUB, how could we represent such sets in Simscript?

7.2.2. Simscript and Simulation

The clock mechanism in Simscript uses the above notions. An event notice is a temporary entity with one attribute specifying the simulated time at which the corresponding event is to occur; pointers enable the event notice to be filed in a special events set. In fact each possible class of events is held in a separate set, as shown in Figure 7.2.1.

In this example event notices A and B refer to the same type of event, C, D, and E to another type, and F to a third. The sets are ordered by event time, so that A is due to occur before B, and so on. Besides those necessary for the clock mechanism, an event notice of a particular type may also include whatever other attributes the programmer wishes to give it. When an event is due to occur, its event notice is removed from the events set, and a routine corresponding to the required class of event is called. If the event notice contains additional attributes, these are passed as parameters to the event

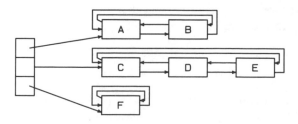

Figure 7.2.1. A Simscript schedule with three classes of event.

routine. When the routine ends, the system must look at the first event notice in each of the possible lists of events to determine which is the next to occur; the simulation clock is advanced, the relevant event notice is taken out of its event set, and so on.

The system distinguishes between internal and external events. An internal event is explicitly scheduled by the program; the event notice includes whatever parameters are necessary. Details of external events, including their type, the time at which they are to occur, and any necessary parameters, are read from designated input devices. The system inserts the corresponding event notices in the schedule.

Instructions for adding internal events to the schedule are adequate. An event is simply a routine with a slightly different declaration; parameters are passed using the standard mechanism. If ARRIVAL is the name of an event, we can schedule it in any of the following ways:

```
SCHEDULE AN ARRIVAL GIVEN parameters AT time expression

SCHEDULE AN ARRIVAL GIVEN parameters IN expression UNITS

SCHEDULE AN ARRIVAL GIVEN parameters NOW .
```

To keep a pointer to an event notice in the schedule, use the option CALLED:

```
SCHEDULE AN ARRIVAL CALLED NEWMAN ...
```

This enables us subsequently to write

```
CANCEL THE ARRIVAL CALLED NEWMAN .
```

(Once again the particular choice of an optional word may conceal important details:

```
SCHEDULE AN ARRIVAL CALLED NEWMAN ...
```

and

```
SCHEDULE THE ARRIVAL CALLED NEWMAN ...
```

are different. The former creates a new event notice, while the latter assumes that the event notice already exists.)

Simscript provides various random number generators, including uniform, normal, binomial, Poisson, exponential, beta, gamma, Erlang, and Weibull. Facilities are also provided for sampling from empirical distributions. The basic uniform generator is of the multiplicative type. Details of the method of implementation are given in Payne *et al.* (1969). Ten streams of random numbers are provided, all using the same generator but with starting seeds set 100,000 values apart.

One convenient feature of Simscript is its capacity to accumulate automatically certain simple statistics. For instance, if X is a real variable, then the instruction

```
TALLY M AS THE MEAN AND V AS THE VARIANCE OF X
```

sets up a mechanism to track the value of X and to accumulate the necessary sum and sum of squares to calculate the required mean and variance. Any subsequent reference to M or V provides the up-to-date values of these statistics. However, unless care is taken, these estimates may be badly biased; see Chapter 3. Besides the mean and variance, the maximum and minimum values of X can be tracked in this way. The mechanism can be used for entities as well as for simple variables:

```
PERMANENT ENTITIES

    EVERY MAN HAS SOME CASH AND OWNS A FAMILY
  .
  .
  .
TALLY AV.CASH AS THE MEAN AND MAX.CASH AS THE MAXIMUM OF CASH
  .
  .
  .
FOR EACH MAN, LIST AV.CASH(MAN) AND MAX.CASH(MAN) .
```

When the verb TALLY is used, means, variances, and so on are calculated without reference to the passage of simulated time. To calculate these same statistics, but now weighting the observations by the length of simulated time for which the tracked variable has had its current value, use ACCUMULATE instead of TALLY. Histograms can also be generated using either TALLY (the number of times the tracked variable is in each class) or ACCUMULATE (the length of simulated time the tracked variable is in each class):

```
TALLY FREQ(0 TO 100 BY 5) AS THE HISTOGRAM OF X .
```

If more sophistication is required, Simscript II.5 also makes available library routines for experimental design and for calculation of confidence intervals: see Law (1979). However, avoid blind use of any statistical methods that you do not understand.

A graphics package is also available so that results can be presented in a more perspicuous form: see West (1979).

7.2.3. A Complete Simulation in Simscript

EXAMPLE 7.2.1 [adapted from Bobillier *et al.* (1976), p. 103, where a similar example is programmed in GPSS]. The take-out counter in a fast-food restaurant is served by two employees, one making hamburgers, and the other wrapping and selling them. Customers are served in the order of their arrival. The cook produces hamburgers at a steady rate; hamburgers which are ready but not yet needed can be kept in stock on the counter.

Customers arrive on average every 2.5 min, the distribution of time between successive arrivals being exponential. The number of hamburgers a customer orders has the following distribution:

Probability	0.1	0.4	0.3	0.2
Number of hamburgers	4	5	6	7

Competition in the neighborhood is strong: if a customer sees three people already waiting in line, he goes elsewhere.

The cook produces one hamburger in a time uniformly distributed between 30 and 50 s. It takes a fixed time of 45 s plus 25 s per hamburger to wrap a customer's order and take his money. This process begins only when all the required hamburgers are available.

We intend to simulate this system for a period of 4 hours, to determine

(a) the total number of customers arriving and the number of customers lost;
(b) the average time a customer has to wait for his order; and
(c) the percentage of the server's time which is usefully occupied.

Figure X.7.5 in the appendix gives a simple program to determine these values. Five event types are used: the arrival of a customer, the start of a service, the preparation of another hamburger, the end of service to a customer, and the end of the simulation. Of these events, only the end of service needs an extra parameter: we need to know the customer concerned to measure his time in the system.

ORDER.SIZE is an example of an empirical distribution. The name is declared in the preamble. Necessary values are read from data cards as indicated in the main program section. (The data will be 0.1 4 0.4 5 0.3 6 0.2 7 *) The declaration of the set QUEUE automatically generates a number of other variables, including in this case N.QUEUE and F.QUEUE: the former gives the number of items in the set, and the latter points to its first member.

Control is given to the timing mechanism and the event routines by the instruction START SIMULATION. If ever no events remain in the schedule to be simulated, control comes back to the main program. In our example, however, the STOP.SIMULATION event prints the results and stops the program. Notice the simple way in which formats are specified. The variable TIME.V used in the event routines is a system variable representing the simulated time.

Notice next the use of DEFINE ... TO MEAN ... This defines a simple pattern of text replacement, so that in the example with

```
DEFINE SECONDS TO MEAN UNITS
DEFINE MINUTES TO MEAN *60 UNITS
```

when we subsequently encounter a statement such as

```
SCHEDULE A NEW.HAMBURGER IN 35 SECONDS
```

it is treated as

```
SCHEDULE A NEW.HAMBURGER IN 35 UNITS
```

while a time such as 3 MINUTES is compiled as 3 *60 UNITS. This again illustrates the basic dilemma of Simscript. Used elegantly, this facility makes it easier to write clear programs; used foolishly (for example, DEFINE A TO MEAN THE), it renders programs totally opaque.

PROBLEM 7.2.4. Program and run the same example in Fortran or some other general-purpose language. Compare critically the Fortran and the Simscript models.

PROBLEM 7.2.5. Suppose the quantity of most interest is L, the average number of customers lost. Estimate this quantity without using simulation.

PROBLEM 7.2.6. Which variance reduction techniques can conveniently be used if L is to be estimated using simulation? How easy is it to incorporate each possible technique into a model (a) in Simscript? (b) in Fortran?

PROBLEM 7.2.7. Estimate L using simulation. Compare your result, and the effort involved, with that of Problem 7.2.5.

PROBLEM 7.2.8. Once again, the model presented is based on an imaginary situation, drastically simplified in the interest of clarity. In a comparable real-life project, which of our simplifying assumptions would still be tenable, and which would probably have to be abandoned? How would you gather data? How much would it cost?

7.2.4. The Standard Example in Simscript

To allow a direct comparison between languages, Figure X.7.6 shows how Example 1.4.1 might be programmed in Simscript. The program given performs only one run of the simulation. As well as printing the number of customers arriving at the bank and the number of services, we accumulate and print the average length of the queue and the average waiting time.

The event routines are modeled closely on the corresponding sections of code in the Fortran version of the model. As a minor variation, we stop the simulation by canceling the only arrival in the schedule when the door of the bank is closed at 3 P.M. Customers in the bank are allowed to complete their services, so that the value of average waiting time will be correct. When all services end there are no events left in the schedule, the main program is resumed, and the results are printed.

Variables defined in the preamble are global, while those defined in individual routines are local. All local variables are initialized to zero on first entry to a routine. We take advantage of this in NEXT.ARRIVAL to set up the necessary tables. The declaration of an array is performed statically, while the allocation of space is performed dynamically (using RESERVE). Routines in Simscript are in general recursive, and space for local variables is reallocated and reinitialized every time the routine is entered. In the routine NEXT.ARRIVAL, however, we want the arrays and the local variables CUMUL and SEG to be saved between calls: we therefore declare this explicitly.

Four separate random number streams determine the number of tellers and generate arrivals, services, and balks. As noted above, the four streams all use the same underlying generator, with seeds set 100,000 values apart.

Do not be misled by the apparent liberality of the Simscript naming conventions. On the IBM 370 implementation used to test the example names were truncated to at most seven characters (often less). An earlier version of the program using the innocuous-looking variable names TELLERS.FREE and TELLERS.AT.WORK failed to run for this reason, with no warning message from the compiler. (This problem may be less acute on other machines or with later releases of the Simscript compiler. Check the system you are using.)

7.2.5. Processes and Resources

Our brief description of Simscript suggests that the language is essentially event-oriented, with event routines containing the instructions necessary to change the state of the system. More recently, new statements have been added to the language (!) to allow models to be described in terms of processes and resources. This seems to be the approach now favored by CACI Inc.

In essence, a process is a special kind of temporary entity, and a resource is a special kind of permanent entity. Processes are initiated by an ACTIVATE statement, they can WAIT, and they can be interrupted and resumed. Processes can REQUEST and RELINQUISH resources; they are made to wait if they request more of a particular resource than is currently available.

Russell (1983) emphasizes this approach to model-building in Simscript.

7.2.6. Summing-up

There are undoubtedly many good ideas in Simscript struggling to get out. The basic concepts of temporary and permanent entities, of attributes, and of sets are sound and useful, though some details of implementation are poorly handled. The event list and event notices provide simple and efficient control of simulations. Facilities for generating random variables are good. The mechanisms for automatically accumulating means, variances, and some other statistics are a great convenience when used properly. Externally generated events are easily handled. It is not hard to include adequate tracing in a program. Finally, some useful facilities are provided for printing formatted reports.

Of more doubtful value to the average programmer are the different facilities provided for storage management, which we have not mentioned above. It can certainly be argued that the ability to release the space occupied by arrays or routines to free storage, or to declare sets without all the standard attributes (with only one-way linking, say), enables an experienced user to fit large jobs into core. However, such practices are a possible source of many errors, especially when large programs are maintained by a number of different people.

PROBLEM 7.2.9. Adapt the program to perform several runs, paying special attention to the problem of reallocating (or not) the arrays in NEXT.ARRIVAL.

The most serious drawback of Simscript is its uncontrolled syntax: the language has essentially no formal structure. Bad control mechanisms abound, the use of optional words and keywords is inconsistent, and the eclectic selection of features to be included makes this confusion total. Half an hour with a Simscript manual should, we hope, convince any reader who finds this criticism too severe.

The problems this causes can be alleviated by insisting on rigorous coding conventions. However, this often comes down to deliberately avoiding options and defaults offered by the system, in which case there is little point in having such latitude to begin with. Despite the argument in favor of perpetual compatibility with earlier compilers, a major overhaul of the Simscript language is overdue.

7.3. GPSS

GPSS is a language *sui generis*, first introduced in 1961. It has undergone the usual series of changes and developments. The current IBM version is GPSS V [see, for instance, IBM (1971)]. An excellent introduction, with many worked examples, is Bobillier *et al.* (1976). Schriber (1974) is an older but still popular introduction. Manufacturers other than IBM may give GPSS slightly different names, presumably for copyright reasons, but the language is in general well standardized; see, however, Section 7.3.6.

7.3.1. The Basic Concepts

A model in GPSS takes the form of a flow-diagram. Some 50 or so different *blocks* are available for constructing flow-diagrams: each has its own name, a special symbol, and a predefined purpose. To construct a model, the programmer selects and connects together a certain number of blocks. When a

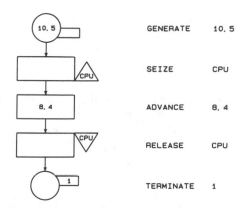

Figure 7.3.1. GPSS flowchart and program.

program is written out on paper, or punched on cards, each line specifies one block. Although a model in GPSS may therefore look quite like a program in an ordinary high-level language, the distinction between a GPSS block and an ordinary language's instructions is important. Figure 7.3.1 shows part of a GPSS model, first using the standard GPSS symbols for the blocks, and then using the more convenient notation with the names of the blocks.

Moving through these blocks are entities called *transactions*. Each transaction, besides occupying a particular block of the flow diagram, may also carry around with it a set of parameters. At any given instant, there may be many transactions at different places in the flow diagram.

Transactions provide a natural means of modeling, say, the movement of customers through a service facility. Just as customers can claim the attention of servers, so, in a GPSS model, transactions can reserve for their own use the different resources in a model. As we shall see, a transaction to some extent resembles a process in a language like Simula, discussed in Section 7.4. In Simscript the closest analog to a transaction is again a process (§7.2.5) or a temporary entity (§7.2.1). However, in Simula the resource being seized may also be represented by a process, and the interaction between, say, customers and servers may be initiated and controlled by either of the processes involved. In GPSS, resources are essentially passive. As a result, it is often easier to model simple interactions in GPSS, but more complicated situations may be difficult to describe.

As simulation of a model progresses, transactions move from block to block of the flow diagram. Some blocks create transactions; others delay their progress for a certain period of simulated time; others destroy them. More generally, the entry of a transaction into some blocks causes changes to occur in the state of the model. Finally, there are blocks which can refuse entry to transactions. They either go elsewhere or wait until the block subsequently allows them to enter.

Generally speaking, a transaction is held on one of two lists, the current events chain (CEC) or the future events chain (FEC). A transaction on the CEC is trying to make progress through the block diagram; it can, however, be held up if a block refuses to let it enter. A transaction on the FEC is not ready to move; only later, as the simulated clock moves on, is it transferred to the CEC to continue its progress.

At any simulated time instant, GPSS scans all the transactions on the CEC. Any transaction which can move further through the flow diagram does so. This movement ends for one of three reasons: the transaction is destroyed, a block refuses to let it enter, or it enters a block which delays it for a period of simulated time. In the last case the transaction is removed from the CEC and added to the FEC. If a moving transaction has caused a change in the state of the model (if it has acquired or released a resource, for instance), the GPSS scan begins again at the beginning of the CEC; otherwise the next transaction on the CEC is scanned. Sooner or later none of the transactions on the CEC can move further.

At this point the FEC is examined. This chain is ordered so that the transactions which are free to move the earliest are at the beginning of the chain, while those which have the longest to wait are at the end. When no further transactions on the CEC can move, GPSS advances the simulated clock in one step to the time when the first transaction on the FEC is ready. All those transactions now free to continue their progress are copied from the FEC to the CEC, and the scan through the CEC beings once again.

Each transaction has a priority, which can be changed as it progresses through the block diagram. The CEC is held in descending order of priority, so that high-priority transactions move before low-priority ones at any simulated instant. Transactions also have a variable number of parameters which can be used to carry information. Essentially all values in GPSS, including the clock, are represented as integers: this can be inconvenient for the experienced user and catastrophic for the unwary.

Transactions are created by **GENERATE** blocks, and destroyed by **TERMINATE** blocks. A model can contain several **GENERATE** or **TERMINATE** blocks if necessary, each operating independently of the others. **TERMINATE** blocks also determine the end of the simulation. The run is started by a control card such as

START 50

which initializes a counter called the termination counter (to the value 50 in our example). Each time a transaction enters a block

TERMINATE 1

say, the termination counter is decremented (by 1 in this case, or in general by the parameter of the TERMINATE block). When the counter reaches zero, the simulation is stopped.

Transactions are delayed by the **ADVANCE** block. Whenever a transaction enters a block such as

ADVANCE 10

it is removed from the CEC and added to the FEC. When the simulated clock has advanced by 10 units, the transaction returns to the CEC to continue its progress.

7.3.2. Resources in GPSS

GPSS offers several kinds of objects for modeling resources. In GPSS books and manuals these are often called entities; they should not be confused with the entities of Simscript. For instance, a resource which can be of service to only one user at a time is called a *facility*. Once a transaction seizes a facility, any other transaction trying to seize the same facility is delayed until the first transaction releases the resource. Suppose, for example, that our model includes a facility called CPU. Then the sequence of blocks

```
SEIZE     CPU

ADVANCE   7

RELEASE   CPU
```

represents a situation where a transaction needs to use the resource for 7 time units. Any other transaction arriving at the block SEIZE is refused entry until the former transaction has entered RELEASE.

This simple sequence conceals a surprisingly powerful mechanism. First, GPSS automatically maintains a queue for each facility in the model, should several transactions contend for the same resource. Next, statistics of facility usage (number of times used, average length of use, percentage of time in use, and so on) are automatically collected and printed for each facility in the model. Features such as preemption or periodic unavailability of resources can be modeled with equal simplicity. It is not necessary to declare facilities: the mere appearance of the name in a SEIZE block suffices.

Resources which can be shared by several transactions are modeled using *storages*. Suppose we want to model a computer system which has 64 K bytes of memory. Then we might declare

```
MEMRY STORAGE 64
```

if memory is allocated in kilobyte units. If transactions represent programs, then a request for 16 K bytes of memory might be represented by the sequence of blocks

```
ENTER MEMRY,16
        .
        .
        .
LEAVE MEMRY,16 .
```

As with facilities, a transaction arriving at the ENTER block at a time when 16 units of the resource are not available is automatically delayed by GPSS until other transactions release the necessary units. Again, as with facilities, statistics on the usage of the resource are automatically compiled and printed.

The statistics automatically gathered by GPSS can be augmented using entities called *queues*. The pair of blocks

```
                    QUEUE INQ
                       .
                       .
                       .
                    DEPART INQ
```

show a transaction entering and leaving a queue called INQ. At the end of the simulation, GPSS automatically prints the number of entries to each queue, its mean and maximum length, the average time spent in the queue, and so on. A "queue" in this sense has nothing to do with contention for a resource: it is simply a device for measuring the flow of transactions through different parts of a model.

Further statistics can be gathered using *tables*, which are essentially histograms of observed values in the model. In particular, a table can be associated with a queue to obtain a histogram of the times spent in the queue by different transactions. Remember, however, the warning given in Section 3.3.5. The times spent in a queue by successive transactions are usually highly correlated, so that the sample variance calculated automatically by GPSS is worthless. Ignore it.

It is rarely necessary to write output instructions in a GPSS model, as many values of interest are printed automatically by the system, either at the end of the simulation or at regular intervals, as required. If this standard output is not adequate (or, more likely, if there is a great deal too much of it), the GPSS report editor can be used to produce selected statistics, including graphical displays of histograms.

As we shall see, however, arithmetic capabilities in GPSS are poor. As one consequence, extensions to the statistics automatically gathered by GPSS may require rather tortuous programming.

PROBLEM 7.3.1. If a GPSS model contains a queue called FILE, the number of items in this queue is referred to as Q$FILE. Define

$$X(t) = \begin{cases} 1, & \text{when Q\$FILE} > 20, \\ 0, & \text{otherwise.} \end{cases}$$

Suggest how the model might be made to compute the mean of $X(t)$.

7.3.3. Generating Random Variates

Facilities for generating random variates in GPSS are poor. Apart from some simple mechanisms incorporated into the language, most distributions have to be supplied in tabulated form.

At the simplest level, certain blocks include a "spread." Thus

GENERATE 10,5

creates a new transaction every 10 time units on average, with a spread of 5: in GPSS terms, this means that inter-generation time is chosen with equal probability to be 5, 6, 7, ..., 13, 14 or 15 time units. Similarly

ADVANCE 15,3

delays a transaction by 12, 13, ..., 17 or 18 time units. At the same level of sophistication

TRANSFER   .123,BLOKA,BLOKB

sends an incoming transaction on to the block called BLOKA with probability 0.877, and to the block BLOKB with probability 0.123.

The GPSS user has access to eight (or more) identical uniform random number generators called RN1, ..., RN8 (described in Example 6.3.1). The standard way of using these to model, for instance, Poisson arrivals, follows:

```
EXPON           FUNCTION          RN1,C24
0.0,0/0.1,0.104/0.2,0.222/0.3,0.355/0.4,0.509/0.5,0.69
0.6,0.915/0.7,1.2/0.75,1.38/0.8,1.6/0.84,1.83/0.88,2.12
0.9,2.3/0.92,2.52/0.94,2.81/0.95,2.99/0.96,3.2/0.97,3.5
0.98,3.9/0.99,4.6/0.995,5.3/0.998,6.2/0.999,7.0/0.9997,8.0
                GENERATE          50,FN$EXPON
```

The function EXPON takes the random number RN1 as argument. C24 indicates that 24 points are to be tabulated, and that linear interpolation is to be used (C means continuous). The tabulated points specify the inverse of an exponential distribution with mean 1. When the "spread" of a GENERATE or ADVANCE block is a function (FN), then the given mean is multiplied by the function to obtain the required value. (This is different from the meaning of a constant "spread," shown above.) Thus in our example intervals between transactions produced by the GENERATE block follow an exponential distribution with a mean of 50.

This method of generating random variates is both clumsy and inaccurate. First, except for the normal and exponential distributions, tabulated in most GPSS manuals, it may not be at all clear how many points need be provided to obtain a given accuracy (see Problem 7.3.11). Second, since GPSS uses integer values for the simulated clock, the fractional part of generated values is truncated. In the example above, the mean simulated time between generated transactions may be more like 49.5 than 50 units. For a small mean, this bias can be considerable.

PROBLEM 7.3.2. What distribution is obtained if random values from an exponential distribution with mean λ are truncated to the nearest lower integer?

Other pitfalls lie in wait for the unwary user. Suppose in the simulation of a computer system we assign random core and time requirements to transactions:

```
                    ASSIGN    1,FN$CORE,PF
                    ASSIGN    2,FN$TIME,PF
                       .
                       .
                       .
        CORE    FUNCTION RN2,D4
        0.2,16/0.4,32/0.75,64/1.0,128
        TIME    FUNCTION RN3,D3
        0.25,1/0.5,10/1.0,50
```

In this case parameter 1 of each transaction holds the core requirement. This is a function which takes the value 16 with probability 0.2, the value 32 with probability 0.2, the value 64 with probability 0.35, and the value 128 with the probability 0.25. Similarly parameter 2 holds the required time.

The two functions CORE and TIME have been assigned to separate random number streams: this is often convenient when replicating simulations under different conditions. However, RN2 and RN3 (and all the other generators) are *identical* in GPSS. Unless the seeds are deliberately reset, they give identical sequences of values. In the above example, if the programmer forgets this, then whenever RN2 is small, so is RN3: small values of the function CORE always accompany small values of TIME, and the results of the model are badly biased.

Finally, arithmetic in GPSS is difficult, though GPSS V is a little better in this respect than its precursors. Although some facilities for using floating-point values exist, most variables are truncated to integer form once they are calculated. Except as function arguments, the random number generators return integer values between 0 and 999 instead of fractional values between 0 and 1. Functions have only one argument, and the only way to give their values is by tabulation. One result is that it is often difficult to see how to program even the simplest variance reduction techniques in GPSS. If you doubt this, try the following problem. Check your answer on a machine.

PROBLEM 7.3.3. A GPSS program generates Poisson arrivals using the method shown above. Explain how to change the program to generate an antithetic series of arrivals.

PROBLEM 7.3.4. Find a neat way to implement the alias method of Section 5.2.8 in GPSS.

7.3.4. A Complete Simulation in GPSS

EXAMPLE 7.3.1. A computer controls 20 identical terminals. The sequence of events is:

(a) The user types a command. The total thinking and typing time follows an exponential distribution with a mean of 10 s.

(b) The terminal transmits the command instantaneously to the computer. 60% of all commands are in class 1, 30% are in class 2, and 10% are in class 3, the time and core requirements for each class being as follows:

Class	1	2	3
Time (ms)	100	500	2000
Core (K bytes)	16	32	64

(c) The computer has 256 K of core available. As commands are received, they contend for memory space. Once space is available, processing can begin.
(d) CPU time is allocated on a simple round-robin basis to each command in the memory in quanta of 100 ms. Once a command has received all the CPU time it needs, it releases its memory allocation. A reply is sent instantaneously to the terminal, and the cycle of events begins again.

Figure X.7.7 of the appendix gives a program to simulate this system for a period of 10 min to determine the distribution of response times for each class of command.

A single transaction represents the stream of commands from each terminal. Imagine it as leaving the terminal, contending for resources, carrying a reply back to the terminal, and so on. Three transaction parameters are used: PF1 gives the class of command being simulated, PF2 gives the remaining requirement of CPU time, and PF3 gives the memory requirement. As can be seen in the declarations of the three functions CLASS, TIME, and CORE, CLASS is chosen randomly, while TIME and CORE are functions of the class in PF1. (In the function declarations, C means "continuous": linear interpolation is to be used; D means "discrete": the function yields only the values tabulated; and L means "list": this is a more efficient form of discrete function if the tabulated arguments are 1, 2, 3, ...)

Four queues gather statistics: queue 1 for class 1 commands, and so on, and queue 4, called ALL, for combined results. A histogram is associated with each queue, giving the distribution of time transactions spend in the queue. Thus, the histogram CLS1 is associated with queue 1; the histogram has 15 columns, beginning with times less than 10 simulated time units (1 s) and going by steps of 10 units.

The two-block timer sequence decrements the termination counter once every minute of simulated time. In conjunction with the START control cards, this determines the simulation run length.

The main body of the simulation is straightforward. The GENERATE creates 20 transactions at the outset of the simulation, and these are never destroyed. The block

```
ASSIGN 2-,1,PF
```

decrements parameter 2 of entering transactions;

TEST G   PF2,0,DONE

lets transactions pass if they have PF2 > 0, and diverts them to DONE otherwise; and, as noted, BUFFER removes the current transaction to the end of the CEC so that it will now be last in the contention for the CPU.

In an attempt to minimize the effect of the start-up transient (the system is empty at time 0), the model is simulated for 1 min with no printing of results. Then all the histograms and so on are reset to zero, and the simulation is continued for 10 min as required. At the end of this period all the relevant (and irrelevant) statistics of the model—including the histograms, CPU and memory utilization, the number of commands processed, and so on—are printed automatically.

PROBLEM 7.3.5. Transactions contending for a GPSS storage are not necessarily served on a FIFO basis. In the example, a transaction requiring 16 units of the storage MEMRY may overtake a transaction requiring 32 or 64 units if only the smaller amount is available. Modify the program to enforce a FIFO discipline on the queue for MEMRY. What effect does this have on response times?

PROBLEM 7.3.6. The periods of 1 min for initialization and 10 min simulated run were chosen quite arbitrarily. Can you do better than this?

PROBLEM 7.3.7. The easy way to replicate runs of the above model adds pairs of control cards

RESET
START   10

for each run required. However, successive runs are now not independent, since the ending state of one is the starting state of the next. Does this matter? If so, what can you do about it? (See also Problem 7.3.12.)

PROBLEM 7.3.8. Reprogram the example in Fortran or in Simscript.

Once again, we cannot stress too strongly that our example is pure fiction, intended solely to illustrate some features of GPSS. None of the assumptions made in the example will survive even cursory examination if it is intended to model real computer systems.

The succinctness of GPSS programs and easy GPSS model manipulation can be especially seductive. An undue emphasis on terse programming may trap the user into imagining that the brevity of the program outweighs possible oversimplification of the model. Searching for even qualitative insight from a simplistic model often leads to delusion: the errors may be gross but perhaps not evident. For instance, a persuasive argument can be made in favor of modifying Example 7.3.1 to show, at least in broad outline, the effect of using different time quanta, or of adding more memory to a system, or of giving different priorities to different classes of command (all trivially easy to program in GPSS). Such arguments should only be accepted after close examination of the sensitivity of the model to gross errors in the underlying assumptions.

The iconographic aspects of GPSS do not lessen the need for careful model validation. Inexperienced users are often blind to this caution; they also replicate runs inadequately or not at all. Even with a perfect model, this is skating on thin ice. Many beginners, faced with quite normal variations between runs of a simulation, refuse to accept them and instinctively look for errors in the program. (This is particularly so when modeling computer systems, where more often than not some of the system resources are operating near capacity.) Of course, this is true for any simulation model: GPSS, however, seems sufficiently "real" that the effect is more pronounced.

PROBLEM 7.3.9. Run the program for Example 7.3.1 just once, and obtain the mean response times for class 1, class 2, and class 3 jobs. Try to guess confidence limits for these values. Now replicate the simulation and check your intuition.

7.3.5. The Standard Example in GPSS

Figure X.7.8 shows how Example 1.4.1 might be programmed in GPSS.

The time unit has been taken to be 1 s, so that GPSS's bad habit of truncating real values to integer should not unduly bias the distribution of inter-arrival times. We are then forced, as outlined in the comments to the program, to work out that the bank is open for 18,000 seconds, and so on.

We admit to fudging the arrival distribution for programming convenience. Whenever a new arrival is to be generated, we calculate the interval simply by multiplying an exponential variate with mean 1 by the mean interarrival time valid at the moment when the calculation is made. If the resulting interval crosses a breakpoint of the arrival rate function, this is wrong. Since the arrival rate function has only two breakpoints, at 11 A.M. and 2 P.M., the resulting error is probably small. In this particular case, the effort to program the distribution correctly seems out of all proportion to the increase in accuracy achieved: but we confess that this is setting foot on the road to ruin.

PROBLEM 7.3.10. Program the arrival distribution correctly, insofar as GPSS's integer clock will let you.

As noted in the comments, the tabulated inverse of the exponential distribution with mean 1 was simply copied from an IBM manual. This implies, among other things, that we have little idea how good the approximation is.

PROBLEM 7.3.11. The IBM manual [IBM (1971), p. 53] claims that "the function gives results which are accurate to within 0.1 percent for $45 < m \le 250$ and 1.0 percent for $m \le 45$," where m is the mean of the exponential distribution to be generated.

What exactly do you think this means? How could you find out if you are right?

The first GENERATE block creates transactions to represent customers. A GPSS queue called BANK gathers statistics about time spent in the system, and a second queue called WLINE represents the line of customers waiting for a teller. The Erlang service time is conveniently modeled as the sum of two exponential variates (multiplied by 60 since the required mean is 2 min, not 2 s).

The second GENERATE block creates a single transaction at time 1 (time 0 does not exist in GPSS) to control the overall timing of the simulation. The model uses three tellers in every case, but sometimes only one or two of them are made available by the FAVAIL blocks. The final TEST ensures that all the customers leave the bank before the simulation ends and the final report is produced.

Without writing a single output instruction, the automatic report gives us all we need to know about the model and more. Among the figures to be found therein are the number of tellers on duty, and the number of arrivals, services (by each teller and in total), and balks. The utilization of each teller is also given, along with the average and maximum numbers of customers in the bank and in the waiting line, their average waiting time, the percentage of customers who did not have to wait at all, and so on. With just one extra instruction (QTABLE) we could obtain the complete distribution of waiting times.

The example illustrates well the strengths and weaknesses of GPSS. The program is considerably shorter than any of the versions in other languages (cf. Figures X.7.2, X.7.6, and X.7.11); its output is much more complete; unfortunately, it is more or less wrong. For another snag, try

PROBLEM 7.3.12. To obtain 50 runs of the model, you could simply punch 49 pairs of cards

```
                    CLEAR
                    START    1
```

and insert them in the deck. This leaves the problem of collating 50 automatic reports. Find a better way to replicate the model.

7.3.6. Summing-up

Just as the original version of Simscript is built around the notion of events, GPSS is centrally concerned with transactions, which move through the block diagram, using and releasing passive resources, and interfering with one another's progress as they go. This is a natural way to view many queueing systems, and accounts to some extent for the ease with which GPSS can be used, especially by beginners.

It seems fair to say that GPSS is entirely composed of *ad hoc* elements. Initially, this is an advantage. Facilities and storages, for instance, are *ad hoc* concepts to model frequently occurring real-life situations, and as such are very useful. It is this intelligent choice of *ad hoc* concepts to include which makes programs such as Figure X.7.8 so succinct and transparent compared to their equivalents in other languages.

Beyond a certain elementary level, however, the *ad hoc* nature of the language becomes a handicap. The more obscure GPSS blocks have bewildering arrays of parameters for coping with special situations: it seems certain that some blocks were initially invented as they were needed, with no thought to making a more unified language. Such GPSS entities as *user chains*, for taking transactions out of the normal GPSS scan, either for efficiency, or for modeling nonstandard queue disciplines, also give the impression of having been put into the language as afterthoughts, rather than in an overall design process.

Most curiously, GPSS has many of the features of a "toy" compiler, produced as an amusing exercise rather than as a major software tool. For instance:

(a) GPSS programs are usually interpreted;
(b) of IBM's versions, only the latest, GPSS V, allows free-format input. The format of identifiers is still rigidly controlled;
(c) almost all numerical values are held in integer mode;
(d) GPSS has no subroutine structure;
(e) functions can only be defined by tabulating them;
(f) other features, such as array referencing or what GPSS calls "indirect addressing," are implemented in the crudest possible way.

The most likely conclusion is that, as for Simscript, a rigid insistance on upward compatibility of successive versions has made it impossible to give GPSS the thorough revision and redesign it badly needs.

Frustration with IBM's ossified implementation of GPSS spurred a number of competing nonstandard versions. Henriksen's (1979) GPSS/H, probably the best known, runs much faster than GPSS V. It also adds

(i) Fortran-like constructs (IF ... ELSE, DO, etc.);
(ii) the ability to call Fortran routines;
(iii) interactive debugging facilities.

Because GPSS and Fortran have quite different world views, mingling these languages via (i) and (ii) above has limited value. Likewise, we are skeptical about the worth of (iii): quick fixes from a terminal seldom produce good software.

Despite its drawbacks, GPSS is quick and easy to use for many models of low or average complexity. When the need is for a crude tool to produce approximate results rapidly, GPSS has no rival. However, many such models yield to analytic or numerical attack. Furthermore, if a simple model programmed in GPSS turns out to be invalid, it may be difficult or impossible to enrich it without reprogramming in another language.

7.4. Simula

Simula was developed at the Norwegian Computing Center from 1963 onwards. Like all widely used languages, it has progressed through several versions. Although earlier manuals bore the subtitle *A Language for Programming and Description of Discrete Event Systems* [Dahl and Nygaard (1967)], it was later realized that the language possesses some powerful facilities useful in a wider context, and the emphasis on simulation is now reduced. Simula remains, however, a popular simulation language.

Simula 67, the official title of the current version [Dahl *et al.* (1970)], contains Algol 60 as a subset (except that Algol *own* variables are not included). We saw that any general purpose language, including Algol, can be used for simulation. To this general base, Simula adds concepts particularly useful in simulation programming.

As with Pascal, reserved words are in bold face, while identifiers mentioned in the text are in italics.

7.4.1. The Class Concept in Simula

A *class* in Simula is a set of objects with similar characteristics. In the simplest case these objects are structured variables which can be created and destroyed dynamically. For instance, we can declare a class of objects called cows by the declaration

```
class cow;
begin
integer age, weight;
real yield
end;
```

Each such object contains two integer fields and one real field. To refer to these objects we need pointer variables, declared for example by

```
ref (cow) daisy, elsie, gertie;
```

To create one of these objects and to set values into the data fields we might use

```
daisy :- new cow;
daisy.age := 5;
daisy.weight := 1234;
daisy.yield := 987.6
```

(The assignment operators :− and := are used for pointers and values, respectively.) A different way of obtaining the same effect declares the fields *age*, *weight*, and *yield* as parameters of the class *cow*, not as local variables. Now the declaration of the class becomes

```
class cow (age, weight, yield);
integer age, weight;
real yield;
begin
end;
```

and an object can be created by

```
elsie :- new cow (4, 1316, 854.2).
```

A class declaration is identical, except for the keyword **class**, to a procedure declaration in Algol.

Besides data fields, objects can also contain instructions, which are normally executed when the object is created. Extending our definition of cows to

```
class cow (age, weight, yield);
integer age, weight;
real yield;
begin
Boolean old;
old := age > 5
end;
```

then when a *cow* is created by

```
gertie :- new cow (6, 1210, 879.0)
```

the instruction in the body of the declaration is executed, setting the local Boolean variable *old* to be **true**. We can test this variable subsequently using statements such as

```
if gertie.old then ...
```

Classes as Co-routines. In the simplest case all the instructions in an object are executed when the object is created. However, two special procedures are provided to allow quasi-parallel execution of the code contained in several objects. The standard procedure *detach* interrupts the execution of an object, returning control to its creator; and if *ptr* is a pointer to an object, then the

```
          begin

          class player (n);
          integer n;
          begin
          ref (player) next;
          detach;
          next :- tab [if n = 4 then 1 else n + 1];

newturn :

                   {insert here program for one move}

          resume (next);
          goto newturn
          end declaration player;

          ref (player) array tab [1 : 4];
          integer i;

          for i := 1 step 1 until 4 do
                    tab [i] :- new player (i);
          resume (tab[1])

          end program;
```

Figure 7.4.1

procedure *resume (ptr)* interrupts the execution of the current object, re-starting the object pointed to by *ptr* at the point it had previously reached.

As an example, Figure 7.4.1 outlines a structure that could be used to simulate a four-person game. Each player is represented by an object of the class *player*. An array of four pointers allows access to each of these objects. The main program creates each object in turn: because of the *detach* at the beginning of each, play does not begin until all four objects exist. At this point the main program gives control to *player* (1), who in turn reactivates *player* (2), and so on, each object playing one move in the game before passing control to its successor.

Prefixed Classes. Suppose we wish to declare two classes of object, called respectively *man* and *woman*, which share many common attributes but which also have some unique elements. One way to do this in Simula declares that both *man* and *woman* are subclasses of a third class, *person*:

```
          class person;
          begin
          real height, weight;
          integer age
          end;

          person class man;
          begin
          ref (woman) wife
          end;

          person class woman;
          begin
          ref (man) husband
          end;
```

When the name of one class (in this case *person*) prefixes the declaration of another (in this case *man* or *woman*), the latter inherits, as well as its own local variables, any local variables declared for the former. The same applies to any parameters appearing in the declarations, and to any instructions. In our example, an object of the class *man* has four local variables: *height*, *weight*, *age*, and *wife*. Not only does this permit a certain abbreviation if several classes of object share common attributes, but also it allows us to declare, for example,

```
ref (person) teacher,
```

a pointer variable which can refer to either a *man* or a *woman*.

7.4.2. Simulation Using System Classes and Procedures

The Simula system provides two predefined classes, *simset* and *simulation*, and a number of predefined procedures. These provide the basic mechanisms for writing simulation models.

The Class simset. The class *simset* contains facilities for the manipulation of two-way lists called sets. These lists have the following structure:

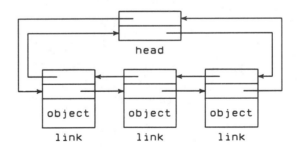

The class *simset* has the following outline:

```
class simset;
begin
class linkage; ...
linkage class head; ...
linkage class link; ...
end;
```

When the name *simset* prefixes a program, three other classes become available. (We mentioned that local variables and parameters are inherited from a prefixing class. In fact all the definitions made in the prefixing class are inherited: other classes, procedures, and so on.) The class *link* can be used as a prefix to add successor and predecessor pointers to a data object; the class *head* allows suitable head cells to be created; and the class *linkage* is simply a convenience so that pointers can go to head cells or link cells indiscriminately.

The class *link* in its turn contains definitions which can be used by objects in this class:

```
procedure out
procedure follow(X); ref (linkage)X;
procedure precede(X); ref (linkage)X;
procedure into(S); ref (head)S.
```

Finally the class *head* contains declarations for

```
ref (link) procedure first;
Boolean procedure empty;
integer procedure cardinal;
procedure clear.
```

To see how all these mechanisms fit together, suppose we are to write a simulation of a queueing system in Simula. Our program might begin

```
simset begin
ref (head) queue;
link class person;
     begin
     real height, weight;
     integer age;
     end;
```

and so on, using the definitions of the classes *man* and *woman* already seen. Because *link* prefixes *person*, objects of this class now include the necessary pointers as members of two-way linked lists. The variable *queue* can point to the head cells of such lists. If *John* and *Mary* are declared by

```
ref (man) John;
ref (woman) Mary;
```

then we might create such a list by, for instance

```
queue :- new head;
John :- new man;
Mary :- new woman;
John.into (queue);
Mary.precede (John);
```

Just as *John.height* refers to the local variable of the object pointed to by *John*, so *John.into* (*queue*) indicates that the procedure *into* operates on this same object. The procedures belonging to an object of the class *head* can be used similarly, so that we might test, for instance

```
if queue.empty then ...
```

Because our queue can contain two kinds of objects, men and women, we sometimes need to know which is which. This can be accomplished using an **inspect** statement:

```
inspect queue.first
        when man do ...
        otherwise do ...
```

The Class simulation. The system class *simulation*, a subclass of *simset* (i.e., the declaration is *simset* **class** *simulation*), introduces simulated time. Objects representing events are scheduled and executed using the declarations in this class.

Consider first the following group of declarations:

```
link class eventnotice (evtime, proc);
          real evtime; ref (process) proc; ...
link class process;
          begin ref (eventnotice) event; ...; end;
ref (head) SQS;
ref (eventnotice) procedure firstev;
          first.ev :- SQS.first;
ref (process) procedure current;
          current :- firstev.proc;
real procedure time; time := firstev.evtime;
```

Two classes of objects have been declared, *event notices* and *processes*. An event notice points to a process, and vice versa. A special list called *sqs*, the sequencing set, holds event notices. This list is ordered by event time, so that the first event notice in *sqs* points to the process currently being simulated. The simulated clock is set to the scheduled time of the current process.

Figure 7.4.2 shows this structure. In this example *firstev* points to the left-hand event notice, *current* points to process *A*, and *time* equals 1.3.

Although this structure resembles the event lists we have already seen in Fortran or Simscript, an event in Simscript and a process in Simula are different concepts. Generally speaking, an event takes place at a given instant of simulated time; its activation causes the values of variables in the model to change. A process may exist over a period of simulated time; it carries within itself variables to describe its own state, and also instructions to simulate all those events in which the process may take part. We return to this distinction in Section 5 of this chapter.

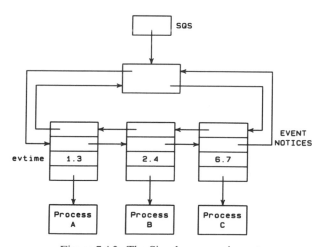

Figure 7.4.2. The Simula sequencing set.

Processes are a sub-class of *link*, so they can be members of two-way lists. However, in the sequencing set event notices are linked, not processes. The definition of the class *process* includes in its turn several procedures:

```
Boolean procedure idle;
Boolean procedure terminated;
real procedure evtime;
ref (process) procedure nextev
```

A process is idle if it does not have a corresponding event notice. It is terminated if all its instructions have been executed. If it has an event notice in the *sqs*, then *evtime* is the time when the process is scheduled to occur, and *nextev* points to the process which is scheduled to follow it.

Finally the class *simulation* includes another group of declarations:

```
procedure hold (t); real t;
procedure passivate;
procedure wait (S); ref (head) S;
procedure cancel (X); ref (process) X;
procedure activate ...;
```

The procedure *hold* reschedules the current process t time units later; *passivate* renders the current process idle (i.e., takes it out of *sqs*) and starts the next; *wait* puts the current process into a list S, and then "passivates" it; and *cancel* removes a process from the *sqs*. The procedure *activate* is not called directly. Instead, Simula provides a number of scheduling statements:

```
activate X at t;
activate new A before X;
reactivate proc delay 15.0;
```

These statements can be used only within a block prefixed by the system class *simulation*.

System Procedures. Simula provides procedures for operating on special kinds of objects called *texts*. In particular, standard input and output procedures are defined in this way. (In Algol, the language underlying Simula, input and output procedures are not defined.)

Simula also provides standard procedures for generating random numbers. All the usual generators are there, including procedures for drawing from empirical distributions. On the other hand only two rudimentary procedures, for calculating time-integrated values and for accumulating histograms, are provided for gathering simulation statistics.

7.4.3. A Complete Simulation in Simula

EXAMPLE 7.4.1. An airport check-in counter is manned by a number of clerks. Passengers arrive at an average interval of 1 min, the distribution of inter-arrival times being exponential, and form a single queue in front of the counter. The time taken by a clerk to serve a passenger is normally distributed, truncated at 0, with a mean of 3 min and a standard deviation of 0.5 min before truncation; 25 % of arriving passengers have not paid the departure tax and are directed by the clerk to the tax counter.

This counter is also manned by a number of tax collectors. Service time has an exponential distribution with a mean of 2 min. After paying his tax, a passenger visits the check-in counter again. This time, however, it takes only 1 min to serve him.

After one or two visits to the check-in counter, as the case may be, passengers go to the departure lounge.

We simulate this system, shown in Figure 7.4.3, for a period of 2 hours, to determine the average time a passenger takes to reach the departure lounge. A simple program to determine this is given in Figure X.7.9 of the appendix.

The only mechanism of any importance which we have not previously seen is the use of **inspect** to extend the scope of the names in an object. The statement

```
inspect checkinq.first when passenger do ...
```

not only tests the class of the first object in the check-in queue (this is unnecessary, since it can only be a passenger), but also makes its local variables accessible: we can access the variables *repeat*, *taxpaid*, and *timein* inside the **when** clause without using the dot notation. In the same way, in the **when** clause we may reference an object as **this** *passenger*.

The **otherwise** clause is executed if none of the **when** clauses applies. In the example, this is only possible when *checkinq. first* is equal to **none**, i.e., the queue is empty.

In Simula, variables are initialized automatically. It is thus not necessary to set *repeat* to **false**, nor *totwait* and *totpass* to zero. *Inint*, *outint*, *outtext*, and so on are standard input and output procedures.

PROBLEM 7.4.1. Reprogram Example 7.2.1 in Simula.

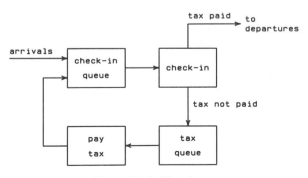

Figure 7.4.3. The airport.

PROBLEM 7.4.2. Define a class of objects called nodes, each with two links *left* and *right*, and an information field. Write a program to read a series of numbers and construct a binary tree. The information fields of the nodes in postorder [Knuth (1973), p. 316] should give the numbers which were read in ascending order.

7.4.4. Demos

The complexity of Simula is offputting. To write even the simplest simulation using the predefined classes *simset* and *simulation*, the user needs a working knowledge of Algol, not currently fashionable; he must also wade through a sea of preliminary definitions: links, linkages, heads, processes, and so on.

One promising way to reduce this complexity is to provide in Simula some of the *ad hoc* constructs for simulation which make GPSS so easy to use. It is not immediately evident that there is a close resemblance between the process-based approach of the former, and the transaction and flow-diagram approach of the latter. However, consider how processes are implemented in Simula. A process is an object of a certain class: each such object contains local variables, and also (at least conceptually) a local copy of a sequence of instructions. As a process is executed, Simula maintains a local program counter (LPC) to keep track of which instruction is next to be performed. If the LPC is envisaged as moving through the instructions carrying its local variables with it, the resemblance to transactions moving through a flow diagram carrying their parameters with them is obvious.

A GPSS block can refuse entry to a transaction; there seems no way, however, in which a Simula instruction can "refuse entry" to the LPC. Nevertheless, if the Simula instruction is a procedure call, the procedure can deactivate the calling process and make it wait until certain conditions are met. In this way operations like the GPSS SEIZE can be implemented.

Vaucher (1973) exploits this resemblance to define a class in Simula containing definitions for Simula objects closely resembling the familiar GPSS entities. This basic idea was taken up and much improved by Birtwhistle in a system called Demos [Birtwhistle (1979a, b)]. Demos is a separately-compiled class in Simula. It can be prefixed to a user's program, which then has access to all the declarations of Demos in the usual way. These declarations include Simula objects, strongly resembling corresponding GPSS entities, and procedures for operating on them.

Thus a Demos entity resembles a GPSS transaction; a Demos resource is like a GPSS storage (a facility is simply a resource with capacity 1); a Demos histogram corresponds to a GPSS table; and so on. Instead of ADVANCE, we find the procedure *hold*; instead of SEIZE and RELEASE, there are the procedures *acquire* and *release*; and instead of TABULATE there is a procedure *update*. This close correspondence holds for many of the features of GPSS. In particular, Demos provides an automatic report much as GPSS does.

For comparison, Figure X.7.10 of the appendix shows Example 7.3.1 recoded using Simula and Demos. As for GPSS, no explicit output statements are needed to produce a report. Parameters in quotes are titles to identify the output. The necessary objects are created using **new** statements, and then (as in §7.3.4) the program simulates a 1 min warm-up period, after which the main simulation models 20 min of operation.

To give the reader some idea of how Demos works, we outline just two elements: Demos "facilities" and the automatic report. In Vaucher's original system, a facility is defined by

```
link class facility (ident);
     text ident;
     begin
     ref (transaction) occupant;
     ref (head) inq;
     integer entries;
     real busy time, tlast;
     inq :- new head;
     into (facilityq)
     end;
```

It is easy to see how this class can be used. The reference to an *occupant* points to the transaction seizing the facility, while the list *inq* holds other transactions which are waiting to use it. Variables such as *entries* and *busytime* accumulate statistics of facility usage, exactly as GPSS does. When a facility is created, it is added to a list of all facilities, *facilityq*, so that when the time arrives to print a standard report, it is possible to find all the facilities which have been created.

The corresponding procedure to seize a facility is (approximately)

```
procedure seize (fac);
ref (facility) fac;
begin
if fac.occupant = = none then begin
       fac.occupant :- this transaction;
       fac.tlast := time end
else begin priority into (fac.inq);
       passivate; out end;

fac.entries := fac.entries + 1
end;
```

The procedure *priorityinto* is like *into*. It adds an item (in this case a transaction) to a list, ordering the list on the value of a given field.

Automatic reporting is achieved using the Simula statement **inner**. If a class A is used as a prefix to a class B, we have seen that B inherits the local variables and the parameters of A, and any procedures defined in A. It also inherits any code in the body of A. If A is defined by

```
class A;
begin
      code A1;
      inner;
      code A2
end;
```

and B by

```
A   class B;
    begin
    code B
    end;
```

then when an object of class B is created, its code body is code A1; code B; code A2. The statement **inner** is replaced by the code of the inner class. Now the class Demos can be defined in outline by

```
class Demos;
begin
class entity ...
class resource ...        declarations
class histogram ...
initialization;
inner;                    replaced by the user program
print final report
end
```

In this way the user is relieved of the necessity to write his own reporting routines. If he also wants intermediate results, he can call the Demos routines directly from his program, since, like all the other definitions, these are inherited from the prefixed class.

Demos provides other useful facilities that GPSS does not, including Simula's random generators enhanced by procedures which automatically generate well-spaced seeds when several random number streams are used or which automatically generate antithetic variates. There is also a simple but useful mechanism for handling the *wait until* problem (see §7.5). Demos does not use either *simset* or *simulation*, and the event list is held as a binary search tree.

PROBLEM 7.4.3. What happens if events tend to be put on the list in the order that they are to occur?

The reference manual of Demos includes a complete listing of the program. Like GASP II (see §7.1.5), and despite a great increase in power, the system is sufficiently succinct and transparent to be easily grasped, even (since Demos does not delve into the obscurer depths of Simula) by a user whose knowledge of Simula does not extend beyond the basic concepts. This makes the tracing of errors and the prediction of system behavior particularly easy. Moreover, since Simula is a block structured language, any feature a user dislikes or finds inadequate can be redefined in the user program.

As a striking instance of this, the Demos reference manual contains an example showing how to replace the entire event list mechanism. If, after reading Problem 7.4.3 and Section 1.7, a user feels that the default clock mechanism is not suitable for his particular problem, Demos makes it easy for him to build any other.

7.4.5. The Standard Example in Simula With Demos

Figure X.7.11 shows one way to program Example 1.4.1 in Simula using the Demos package.

The three variables *arrivals*, *services*, and *balks* point to objects of class *count*. This class provides a simple data-gathering mechanism, activated by a call of the procedure *update* (for example, *arrivals.update* (1) to add 1 to the counter called *arrivals*). The final values of such counters are printed automatically at the end of a simulation.

Demos provides three classes of random variate generators, *rdist* (real), *idist* (integer), and *bdist* (Boolean). In the example *arrdist* is a subclass of *rdist*, and *balkdist* is a subclass of *bdist*. Again, these classes include an automatic report at the end of the simulation. More importantly for us, each distribution in these classes (created by **new** *dist*, and sampled by *dist.sample*) has its own seed chosen so that the sequences of values used do not overlap. The underlying uniform generator is the multiplicative generator

$$X_{k+1} = 8192X_k \bmod 67099547$$

tested by Downham and Roberts (1967). Seeds are spaced 120633 values apart. [See Birtwhistle (1979b) for implementation details.]

The arrival distribution is coded as an object of class *arrdist*. When created, this object initializes the appropriate tables (cf. Figure X.7.4). The procedure *sample* associated with the object uses these tables as global variables.

The distribution for balking, which is a Boolean distribution returning **true** or **false**, is associated through its parameter with a particular resource (an object of class *res*: *tellers* in our example). Resources in Demos possess an attribute called *length*, which is the length of the queue of transactions waiting to use them. This attribute is used in the procedure *sample*. (The programming here is clumsy. We would like the balking distribution to be part of a resource, since that is where it belongs logically, and also to be a Demos distribution, to take advantage of the automatic generation of seeds. Unfortunately we cannot have a class prefixed by both *res* and *bdist*.) The same attribute is used at the end of the main program when we add the number of customers remaining in the queue to the counter *services*.

The rest of the program is straightforward. The routine *trace* called at the beginning of the main program sets a flag so that an ample record of the events simulated is automatically printed. The final report includes the values of arrivals, services, and balks.

PROBLEM 7.4.4. The automatic report includes a figure for teller utilization. However the resource called *tellers* is created at $t = 0$ and immediately "occupied" until $t = 60$ to represent the fact that the bank is not yet open. The reported utilization is therefore incorrect. Find a way to fix this.

PROBLEM 7.4.5. Find a way to make 50 runs of the model, putting the useful results out onto a file for later analysis.

7.4.6. Summing-up

We showed that Simula is complex, inordinately so we believe. Worse, as one delves deeper into the language this complexity does not disappear: there is no point, at least in our experience, at which the underlying structure becomes evident and self-explanatory. If you doubt this, try to read in Dahl *et al.* (1970) the details of the *detach* procedure (in particular the description of detaching an unattached object). Part of the complexity may be due to poor documentation; the major part, however, is real. Demos manages without most of this complexity, but some users will be obliged to plumb the depths beneath Demos.

Although Simula has good features, there are blemishes. For instance, the procedure *activate*, apparently declared normally in the class *simulation*, is only accessible through the special statement **activate** (and this statement is illegal unless the prefix *simulation* is present). As with most languages of the Algol family, separate compilation of procedures or classes is difficult. Accumulating statistics in a simulation (unless Demos is used) is as tedious as in Fortran, and producing formatted reports is worse. The combination of wellnigh impenetrable complexity in certain areas with a tedious triviality in others makes Simula a curiously ill-balanced tool.

On the positive side, the basic concepts of classes of objects are elegant and clever. Classes provide powerful facilities for structuring data. They can be used to implement quasi-parallel processes, useful for example in operating system design. Vaucher (1975) uses prefixing to formalize concisely and lucidly certain ideas in structured programming. Finally, the fact that a powerful extension like Demos can be built onto the language with relative ease argues that the basic mechanisms are adequate as the foundation for further improvements.

In sum, Simula is not as good a simulation language as it might be. It contains excellent ideas, some rather badly carried out. When extended by the Demos package, it meets most of our criteria for transparency and relative conciseness. Our hypothetical consumers' guide to simulation languages would give this combination qualified approval. Unfortunately not all situations can be modeled using Demos. To give just one example, a transaction in Demos can be in only one queue at once, which can be inconvenient.

PROBLEM 7.4.6. Repeat Problem 7.2.3 using Demos.

PROBLEM 7.4.7. How could you implement GPSS's TRANSFER BOTH and TRANSFER SIM in Demos?

Mastering the complexity of "pure" Simula does not seem to us worthwhile when measured against the extra effort needed to write simulations in a more perspicuous general-purpose language. For the reader interested in pursuing his interest in Simula, we recommend Birtwhistle *et al.* (1973), or, for those whose Swedish can cope, Leringe and Sundblad (1975) or Siklósi (1975). Franta (1977) is also useful. Landwehr (1980) details another way to extend Simula to facilitate the collection of statistics; the approach is more formal than Birtwhistle's and the end result rather less transparent.

7.5. General Considerations in Simulation Programming

7.5.1. Language Design

Simscript, GPSS, and Simula have been in use for well over a decade. As we remarked earlier, all three have major flaws. Competitors to them abound, some achieving commercial success, but none has yet made a serious impact on language design. A concerted effort to define and implement a modern programming language with adequate simulation capabilities is long overdue. Whether there should be a language designed especially for simulation is less clear. Maintaining a program written in a specialized language is hard after the programmers who wrote it leave.

Part of the problem is that everyone has his own opinion concerning the desirable features of such a language. Despite the strictures of Sheil (1981): "The study of programming languages has been central to computer science for so long that it comes as a shock to realize how little empirical evidence there is for their importance," we set forth our own ideas below.

First, like any other modern language, a simulation language needs an adequate formal syntactic structure, for efficient compilation and programming ease. Of the languages we have looked at, only Simula meets this first criterion.

Next, the semantics of the language should be clear and well defined. A certain informality in definition seems necessary if clarity is not to suffer. Of the languages we have seen, Simscript is often muddled; Simula's semantics are occasionally so opaque that one can only take on trust that they are well defined; while GPSS sets out well but any structure collapses as more and more *ad hoc* concepts are heaped on.

The facilities offered by the language should include an easy-to-use mechanism for defining and manipulating powerful data structures. A clock mechanism is essential; it should be efficient but easily replaceable. The language should include a battery of random number generators, again as default options which can be replaced; good facilities are needed for collecting statistics and printing reports. If existing program libraries, such as the IMSL package, are easily accessible, so much the better. Finally all these facilities should not be obtained at the price of catastrophic run-time efficiency.

Any statistical mechanisms provided automatically by the language must be clearly defined and (even more important) they must be correct. In the introduction to this chapter we noted that we would usually be reluctant to use an existing simulation language for important programs. Much of this reluctance is due to the difficulty of finding out exactly what such languages *do*. Simscript's TALLY and ACCUMULATE, GPSS's automatic report, and Demos's procedure *update*, when correct, are useful. Determining exactly how they all deal, for example, with periods before the first event or after the last event simulated requires considerable effort. The "correct" statistical mechanism is not the same for all models.

We noted above that Simscript is built around the notion of events, GPSS around the idea of transactions moving through a flow diagram, and Simula around processes. There also exist languages built around the notion of *activities*. The best known is CSL: see Clementson (1973). Clearly the notion of a process includes that of an event: an event is simply a process all of which is completed simultaneously. We have also seen that there can be a close resemblance between processes and the transaction and flow diagram approach.

The Demos package, outlined above, is easy to learn, easy to use, easy to alter if one dislikes the standard procedures, and powerful enough to build models which are computationally and statistically sophisticated. If pursued further, this approach seems to promise an elegant way of combining the undoubted usefulness of the kind of entities predefined in GPSS, with the formal structure, computational power, and extensibility of Simula. Although such efforts were hampered at first by some apparently arbitrary restrictions in the Simula compiler, these have largely disappeared, and if Simula can be made more transparent, then this avenue invites further exploration.

Languages such as Pascal Plus (see §7.1.7) and Parallel Pascal [Brinch Hansen (1977)] for controlling parallel processes offer another approach. Constructs useful for controlling parallel or quasi-parallel processes in real time are also useful for performing the same task in simulated time. At present all events in a simulation usually share the same event list and the simulated clock has the same value in each ongoing simulated process. If in fact these simulated processes are noninterfering over a certain period of simulated time, each could have its own local event list and local clock. Only when such a process needs to interact with others must it wait until their simulated clocks catch up. One advantage of such an approach is that it would tend to keep event lists short and efficient.

Finally, there are a number of simple-seeming problems in simulation language design to which no good solution is yet available. The classic is the "wait-until" problem: if a language includes some such construct as

```
wait until (Boolean expression)
```

the simple way to implement this checks the value of the Boolean expression after every simulated event. This may be both inefficient and wrong. For instance, if the Boolean expression is simply

```
time = 100
```

and no event is scheduled to occur at this time, the condition will never be satisfied. None of the known ways to alleviate such problems is entirely satisfactory.

Again, the Demos extension to Simula offers a simple but useful facility. An entity can put itself into a "condition queue" by some such procedure call as

```
holdqueue.waituntil (boxes.avail > 0 and lids.avail > 0)
```

If the condition is true, the entity does not go into the queue; otherwise, it joins the queue and waits. The queue is only re-examined when some other entity executes the call

```
holdqueue.signal
```

At this time one or more blocked entities may find that they can proceed. This technique, as mentioned, is both useful and easy to understand. However it does not solve the general problem.

PROBLEM 7.5.1. Customers queue for a resource. If they are still in the queue after 100 time units, they leave and go elsewhere. Find a way to model this situation correctly in any of the languages discussed in this chapter.

PROBLEM 7.5.2. You hope to implement the general construct

```
wait until (Boolean expression)
```

The Boolean expression may in particular depend on simulated time. The waiting process is to be reactivated as soon as the Boolean expression becomes true, the important point being that this may happen at an instant not an event epoch. Define an *efficient* implementation as one that does not evaluate the expression at every instant of simulated time (necessarily discrete on a machine with finite accuracy). Show that developing such an implementation is NP-hard, as defined for example in Garey and Johnson (1979).

Hint: The satisfiability problem, known to be NP-hard, is to find values (true or false) for the n logical variables in a Boolean expression that make the whole expression true. Show that the satisfiability problem can be made equivalent to a *wait until*. To do this, let the kth bit of the number representing simulated time correspond to the kth logical variable for $k = 1, 2, \ldots, n$.

7.5.2. System Considerations in Simulation

We repeat briefly here some of the warnings given in Section 1.8.

Large simulations, like other large application programs, sooner or later experience problems, whatever the language, whatever the machine, and whatever the operating system being used. Such problems are best avoided by good design of simulation models. Sufficient data is rarely available to justify large, minutely detailed models. If a program is too big to run, first try, without oversimplifying, to conceive a less complex but equally valid model. If runs are uneconomically long, think about using statistical techniques to reduce them and also question whether the input data can support the weight being put on it.

Large simulations often cause serious memory management problems. Almost always data objects are created and destroyed randomly. When using a general-purpose language, it may be possible to group data objects in storage to minimize overhead if they cannot all be held in main memory. When using a simulation language, however, the user usually has little control over memory allocation. Simscript offers some facilities for explicitly creating and destroying arrays and other data objects. GPSS systems may allow the block diagram to be segmented, so that only required blocks are in core; it is sometimes possible to declare which transaction parameters, or what proportion of a particular kind of object, can conveniently be held on secondary storage. The Simula user can only rely on the garbage-collector; it is also notoriously difficult to segment Algol (and hence Simula) programs.

Checkpointing long-running simulations and restarting in the case of machine error is usually a minor problem, at least for finite-horizon models. Although a series of runs may be long, each individual run is usually short enough not to need checkpoints. Thus we never need save the whole state of a model, but just the accumulated statistics at regular intervals, together, perhaps, with current values of the random number seeds. If individual runs are long enough to need checkpoints, re-examine the underlying model. In the case of infinite-horizon simulations where there is just one long run, the need to take checkpoints at intervals adds one more problem to those already seen in Section 3.3.

Finally, for a simulation to be portable, the language used should be stable and well-standardized. Over the last 10 or 15 years, Fortran has been more stable than any of the other languages we have described. Even for Fortran, however, a revised standard has appeared. Disappointingly, it does little to increase the power of the language for simulation. Other revisions of the standard are expected during the 1980s. For any other language, different installations are likely to use different versions, or at best different implementations of the same version. If a program is to be portable, settle these points before writing the simulation; afterwards is too late. For an excellent guide to writing portable Fortran programs, see Larmouth (1981).

7.5.3. Statistical Considerations

Chapter 8 shows how some of the statistical techniques previously discussed can be programmed in practice. It is worth mentioning here, however, that this may be significantly easier in one language than in another.

In the worst case, some languages may produce wrong answers: GPSS's bad habit of truncating all variables to integer form is an example of this. We have also seen (§7.5.1) that any language which automatically tallies means and variances should make the cautious user pause: have initialization and end effects, for instance, been properly taken into account, and if not, can anything be done about it? The tally process must be precisely defined in the language manual. Ideally, the user should be allowed to specify random variables whose means and variances are then tallied.

Such considerations may induce us to revise our initial estimate of the usefulness of a language. The price of automatic collection of statistics is always a loss of control over what is going on. If the language is well designed and gives us exactly what we need in a convenient way, this may not matter. In other cases we may tolerate minor errors in our results (for example, failing to tally the last value of a variable which changes frequently) in return for the convenience of using the built-in mechanism. However this is not always acceptable. As usual, the ideal system offers predefined, simple mechanisms for the situations most commonly encountered and permits detailed and correct handling of special cases. No such system yet exists.

CHAPTER 8
Programming to Reduce the Variance

The chapter begins with a section showing different ways of programming an input distribution for a simulation. This illustrates parts of Chapter 4. We continue with practical examples of how to program some of the variance reduction techniques presented in Chapter 2. In particular, several techniques are applied in turn to the bank teller model of Example 1.4.1. A simple program implementing this model was given in Section 7.1.6. Now we improve this simple program to obtain better results at little or no extra cost in computer time. The examples are programmed in Fortran; however, recasting them in terms of another language (except perhaps GPSS) is not hard.

The examples are intended to illustrate how to program the variance reduction techniques, not to compare their relative efficiencies (which is why they follow Chapter 7 rather than Chapter 2). The basic model of Example 1.4.1 was chosen with no particular technique in mind to provide a testbed for several methods. It is likely that it accidentally serves to make one technique look rather good, and another comparatively poor. We emphasize again that this *is* an accident: do not generalize from the results of a particular technique in this chapter to a conclusion about its power in a wider context.

PROBLEM 8.1. Get our standard example working on your local machine. (In later problems we shall assume this has been done.) All our experiments were carried out using the starting seed 1234567890, other seeds being spaced 200,000 values apart; see Section 7.1.6. Choose a different starting seed, and re-run the program. Compare your results to those of Figures 7.1.5 and 7.1.6.

As usual, we compare techniques using the observed sample variances. To avoid cluttering the discussion, we assume for the moment that the variance of these sample variances is small: that is, we provisionally assume that our sample variances accurately estimate the true variances. Though we do not go deeply into the question, we give some evidence of the truth or falsity of this assumption in Section 8.8.

8.1. Choosing an Input Distribution

Chapter 4 discusses ways of choosing an input distribution for a simulation. In this section we compare an exact method for generating random variates from a given distribution to an approximate technique which generates variates by interpolation between tabulated values. We also give an example illustrating the quasi-empirical distribution presented in Section 4.6. Finally we test the robustness of our results by fitting two "wrong" distributions to our input data.

8.1.1. Using an Exact Method

As in the rest of the chapter, we use the example presented in Sections 1.4 and 7.1.6. For simplicity, however, we assume until further notice that there are always two tellers at work in the bank. We also assume that we can afford to make exactly 50 runs.

Making the necessary trivial changes in the program of Figure X.7.2, and using the exact routine for generating Erlang variates shown in Figure X.7.3, we estimate the expected number of customers served in the straightforward way, averaging over the 50 runs. Calculated thus, our estimate is 233.34 with a sample variance of 1.74. Getting the exact expectation seems intractable.

As we remarked in Section 7.1.6, this exact Erlang generator transforms the underlying uniform random variates in a monotone way, offering hope that techniques such as antithetic variates and common random numbers will work. However, unlike inversion, the method uses more than one uniform variate per Erlang variate generated, which could cause synchronization problems later should we wish to use common random numbers but a different service-time distribution. Problem 7.1.11 suggests one way Erlang variates can be generated by inversion. The following section suggests another.

8.1.2. Inverting a Tabulated Distribution

For this experiment, we replace the exact Erlang generator by an approximate generator based on a number of tabulated values. The Erlang distribution with mean 2 and shape parameter 2 has

$$F(x) = 1 - (1 + x)e^{-x}$$

and we tabulate this function for $x = 0$ to 4 in steps of 0.2 and for $x = 5, 6,$ and 7. We choose a final point so that the mean of the interpolated distribution is exactly 2. This final point falls at about $x = 7.8$: our generator therefore does not provide values greater than this limit.

PROBLEM 8.1.1. Try tabulating $F(x)$ at $x = 0$ to 4 in steps of 0.2 and for $x = 5, 6, 7,$ and 8, adding a final point to make the mean of the interpolated distribution exactly 2. What goes wrong?

PROBLEM 8.1.2. In this example we chose the points at which to tabulate F arbitrarily using a rough sketch of the function. In general, given a cdf F and a set of interpolation points, how would you define the resulting "interpolation error"? Remember that the tabulated values are used to calculate F^{-1}, not F, and that the distribution may have one or two infinite tails. (See also Problem 5.2.5 and the introductory remarks to Chapter 4.)

PROBLEM 8.1.3 (continuation). Given a cdf F and a maximum permissible interpolation error, defined in whatever way you chose in Problem 8.1.2, design an algorithm to produce a suitable set of interpolation points. Bear in mind that if you are using a table to invert F, it may well be that F^{-1} is hard to calculate directly: your algorithm should take account of this.

To generate random variates, we generate a uniform random variable U and then approximate $F^{-1}(U)$ using linear interpolation. If F is tabulated at $x_0, x_1, x_2, \ldots, x_n$, with $F(x_0) = 0$ and $F(x_n) = 1$, we first find i such that $F(x_i) \leq U < F(x_{i+1})$. This can be done rapidly using binary search since the tabulated values are ordered. Then the value returned is

$$X = x_i + (U - F(x_i)) \cdot \frac{x_{i+1} - x_i}{F(x_{i+1}) - F(x_i)}.$$

A suitable program for our example is shown in Figure X.8.1. A bucket search scheme (see §5.2.1) is faster but slightly more complicated.

Using this technique for generating Erlang variates, we again made 50 runs of the model with two servers. The average number of customers served was 234.50 with a sample variance of 1.82. These are essentially the same results we obtained using the exact Erlang generator (although this agreement might not happen with a different number of runs or on a different model). The total computer time required is about 10% less using the approximate method, with negligible extra programming. One can infer that variate generation is significantly more than 10% faster. This shows that one need not sacrifice variate generation speed to use a method (inversion) compatible with synchronization. The latter is our primary consideration.

PROBLEM 8.1.4. See how fast the model runs using VCHI2 or VCHISQ to generate Erlang variates (cf. Problem 7.1.11).

8.1.3. Using a Mixed Empirical and Exponential Distribution

In this section and the next we suppose that we do not know the exact form of the service-time distribution for our model of a bank. Instead, we suppose that 25 observations of the service time are available. These are listed in ascending order in Table 8.1.1. In fact, these "observations" were generated using an exact Erlang generator, so that if our methods are sound we expect to obtain similar results to those of the preceding sections. Figure 8.1.1(a) shows the fit between the "observations" and the true distribution.

Table 8.1.1. 25 "Observations" of Service Time

0.36	0.82	1.92	2.51	3.60
0.41	0.97	1.93	2.66	3.74
0.50	1.11	2.24	2.72	4.19
0.68	1.77	2.26	3.44	7.12
0.70	1.90	2.27	3.45	7.54

We first try fitting a quasi-empirical distribution, discussed in Sections 4.6 and 5.2.4, with parameter k arbitrarily set to 5. The mean of our observations is 2.43 and their variance 3.43. As a rough check, we generated 5000 values from the fitted distribution. The observed mean and variance were 2.44 and 4.24. The mean agrees well with the mean of our sample, although the variance is rather high. The underlying exact Erlang distribution has mean and variance 2, but we are not supposed to know this.

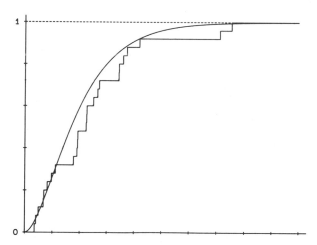

Figure 8.1.1(a). The true Erlang and the "observations."

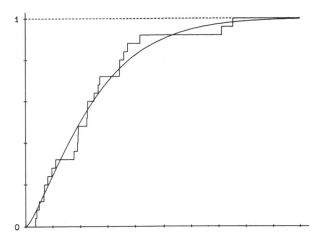

Figure 8.1.1(b). Fitted Weibull and the "observations."

We then used this distribution in a simulation of the bank model. Making 50 runs as before, the average number of customers served was 214.88, with a sample variance of 2.30. This estimate is naturally lower than our previous ones, since our estimate of mean service time based on the sample of 25 observations is too high. It seems likely that this error swamps inaccuracies introduced by differences in the shape of the distribution. Our experimental results below bear this out. With more observations the mean of the underlying distribution is more accurately estimated so the balance could shift. Total computing time in this experiment was 20% less than for an exact Erlang generator and 10% less than for a tabulated generator.

8.1.4. Testing Robustness

In the previous section we fitted a quasi-empirical distribution to our 25 observations, making only minimal assumptions about the true underlying distribution. In this section we fit two well-known—but incorrect—theoretical distributions, to test the robustness of our model.

The first distribution fitted is the Weibull (see §5.3.14). Its mean and variance are

$$\mu = \Gamma(1 + 1/c)/\lambda$$

and

$$\sigma^2 = \{\Gamma(1 + 2/c) - \Gamma(1 + 1/c)^2\}/\lambda^2,$$

respectively. We fit this to our observations by matching moments, finding $1/c = 0.76$, $\lambda = 0.379$ approximately. Figure 8.1.1(b) illustrates the fit obtained. When we simulate the bank using this Weibull distribution instead of an Erlang distribution, the average number of customers served is 213.48, with a sample variance of 2.13 (still using 50 runs). Total computing time is somewhat longer using the Weibull distribution than with the exact Erlang distribution.

Finally, we simulated the bank with the Erlang distribution replaced by an exponential distribution fitted to the observations. Figure 8.1.2(a) illustrates the fit obtained. In this case the average number of customers served is 212.58, with a sample variance of 2.74.

PROBLEM 8.1.5. We did not carry out any goodness-of-fit tests with either the Weibull or the exponential distributions adjusted to our "observations." Would you expect such a test to warn us that we are fitting the "wrong" distribution? (Figures 8.1.1 and 8.1.2 may help.) Try it and see.

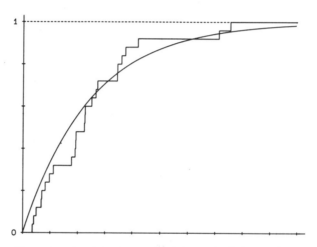

Figure 8.1.2(a). Fitted exponential and the "observations."

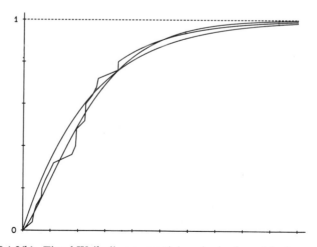

Figure 8.1.2(b). Fitted Weibull, exponential, and mixed empirical-exponential.

PROBLEM 8.1.6. Had our results been less clear-cut, we would have been able to evaluate possible differences more exactly by using common random numbers to compare the results obtained using different distributions. How do you suggest programming this? (Read §8.2 before answering.)

The fitted Weibull and exponential and the mixed empirical-exponential distribution are compared in Figure 8.1.2(b). Their means, obtained in every case by matching the empirical mean, are all the same. In our example this dominates shape differences. Using our 25 "observations," we get essentially the same estimate of the number of customers served whether we use a mixed empirical-exponential, a Weibull, or a pure exponential generator. This being so, the quasi-empirical generator has much to recommend it: it will cope with distributions of many different forms, it requires almost no preliminary calculations, and it is the fastest of the generators we used.

As usual, however, we warn the reader against extrapolating uncritically from a small number of tests on a particular model.

PROBLEM 8.1.7. Use your model of the bank to try a similar series of experiments, this time changing the arrival distribution.

PROBLEM 8.1.8. Write a program (not necessarily in GPSS) to implement the model given in Example 7.3.1. Use it to estimate response times for the three classes of job. Now change your program so that the time required by each class of jobs is not constant, but has an exponential distribution with the given mean. For instance, class 1 jobs have exponentially distributed service times with a mean of 100 ms, and so on. Do the estimated response times change? Keep the second version of your program for later experiments.

8.2. Common Random Numbers

Suppose the manager of our bank has the opportunity to install new terminals which will speed up customer service. More precisely, service times will still have an Erlang distribution with the same shape parameter, but the mean will be 1.8 min instead of 2.0. Before installing the new terminals, he would like to estimate how many extra customers on average this will enable him to serve.

The naive way to estimate this improvement is to simulate the model with a 2.0 min mean, simulate again with a fresh random number stream and a 1.8 min mean, and observe the difference. The first part of Figure 8.2.1 shows the results obtained in this way. For a mean of 2.0 min, 50 runs were already available (see §8.1), and these figures were used; 50 runs were made for a mean of 1.8 to provide a comparable result. Our estimate of the improvement obtained is simply the difference: there are 6.68 extra customers served. To obtain an approximate idea of the accuracy of this estimate, we sum the sample variances of the two individual estimates, getting 3.45.

A better way to estimate the improvement uses common random numbers. Make pairs of runs. At the beginning of an odd-numbered run save all the random number seeds; then carry out the run with mean service time 2.0. At the beginning of an even-numbered run reset the random number seeds to the saved values and then carry out the run with mean service time 1.8. (If we had saved the service times corresponding to mean 2.0, we could have generated the times corresponding to mean 1.8 simply by multiplying by 1.8/2.0, i.e., by 0.9.) Each pair of runs provides *one* observation, the difference in the number of customers served.

Programming this is straightforward. To the basic program we add three integer variables to keep the initial values of the random number seeds so we can use them again (the number of tellers is fixed at two), a logical variable EVEN initialized to .FALSE. to tell us which member of a pair we are on, and a variable AVGSRV used in calls to ERLANG. The short sequence of code shown in Figure X.8.2 must be executed at the beginning of each run.

In general, different runs have a different number of arrivals, services, and so on, so that starting from the same point in a random number stream they make a different number of calls on the generator. Suppose run 1 has 250 services but run 2 only has 240. Then if run 3 starts off where run 2 stopped, there is a dependency between runs 1 and 3: the last 10 random numbers of the former are the first 10 of the latter. To avoid this situation elegantly, we could count how many calls are made by run 1 and how many by run 2 and then take steps accordingly. Though cruder, it is easier and just as effective to throw away some appropriate number of random values between each pair of runs, so that no overlap is possible: this is what we do here.

Both the versions being compared use two uniform variates to generate an Erlang variate, so synchronization is maintained. Using inversion, we would not have to worry about this (cf. Problem 8.2.3).

In our example we carried out 50 pairs of runs. This used about the same computing time as the simple technique, where we carried out 50 runs for each case separately. As Figure 8.2.1 shows, the improvement is dramatic: using common random numbers reduces the sample variance of our estimate by a factor of 12, with the same amount of work.

PROBLEM 8.2.1. In our program we generate service times as required. An alternative generates a service time for each customer as he arrives, to be used later if and when required. This is slightly harder to program, as we have to store these times somewhere. Would you expect it to repay us by providing better synchronization? (Cf. Problems 2.1.1 and 2.1.8.)

PROBLEM 8.2.2. The change in service-time distribution in this example involves only a change of scale, not a change of shape. This being so, we might expect common random numbers to work exceptionally well. Why?

Technique	Mean number served	Sample variance
Mean service time 2.0	233.34	1.74
Mean service time 1.8	240.02	1.71
Difference	6.68	3.45
Direct estimate of difference using common random numbers	6.48	0.28

Figure 8.2.1. Estimating an improvement.

PROBLEM 8.2.3. Suppose instead that the bank's new terminals do change the shape of the service-time distribution: mean service time still falls to 1.8 min, but the service-time distribution will now be exponential. Using elementary queueing theory, guess whether the new terminals will be an improvement or not. Check your guess using simulation with common random numbers, taking whatever steps seem appropriate to keep the Erlang and the exponential generators synchronized.

PROBLEM 8.2.4. Devise several possible priority scheduling rules for the computer model of Problem 8.1.8 and compare them using common random numbers. (Reread §3.2.2 first.)

8.3. Antithetic Variates

It is plausible that the quantity of interest to us—the number of customers served—is a monotone function of the random numbers used to generate service times in our model. Remembering how we programmed ERLANG (see Figure X.7.3), increasing one of the uniform random numbers decreases a simulated service time, which in turn should increase the number of customers served.

It also seems likely that the number of customers served is a "nearly" monotone function of the random numbers used to generate the arrival events. Looking at NXTARR (Figure X.7.4), we see that increasing one of the uniform random numbers decreases a simulated interarrival time; although the situation is complicated by customers refusing to join the queue under heavy loading conditions, decreased interarrival times probably lead to an increased number of customers served. See Problem 2.1.8, however.

In view of this, it is worth trying antithetic variates for the service times, or for the arrival times, or perhaps both, in an attempt to obtain better estimates of the mean number of customers served (see §2.2).

We first consider the service times. Instead of 50 independent runs, we make 25 independent pairs of runs. Within each pair the random number streams used to simulate arrival events and balking are independent; however, if a uniform random number U is used in the first member of the pair to derive a service time, then we use $1 - U$ instead in the second member. Each pair of runs gives us *one* estimate of the number of customers served, namely the average of the two runs.

The programming required is essentially the same as for common random numbers, with only trivial changes. At the beginning of an odd-numbered run we throw away some random numbers from the appropriate series to avoid problems of overlap (cf. §8.2), saving the required random number seeds. At the beginning of an even-numbered run we reset the seeds for the series of interest and repeat the simulation using antithetic variates. With a pure multiplicative generator the method of Section 6.8.1 can be used. If we use antithetic variates on the service time, for instance, then we neither save nor restore the seeds for the other random variates to be generated, nor do we throw any of these other variates away.

Besides these changes in the model, we must adapt the program which works out our results to accept 25 pairs of observations instead of 50 independent estimates. This is trivial. Finally, it is worthwhile to make one or two pairs of runs of the modified model with the trace turned on, to verify that it is behaving as expected.

In Section 8.1 we fixed a base point for our experiments by carrying out 50 independent runs; the average number of customers served was 233.34, with a sample variance of 1.74. When we carried out 25 pairs of runs using antithetic variates on the service time, we obtained an estimate of 234.68, with a sample variance of 2.06. In this case we have seemingly lost a little, rather than gained, by using antithetic variates. Next, we tried the same technique, but using antithetic variates for the arrival times. In this case our estimate turned out to be 234.10 with a sample variance of 1.29, an improvement of about 25% over the straightforward estimate. Finally, we used antithetic variates for both the arrival and the service times. In this case our estimate was 234.80, with a sample variance of 1.60, slightly better than the straightforward estimate.

PROBLEM 8.3.1. Explain how using antithetic variates on two distributions in the model can give worse results than using them on just one of the distributions.

It is perhaps not surprising that antithetic variates do not work well in this particular case, at least as far as service times are concerned. First, we have already seen that our results depend strongly on the mean of the service-time distribution, but only weakly on its shape. This is not true for all models. Second, our exact Erlang generator uses two uniform random numbers for each variate supplied. Theorem 2.1.2 suggests that inversion should make antithetic variates work better.

To test this, we repeated our experiment using the table-driven Erlang generator of Section 8.1.2, which only requires one uniform random number per variate. Recall that in this case our straightforward estimate was 234.50 customers served, with a sample variance of 1.82. Using antithetic values for the service times, and still with 25 pairs of runs, we obtain an estimate of 234.32, with a sample variance of 1.93. Again, antithetic variates give us no improvement in the case of our particular model. As before, the model using the table-driven generator takes 10 % less time than using an exact Erlang.

Instead of calculating sample variances based on 25 independent pairs, we could simply treat the 50 observations as if they were independent and calculate a sample variance in the straightforward way. Since the observations are in fact negatively correlated, the calculated sample variance will be too high, which is a safe error to make. Though this technique may sometimes be useful, in our case the use of 25 pairs is consistently better.

Antithetic variates can be used in conjunction with other techniques. For instance, we might generate pairs of runs with antithetic service times, and then use regression with splitting (see §8.4) between the mean number of customers served and the mean number of arrivals for each pair. The reader can no doubt invent many other possible experimental designs. Do not conclude from our results that antithetic variates are useless in general: this is certainly not the case.

PROBLEM 8.3.2. Use the computer model of Problem 8.1.8 to experiment with antithetic variates. You have at least three candidates to think about: the users' thinking and typing time, the assignment of jobs to classes, and the service times.

8.4. Control Variates

We illustrate first a simple approach, and then two approaches through regression.

8.4.1. The Simple Approach

We return now to the full model. Suppose we can afford 200 runs. Looking at the output from a few runs—indeed simply by thinking about it—we can guess that when three tellers are at work the number of customers served almost equals the number of arrivals: the system has ample capacity and the queue is rarely full. On the other hand, with only one teller at work the number of customers served is hardly affected by the number of arrivals: the teller is busy all day, the queue is usually full, and extra arrivals are lost.

However, since the former case is more frequent than the latter, it may be advantageous to use the number of arrivals as a control variate, and to estimate not the number of customers served, but rather the difference between this and the number of arrivals, i.e., the number of customers lost. We can calculate the expected number of arrivals—247.5—so knowing the difference lets us estimate the mean number of customers served.

This can be seen as a form of indirect estimation, as discussed in Section 3.5. That is, we estimate some secondary performance measure (customers lost) and then use a relationship between this and the primary performance measure (customers served) to estimate the latter.

One sensible way to begin is to carry out a short pilot experiment. Fortunately, we included the number of arrivals as one of our outputs. It remains to alter the program slightly, so that the number of tellers on each run, instead of being random, is predefined (cf. Problem 2.4.2). The changes required in the program of Figure X.7.2 are trivial: add an array INTEGER TELTAB (200), with one entry for each required run; initialize it to the desired values, by a statement such as

```
DATA TELTAB /30*1, 60*2, 110*3/
```

and replace the random choice of TLATWK by

```
TLATWK = TELTAB (REPL)
```

(also suppress all mention of TELLSD if you are tidy-minded).

We carried out 60 runs with this modified program, 20 each with one, two, and three tellers. Figure 8.4.1 summarizes the information thus obtained, with x denoting the number of customers lost.

Anticipating the discussion of stratified sampling in Section 8.5, we now use these pilot runs to select sample sizes for the three strata to minimize variance. Our estimate of θ, the mean number of customers lost, is the weighted average

$$\hat{\theta} = 0.05 \times 97.70 + 0.15 \times 12.50 + 0.80 \times 1.40$$
$$= 7.88.$$

Now suppose, possibly incorrectly, that s_i^2 is a good estimate of the true variance of $\bar{x}_i$ for $i = 1, 2, 3$. If we carry out an experiment with n_1 runs with one teller, etc., the variance of our estimate of θ will be approximately

$$0.05^2 \times \frac{115.38}{n_1} + 0.15^2 \times \frac{62.26}{n_2} + 0.80^2 \times \frac{4.36}{n_3}$$

since the probability of having one teller at work is 0.05, and so on.

Number of tellers	Number of runs	$\bar{x}_i$	s_i^2
1	20	97.70	115.38
2	20	12.50	62.26
3	20	1.40	4.36

Figure 8.4.1. Pilot runs—customers lost.

Number of tellers	Number of runs	$\bar{x}_i$	s_i^2
1	32	100.34	153.91
2	70	14.21	86.75
3	98	1.71	4.04

$$\hat{\theta} = 247.5 - \sum p_i \bar{x}_i = 238.98.$$

Estimated variance $[\hat{\theta}] = \sum p_i^2 s_i^2 / n_i = 0.07.$

Figure 8.4.2. Results using a control variate combined with stratification.

PROBLEM 8.4.1. Show that $\sum (a_i/n_i)$ subject to $\sum n_i = c$ is minimized by

$$n_i = \frac{c\sqrt{a_i}}{\sum (\sqrt{a_i})} \quad \text{(cf. Problem 2.4.1)}.$$

Using this result, we see that if a total of 200 runs are available, we should carry out 32 runs with one teller, 70 with two, and 98 with three (that is, since the pilot runs already exist, we add 12 runs with one teller, and so on). Figure 8.4.2 shows the final results.

Compared with the naive approaches of Section 7.1.6, our final result is much improved. The estimated variance of our estimator has been cut by a factor of at least 25, with no extra computing.

PROBLEM 8.4.2. It would be simpler just to make 200 uncontrolled runs, but then to calculate an estimate using the number of customers lost instead of the number of customers served. Try this using both the very naive and the rather less naive estimators of Section 7.1.6.

8.4.2. Regression With Splitting

In the above paragraphs we used the number of arrivals as a control variate, assuming simply that there was *some* relationship between this and the number of customers served. No attempt was made to explore the form of the relationship. To improve our estimates further, we can employ linear regression to approximate the relationship between the two variables.

To keep the examples simple, we return to the case where two tellers are at work. As before, we assume that we can afford 50 runs. No changes are necessary in the program implementing our model: we simply make 50 runs with the number of tellers held constant at two. A baseline for what follows was established in Section 8.1, where we calculated the expected number of customers served in the straightforward way, averaging over the 50 runs. Calculated in this way our estimate is 233.34 with a sample variance of 1.74.

To use regression with splitting, we follow the technique outlined in Section 2.3, particularly Problems 2.3.6 and 2.3.7. Let A_i and S_i, $i = 1, 2, \ldots,$ 50, be the number of arrivals and the number of customers served at each run. Since $E[A] = 247.5$, if $A'_i = A_i - 247.5$ then $E[A'] = 0$. For each i, we calculate $\hat{k}_i$, the slope of the regression line through the *other* 49 observations, regressing S against A'. Now $S_i - \hat{k}_i A'_i$ is one estimate of the average number of customers served. We now treat these 50 estimates as though they were iid. The resulting overall estimate is unbiased, but our sample variance generally underestimates the real variance because the individual estimates are correlated.

Figure X.8.3 shows a program for carrying out this procedure. It does *not* compute each $\hat{k}_i$ from scratch. When we tried regression with splitting on our data, we found that the expected number of customers served is estimated to be 233.44 with sample variance (biased) 0.76. This estimate appears to be more than twice as good as the naive estimate.

8.4.3. Regression With Jackknifing

As a last variation on the theme of control variates, we tried using regression with jackknifing, which is closely akin to splitting. The technique is described in Section 2.7.

Define A'_i and S_i, $i = 1, 2, \ldots, 50$, as above. Our estimates are calculated by regressing S against A', and noting where the regression line cuts the y-axis (i.e., we calculate the constant term in the regression). To find $\hat{\theta}$—using the notation of Section 2.7—we use all 50 observations, while to find $\hat{\theta}_{-i}$ we leave out observation i and use the other 49. Then the pseudovalues J_i are calculated. We treat them as 50 iid estimates, although again, in fact, they are dependent.

Figure X.8.4 shows a program for carrying out this procedure. Comparison with Figure X.8.3 brings out the similarity between splitting and jackknifing. When we tried the latter on our data, we found that the mean number of customers served is estimated to be 233.44, with sample variance (biased) 0.77. There is nothing to choose between jackknifing and splitting. For comparison $\hat{\theta}$, the estimate using regression through all 50 observations, turned out to be 233.46.

PROBLEM 8.4.3. Use the computer model of Problem 8.1.8 to experiment with control variates. It seems likely, for instance, that the average response time over a given period will depend on the average service time of the tasks executed, and on their average core requirements, among other factors.

PROBLEM 8.4.4 (continuation). Suppose you find that average response time depends on *both* average service time required and average core required. Would using two control variates be better than using just one? See Problem 2.3.8 before you answer.

8.5. Stratified Sampling

One important parameter of the model of Example 1.4.1 is the number of tellers at work at the bank on any given day. Remember that there is one teller with probability 0.05, there are two with probability 0.15, and three with probability 0.80. The naive program of Section 7.1.6 chooses the number of tellers for each run randomly with the required probability. As we saw (Figure 7.1.5) this led, in 200 runs, to 8 runs with one teller, 28 with two tellers, and 164 with three.

Using stratified sampling, we fix the number of runs of each type in advance, hoping to improve our estimate. The necessary changes to the program were given in Section 8.4.1. As there, we begin by carrying out a short pilot experiment with 60 runs, 20 each with one, two, and three tellers. (In fact, of course, we can use the same pilot experiment.) The results are shown in Figure 8.5.1, where x is now the number of customers served.

On the basis of these pilot runs, we may expect that with n_1 runs with one teller, etc., the variance of our estimate of θ, the mean number of customers served, will be approximately

$$0.05^2 \times \frac{61.52}{n_1} + 0.15^2 \times \frac{84.46}{n_2} + 0.80^2 \times \frac{122.68}{n_3}.$$

If, as in the naive approach, we can afford 200 runs, using the result of Problem 8.4.1 we may hope to minimize the variance of our estimate of θ by taking $n_1 = 7$, $n_2 = 26$, $n_3 = 167$. Since the pilot experiment has already been done, however, this is not possible; n_1 cannot be less than 20. Using the same technique to divide the remaining 180 runs, we finally settle on $n_1 = 20$, $n_2 = 24$, $n_3 = 156$. The results are shown in Figure 8.5.2.

In this particular case we have gained nothing (indeed we have lost a fraction) compared to the naive approach of Section 7.1.6. The reason is clear: the naive approach used sample sizes which were nearer to the estimated optimum than those we were able to use in the more controlled experiment. However, this is not the case in general (§8.4.1 illustrates this well).

Number of tellers	Number of runs	$\bar{x}_i$	s_i^2
1	20	150.60	61.52
2	20	232.60	84.46
3	20	246.50	122.68

Figure 8.5.1. Results of the pilot runs.

Number of tellers	Number of runs	$\bar{x}_i$	s_i^2
1	20	150.60	61.52
2	24	235.29	110.22
3	156	246.44	229.25

$$\hat{\theta} = \sum p_i \bar{x}_i = 239.98.$$

Estimated variance $[\hat{\theta}] - \sum p_i^2 s_i^2 / n_i = 1.05.$

Figure 8.5.2. Results with stratified sampling.

PROBLEM 8.5.1. Looking at these results, we see that the pilot runs considerably under-estimated the variance of the simulation output in the case with three tellers. This in turn led us to choose sample sizes quite far from the optimum for the longer simulation. To avoid this kind of accident, we could perform a short pilot experiment, use this to guide a longer simulation, which in turn could guide one longer still, and so on. In the limit, we could add runs one at a time in the stratum where it seems they will do most good. Discuss this approach.

PROBLEM 8.5.2. One version of stratified sampling is *proportional sampling*. In our case, suppose once again that we have 200 runs available. Then make $0.05 \times 200 = 10$ runs with one teller, $0.15 \times 200 = 30$ runs with two tellers, and $0.80 \times 200 = 160$ runs with three tellers. Explain why this technique *must* be better than nothing. Try it. (See Problem 2.4.1.)

PROBLEM 8.5.3. Still with no extra work, we could stratify the problem and use direct estimation of the number of customers served in the case of one teller, estimation of the number of customers lost in the case of three tellers, and either method for two tellers. Use Figures 8.4.1 and 8.5.1 to choose optimal sample sizes for this approach. Calculate the variance you can expect in the overall estimate.

PROBLEM 8.5.4. In our model there is little to indicate whether direct estimation or estimation via a control variate works better in the case of two tellers. When presented with a set of results, the "obvious" thing to do is to calculate an estimate both ways, and use the better one. This has to be cheating. Doesn't it?

PROBLEM 8.5.5. Show how stratified sampling and related techniques might be applied in the computer model of problem 8.1.8. (Hint: See Problems 2.4.3 and 2.4.5.)

8.6. Importance Sampling

Importance sampling is described in Section 2.5. In particular, Example 2.5.1 suggests how importance sampling may be applied to Example 1.4.1. As suggested there, we modify only the service time distribution. With the notation of Example 2.5.1, the original service time distribution B is 2-Erlang with mean 2. Hence

$$b(x) = xe^{-x}, \qquad x > 0.$$

Since, intuitively, large service times tend to lead to a smaller number of customers served, the output function g of our simulation will tend to decrease as x increases. If we hope to reduce the variance of our model, the modified service time $\tilde{b}$ should mimic gb; it must therefore fall away in the tail faster than b. Furthermore, we would like to choose $\tilde{B}$ so that sampling from it is easy, and so that $b(x)/\tilde{b}(x)$ is easily calculated for any x (cf. Example 2.5.1).

The possibilities for $\tilde{B}$ are legion. We try

$$\tilde{b}(x) = \begin{cases} \dfrac{1}{c} xe^{-x}, & 0 < x < t, \\[2mm] \dfrac{1}{c} te^{-x}, & t \le x, \end{cases}$$

where c is a normalizing constant such that

$$\int_0^\infty \tilde{b}(x)\, dx = 1$$

and t is an arbitrary parameter.

PROBLEM 8.6.1. Show that $c = 1 - e^{-t}$.

Note that

(a) $\tilde{b}$ has a lighter tail than b, and

(b) $b(x)/\tilde{b}(x) = \begin{cases} c, & 0 < x < t, \\[2mm] \dfrac{cx}{t}, & t \le x. \end{cases}$

We sample from $\tilde{B}$ by composition. First we pick one of the intervals $(0, t)$ and (t, ∞) with the correct probabilities. On $(0, t)$ the distribution still looks like the original one, namely, 2-Erlang with mean 2. For values of t not too small, generate variates from the original distribution and reject any which exceed t. On (t, ∞) the density is proportional to an exponential pdf.

PROBLEM 8.6.2. Show that the probability of being on $(0, t)$ is

$$1 - \frac{t}{e^t - 1}.$$

PROBLEM 8.6.3. On (t, ∞) we could generate exponential variates with mean 1 and reject any less than t. If t is moderately large, this is hopelessly inefficient. Give a better way.

Figure X.8.5 shows a suitable function for implementing this technique. The common variables BT and BC are t and c of the argument above; BLIM is the probability of being on $(0, t)$; and BMULT accumulates the product of terms $b(x)/\tilde{b}(x)$ as each x is generated. The necessary modifications to the main program are:

(a) declare **B, BT, BC, BLIM** and **BMULT** appropriately;
(b) at the beginning of the program, set **BT** to some appropriate value and calculate **BC** and **BLIM** in terms of **BT**;
(c) at the beginning of each run, set **BMULT** to 1;
(d) replace all three calls of ERLANG by calls of **B**;
(e) at the end of each run, output **NSERV * BMULT** instead of **NSERV**.

PROBLEM 8.6.4. Before you read on, think critically about the example. Do you expect the altered sampling to be better or worse than the original? By how much?

Figure 8.6.1 summarizes the results of 50 runs for the case where two tellers are at work. When t is large, the probability of generating values in the modified tail of $\tilde{B}$ is so small that we have essentially returned to the original model. Should we be surprised that the method is catastrophically bad for small to moderate values of t? No. Our output function g, the number of customers served, is little influenced by any one service time; put another way, $g(\ldots, x, \ldots)$ is quite flat as x varies over a wide range. If g is approximately constant, then the distribution which best "follows" gb is b itself; any modification can only make things worse, not better.

It is only when g is quite variable (and when it can be guessed or plotted fairly accurately) that importance sampling may help.

PROBLEM 8.6.5. In the general case of Example 1.4.1, g is certainly not constant as the number of tellers varies: see, for instance, Figure 8.5.1. Importance sampling would suggest modifying the probability of carrying out a run with i tellers on the job, $i = 1$, 2, 3, to be proportional to $p_i \bar{x}_i$, where the p_i are the true probabilities and the $\bar{x}_i$ are estimates from pilot runs or elsewhere. How does this compare to: (a) stratified sampling (§8.5); and (b) proportional sampling (Problem 8.5.2)? Try it.

t	Estimate	Sample variance
4	212.36	1427.81
6	221.89	188.71
8	229.03	15.63
10	231.18	2.21

Figure 8.6.1. Results of importance sampling.

PROBLEM 8.6.6. Use the computer model of Problem 8.1.8 to experiment with importance sampling. Consider, for instance, modifying the probability that a request falls in a particular class.

8.7. Conditional Monte Carlo

This technique is explained in Section 2.6; see, in particular, Example 2.6.1. Again we suppose that we are interested in the case where two tellers are at work and that we can afford 50 runs.

The trivial modifications to our program follow directly from Example 2.6.1. We initialize a real variable, ESERV say, to zero. Whenever a customer arrives, we increment ESERV by the probability that he will not balk: this requires the addition of two or three instructions to the sequence beginning at statement 100. At the end of a run, we output ESERV instead of NSERV and then calculate the mean and variance of the observed values in the usual way. The test which decides whether or not the customer does in fact balk is left unchanged.

In our experiments the sample mean and variance were 233.66 and 1.97, respectively, for extended conditional Monte Carlo versus 234.62 and 1.74, respectively, for the naive method. Purely on the basis of this experimental evidence the naive method seems a little better, but we suspect that this ordering of the sample variances does not reflect that of the true variances. Whether or not this apparent anomaly is due to the same kind of sampling fluctuations that Figure 8.8.2 reveals we do not know.

We ought to do better by pushing conditional Monte Carlo further following Problem 4.9.11. Programming this method is reasonably straightforward. We are to accumulate

$$P[\text{a hypothetical arrival would not balk}]$$
$$\times \text{ (interval length)} \times \text{ (arrival rate)}$$

over all intervals in which the number of customers in the queue and the arrival rate remain constant. Each actual arrival and each start of service may change the number of customers in the queue, so we must add code to these events to accumulate the required sum. The same code is used when the arrival rate changes: to detect these moments schedule a fictitious event CHANGE at 11 A.M. and at 2 P.M. Finally, we update the accumulated sum when the bank closes its doors at 3 P.M. Then we output the accumulated sum and calculate the mean and variance of the observed values in the usual way. The extra program required amounts to scarcely more than a dozen lines.

Using this technique we obtained an estimate for the mean number of customers served of 233.12, with a sample variance of 1.12. This is some 35 % better than an estimate obtained in the straightforward way, for the same amount of computing.

8.8. Summary

Figure 8.8.1 brings together results scattered through the preceding sections for the case where two tellers are at work. To a limited extent, this allows us to compare the usefulness of different techniques on this particular example. We emphasize once more that no conclusions should be drawn about the power of the techniques in general.

All the techniques we have seen are trivially easy to program, with the possible exception of importance sampling; none requires more than superficial analysis of the problem, perhaps with the same exception; and none materially increases the running time of the simulation program (in our example even importance sampling, again the extreme case, increased running time by less than 10%). All offer worthwhile gains in problems with the appropriate structure.

PROBLEM 8.8.1. Applied to a problem with an inappropriate structure (see the discussion of §8.6), importance sampling produced disastrously *bad* results. Is this liable to happen using the other techniques in inappropriate situations, or are they "fail-safe"?

PROBLEM 8.8.2. If importance sampling can be disastrously bad, one might hope that it can also be spectacularly good. Can you justify such optimism?

Method	Estimate obtained	Observed variance
Straightforward	233.34	1.74
Using tabulated Erlang	234.50	1.82
Antithetic variates		
Service times	234.68	2.06
Arrival times	234.10	1.29
Both	234.80	1.60
Service times with	234.32	1.93
tabulated Erlang		
Control variate		
with splitting	233.44	0.76*
with jackknifing	233.44	0.77*
Importance sampling		
$t = 6$	221.89	188.71
$t = 8$	229.03	15.63
Conditional Monte Carlo		
Using Example 2.6.1	233.66	1.97
Using Problem 4.9.11	233.12	1.12

Figure 8.8.1. Comparison of techniques for the case of two tellers. (All results based on 50 runs. Starred variances are biased low.)

Set	Estimate	Obs. var.	Set	Estimate	Obs. var.
1	233.34	1.74	6	234.62	2.90
2	233.20	3.43	7	234.26	2.51
3	231.32	2.96	8	232.20	2.78
4	231.34	3.32	9	231.76	2.74
5	232.46	2.94	10	231.62	3.56

Figure 8.8.2. Results from 10 sets of 50 runs.

For the full model (with one, two, or three tellers at work) we gave two examples of stratified sampling. With a straightforward estimator the optimal sample sizes for each stratum were close to those obtained when simulating naively, and no improvement was found. The combination of stratified sampling and the use of a control variate, however, gave appreciably better accuracy with no extra computing.

Finally, for comparing policies using the full model, common random numbers gave an order of magnitude gain over uncontrolled simulation, again at no cost.

Throughout this chapter, and indeed throughout the book, we compared techniques using sample variances. It is interesting to see, for our particular example, how reliable this is. As a simple experiment, we ran the straight-forward simulation of the case of two tellers for 10 sets of 50 runs and calculated results separately for each set. Figure 8.8.2 shows the values obtained. Although the sample variances are not wildly unstable, there is a factor of 2 between the largest and the smallest. By chance, the first set of 50 runs, used above to provide a basis for comparison, has the lowest sample variance of all. This emphasizes yet again the tenuous and local nature of the comparisons in this chapter. It also underlines the need for more (possibly many more) runs to get both good estimates of performance and good estimates of their goodness.

The Shapiro–Wilk Test for Normality

Shapiro and Wilk (1965) introduced the test for normality described below. Monte Carlo comparisons by Shapiro *et al.* (1968) indicate that it has more power than other procedures for checking the goodness-of-fit of normal distributions to data.

Suppose you have n observations $X_1, X_2, \ldots, X_n$. Proceed as follows:

(1) Sort the X_i so that $X_i > X_{i-1}$, $i = 2, 3, \ldots, n$.
(2) Calculate

$$b = \sum_{i=1}^{\lceil n/2 \rceil} (X_{n-i+1} - X_i) a_{in},$$

where the a_{in} are given in Figure A.1.

i \ n	2	3	4	5	6	7	8	9	10
1	0.7071	0.7071	0.6872	0.6646	0.6431	0.6233	0.6052	0.5888	0.5739
2	—	0.0000	0.1677	0.2413	0.2806	0.3031	0.3164	0.3244	0.3291
3	—	—	—	0.0000	0.0875	0.1401	0.1743	0.1976	0.2141
4	—	—	—	—	—	0.0000	0.0561	0.0947	0.1224
5	—	—	—	—	—	—	—	0.0000	0.0399

i \ n	11	12	13	14	15	16	17	18	19	20
1	0.5601	0.5475	0.5359	0.5251	0.5150	0.5056	0.4968	0.4886	0.4808	0.4734
2	0.3315	0.3325	0.3325	0.3318	0.3306	0.3290	0.3273	0.3253	0.3232	0.3211
3	0.2260	0.2347	0.2412	0.2460	0.2495	0.2521	0.2540	0.2553	0.2561	0.2565
4	0.1429	0.1586	0.1707	0.1802	0.1878	0.1939	0.1988	0.2027	0.2059	0.2085
5	0.0695	0.0922	0.1099	0.1240	0.1353	0.1447	0.1524	0.1587	0.1641	0.1686
6	0.0000	0.0303	0.0539	0.0727	0.0880	0.1005	0.1109	0.1197	0.1271	0.1334
7	—	—	0.0000	0.0240	0.0433	0.0593	0.0725	0.0837	0.0932	0.1013
8	—	—	—	—	0.0000	0.0196	0.0359	0.0496	0.0612	0.0711
9	—	—	—	—	—	—	0.0000	0.0163	0.0303	0.0422
10	—	—	—	—	—	—	—	—	0.0000	0.0140

(continued)

i	21	22	23	24	25	26	27	28	29	30
1	0.4643	0.4590	0.4542	0.4493	0.4450	0.4407	0.4366	0.4328	0.4291	0.4254
2	0.3185	0.3156	0.3126	0.3098	0.3069	0.3043	0.3018	0.2992	0.2968	0.2944
3	0.2578	0.2571	0.2563	0.2554	0.2543	0.2533	0.2522	0.2510	0.2499	0.2487
4	0.2119	0.2131	0.2139	0.2145	0.2148	0.2151	0.2152	0.2151	0.2150	0.2148
5	0.1736	0.1764	0.1787	0.1807	0.1822	0.1836	0.1848	0.1857	0.1864	0.1870
6	0.1399	0.1443	0.1480	0.1512	0.1539	0.1563	0.1584	0.1601	0.1616	0.1630
7	0.1092	0.1150	0.1201	0.1245	0.1283	0.1316	0.1346	0.1372	0.1395	0.1415
8	0.0804	0.0878	0.0941	0.0997	0.1046	0.1089	0.1128	0.1162	0.1192	0.1219
9	0.0530	0.0618	0.0696	0.0764	0.0823	0.0876	0.0923	0.0965	0.1002	0.1036
10	0.0263	0.0368	0.0459	0.0539	0.0610	0.0672	0.0728	0.0778	0.0822	0.0862
11	0.0000	0.0122	0.0228	0.0321	0.0403	0.0476	0.0540	0.0598	0.0650	0.0697
12	—	—	0.0000	0.0107	0.0200	0.0284	0.0358	0.0424	0.0483	0.0537
13	—	—	—	—	0.0000	0.0094	0.0178	0.0253	0.0320	0.0381
14	—	—	—	—	—	—	0.0000	0.0084	0.0159	0.0227
15	—	—	—	—	—	—	—	—	0.0000	0.0076

i	31	32	33	34	35	36	37	38	39	40
1	0.4220	0.4188	0.4156	0.4127	0.4096	0.4068	0.4040	0.4015	0.3989	0.3964
2	0.2921	0.2898	0.2876	0.2854	0.2834	0.2813	0.2794	0.2774	0.2755	0.2737
3	0.2475	0.2463	0.2451	0.2439	0.2427	0.2415	0.2403	0.2391	0.2380	0.2368
4	0.2145	0.2141	0.2137	0.2132	0.2127	0.2121	0.2116	0.2110	0.2104	0.2098
5	0.1874	0.1878	0.1880	0.1882	0.1883	0.1883	0.1883	0.1881	0.1880	0.1878
6	0.1641	0.1651	0.1660	0.1667	0.1673	0.1678	0.1683	0.1686	0.1689	0.1691
7	0.1433	0.1449	0.1463	0.1475	0.1487	0.1496	0.1505	0.1513	0.1520	0.1526
8	0.1243	0.1265	0.1284	0.1301	0.1317	0.1331	0.1344	0.1356	0.1366	0.1376
9	0.1066	0.1093	0.1118	0.1140	0.1160	0.1179	0.1196	0.1211	0.1225	0.1237
10	0.0899	0.0931	0.0961	0.0988	0.1013	0.1036	0.1056	0.1075	0.1092	0.1108
11	0.0739	0.0777	0.0812	0.0844	0.0873	0.0900	0.0924	0.0947	0.0967	0.0986
12	0.0585	0.0629	0.0669	0.0706	0.0739	0.0770	0.0798	0.0824	0.0848	0.0870
13	0.0435	0.0485	0.0530	0.0572	0.0610	0.0645	0.0677	0.0706	0.0733	0.0759
14	0.0289	0.0344	0.0395	0.0441	0.0484	0.0523	0.0559	0.0592	0.0622	0.0651
15	0.0144	0.0206	0.0262	0.0314	0.0361	0.0404	0.0444	0.0481	0.0515	0.0546
16	0.0000	0.0068	0.0131	0.0187	0.0239	0.0287	0.0331	0.0372	0.0409	0.0444
17	—	—	0.0000	0.0062	0.0119	0.0172	0.0220	0.0264	0.0305	0.0343
18	—	—	—	—	0.0000	0.0057	0.0110	0.0158	0.0203	0.0244
19	—	—	—	—	—	—	0.0000	0.0053	0.0101	0.0146
20	—	—	—	—	—	—	—	—	0.0000	0.0049

i	41	42	43	44	45	46	47	48	49	50
1	0.3940	0.3917	0.3894	0.3872	0.3850	0.3830	0.3808	0.3789	0.3770	0.3751
2	0.2719	0.2701	0.2684	0.2667	0.2651	0.2635	0.2620	0.2604	0.2589	0.2574
3	0.2357	0.2345	0.2334	0.2323	0.2313	0.2302	0.2291	0.2281	0.2271	0.2260
4	0.2091	0.2085	0.2078	0.2072	0.2065	0.2058	0.2052	0.2045	0.2038	0.2032
5	0.1876	0.1874	0.1871	0.1868	0.1865	0.1862	0.1859	0.1855	0.1851	0.1847
6	0.1693	0.1694	0.1695	0.1695	0.1695	0.1695	0.1695	0.1693	0.1692	0.1691
7	0.1531	0.1535	0.1539	0.1542	0.1545	0.1548	0.1550	0.1551	0.1553	0.1554
8	0.1384	0.1392	0.1398	0.1405	0.1410	0.1415	0.1420	0.1423	0.1427	0.1430
9	0.1249	0.1259	0.1269	0.1278	0.1286	0.1293	0.1300	0.1306	0.1312	0.1317
10	0.1123	0.1136	0.1149	0.1160	0.1170	0.1180	0.1189	0.1197	0.1205	0.1212
11	0.1004	0.1020	0.1035	0.1049	0.1062	0.1073	0.1085	0.1095	0.1105	0.1113
12	0.0891	0.0909	0.0927	0.0943	0.0959	0.0972	0.0986	0.0998	0.1010	0.1020
13	0.0782	0.0804	0.0824	0.0842	0.0860	0.0876	0.0892	0.0906	0.0919	0.0932
14	0.0677	0.0701	0.0724	0.0745	0.0765	0.0783	0.0801	0.0817	0.0832	0.0846
15	0.0575	0.0602	0.0628	0.0651	0.0673	0.0694	0.0713	0.0731	0.0748	0.0764
16	0.0476	0.0506	0.0534	0.0560	0.0584	0.0607	0.0628	0.0648	0.0667	0.0685
17	0.0379	0.0411	0.0442	0.0471	0.0497	0.0522	0.0546	0.0568	0.0588	0.0608
18	0.0283	0.0318	0.0352	0.0383	0.0412	0.0439	0.0465	0.0489	0.0511	0.0532
19	0.0188	0.0227	0.0263	0.0296	0.0328	0.0357	0.0385	0.0411	0.0436	0.0459
20	0.0094	0.0136	0.0175	0.0211	0.0245	0.0277	0.0307	0.0335	0.0361	0.0386
21	0.0000	0.0045	0.0087	0.0126	0.0163	0.0197	0.0229	0.0259	0.0288	0.0314
22	—	—	0.0000	0.0042	0.0081	0.0118	0.0153	0.0185	0.0215	0.0244
23	—	—	—	—	0.0000	0.0039	0.0076	0.0111	0.0143	0.0174
24	—	—	—	—	—	—	0.0000	0.0037	0.0071	0.0104
25	—	—	—	—	—	—	—	—	0.0000	0.0035

Figure A.1. Coefficients a_{in} for the Shapiro–Wilk test. (Reproduced by permission from *Biometrika*.)

(3) Calculate

$$W_n = b^2/[(n-1)s^2],$$

where s^2 is the sample variance of the X_i. Percentage points for W_n are listed in Figure A.2. For normal data W_n should be near 1; reject the assumption of normality if W_n is small.

n					Level				
	0.01	0.02	0.05	0.10	0.50	0.90	0.95	0.98	0.99
3	0.753	0.756	0.767	0.789	0.959	0.998	0.999	1.000	1.000
4	0.687	0.707	0.748	0.792	0.935	0.987	0.992	0.996	0.997
5	0.686	0.715	0.762	0.806	0.927	0.979	0.986	0.991	0.993
6	0.713	0.743	0.788	0.826	0.927	0.974	0.981	0.986	0.989
7	0.730	0.760	0.803	0.838	0.928	0.972	0.979	0.985	0.988
8	0.749	0.778	0.818	0.851	0.932	0.972	0.978	0.984	0.987
9	0.764	0.791	0.829	0.859	0.935	0.972	0.978	0.984	0.986
10	0.781	0.806	0.842	0.869	0.938	0.972	0.978	0.983	0.986
11	0.792	0.817	0.850	0.876	0.940	0.973	0.979	0.984	0.986
12	0.805	0.828	0.859	0.883	0.943	0.973	0.979	0.984	0.986
13	0.814	0.837	0.866	0.889	0.945	0.974	0.979	0.984	0.986
14	0.825	0.846	0.874	0.895	0.947	0.975	0.980	0.984	0.986
15	0.835	0.855	0.881	0.901	0.950	0.975	0.980	0.984	0.987
16	0.844	0.863	0.887	0.906	0.952	0.976	0.981	0.985	0.987
17	0.851	0.869	0.892	0.910	0.954	0.977	0.981	0.985	0.987
18	0.858	0.874	0.897	0.914	0.956	0.978	0.982	0.986	0.988
19	0.863	0.879	0.901	0.917	0.957	0.978	0.982	0.986	0.988
20	0.868	0.884	0.905	0.920	0.959	0.979	0.983	0.986	0.988
21	0.873	0.888	0.908	0.923	0.960	0.980	0.983	0.987	0.989
22	0.878	0.892	0.911	0.926	0.961	0.980	0.984	0.987	0.989
23	0.881	0.895	0.914	0.928	0.962	0.981	0.984	0.987	0.989
24	0.884	0.898	0.916	0.930	0.963	0.981	0.984	0.987	0.989
25	0.888	0.901	0.918	0.931	0.964	0.981	0.985	0.988	0.989
26	0.891	0.904	0.920	0.933	0.965	0.982	0.985	0.988	0.989
27	0.894	0.906	0.923	0.935	0.965	0.982	0.985	0.988	0.990
28	0.896	0.908	0.924	0.936	0.966	0.982	0.985	0.988	0.990
29	0.898	0.910	0.926	0.937	0.966	0.982	0.985	0.988	0.990
30	0.900	0.912	0.927	0.939	0.967	0.983	0.985	0.988	0.900
31	0.902	0.914	0.929	0.940	0.967	0.983	0.986	0.988	0.990
32	0.904	0.915	0.930	0.941	0.968	0.983	0.986	0.988	0.990
33	0.906	0.917	0.931	0.942	0.968	0.983	0.986	0.989	0.990
34	0.908	0.919	0.933	0.943	0.969	0.983	0.986	0.989	0.990
35	0.910	0.920	0.934	0.944	0.969	0.984	0.986	0.989	0.990
36	0.912	0.922	0.935	0.945	0.970	0.984	0.986	0.989	0.990
37	0.914	0.924	0.936	0.946	0.970	0.984	0.987	0.989	0.990
38	0.916	0.925	0.938	0.947	0.971	0.984	0.987	0.989	0.990
39	0.917	0.927	0.939	0.948	0.971	0.984	0.987	0.989	0.991
40	0.919	0.928	0.940	0.949	0.972	0.985	0.987	0.989	0.991
41	0.920	0.929	0.941	0.950	0.972	0.985	0.987	0.989	0.991
42	0.922	0.930	0.942	0.951	0.972	0.985	0.987	0.989	0.991
43	0.923	0.932	0.943	0.951	0.973	0.985	0.987	0.990	0.991
44	0.924	0.933	0.944	0.952	0.973	0.985	0.987	0.990	0.991
45	0.926	0.934	0.945	0.953	0.973	0.985	0.988	0.990	0.991
46	0.927	0.935	0.945	0.953	0.974	0.985	0.988	0.990	0.991
47	0.928	0.936	0.946	0.954	0.974	0.985	0.988	0.990	0.991
48	0.929	0.937	0.947	0.954	0.974	0.985	0.988	0.990	0.991
49	0.929	0.937	0.947	0.955	0.974	0.985	0.988	0.990	0.991
50	0.930	0.938	0.947	0.955	0.974	0.985	0.988	0.990	0.991

Figure A.2. Percentage points for W_n for the Shapiro–Wilk test. (Reproduced by permission from *Biometrika*.)

APPENDIX L
Routines for Random Number Generation

This appendix contains examples of generators, and of routines for calculating and inverting cdfs. Many of the routines presented are discussed in Chapters 5 and 6.

All the programs are written in ANSI Fortran. To make them as useful as possible, we have only used language features which are in both the old standard Fortran IV and the new official subset of Fortran 77 [ANSI (1978)]. All the examples have been tested on at least two machines. Error reporting is summary: in most cases a flag called IFAULT is set to a nonzero value if an error is detected, the code used depending on the routine at fault. Names follow the Fortran type conventions, and only dimensioned variables are explicitly declared. The comment AUXILIARY ROUTINES refers to routines to be found elsewhere in the appendix.

The appendix contains the following routines:

Functions

1. ALOGAM   Evaluates the natural logarithm of $\Gamma(x)$ (used to set up further calls).

2. BETACH   Generates a beta deviate using the method of Cheng (1978)—see Section 5.3.10.

3. BETAFX   Generates a beta deviate using the method of Fox (1963)—see Section 5.3.10.

4. BOXNRM   Generates a standard normal deviate using the Box–Muller method—see Section 5.2.10.

5. D2   Computes $(\exp(x) - 1)/x$ accurately (used as an auxiliary routine).

6. **GAMAIN**   Computes the incomplete gamma ratio (used as an auxiliary routine).

7. **IALIAS**   Generates a discrete random variate by the alias method—see Section 5.2.8. Initialization is carried out by the subroutine ALSETP.

8. **IVDISC**   Inverts a discrete cdf using the method of Ahrens and Kohrt—see Section 5.2.1. Initialization is carried out by the subroutine VDSETP.

9. **IVUNIP**   Generates an integer uniform over an interval $(1, n)$ using inversion—see Section 6.7.1.

10. **LUKBIN**   Generates a discrete random variate using the method of Fox (1978b)—see Section 5.2.7. Initialization is carried out by the subroutine SETLKB.

11. **REMPIR**   Generates a variate from an empirical distribution with an exponential tail—see Sections 4.6 and 5.2.4.

12. **RGKM3**   Generates a gamma variate using rejection—see Section 5.3.9.

13. **RGS**   Generates a gamma variate using rejection—see Section 5.3.9.

14. **RSTAB**   Generates a stable Paretian variate using the method of Chambers *et al.* (1976)—see Section 5.3.3.

15. **SUMNRM**   Generates a standard normal deviate by a not very quick and unnecessarily dirty method: its use is not recommended.

16. **TAN**   Computes $\tan(x)$ (used as an auxiliary routine).

17. **TAN2**   Computes $\tan(x)/x$ (used as an auxiliary routine).

✓18. **TRPNRM**   Generates a standard normal deviate using the method of Ahrens and Dieter—see Example 5.2.2.

19. **UNIF**   Generates uniform random numbers. The code is fast and portable—see Section 6.5.2.

20. **UNIFL**   Generates uniform variates using L'Ecuyer's composite method.

21. **VBETA**   Calculates the inverse of the beta cdf—see Section 5.3.10.

22. **VCAUCH**   Calculates the inverse of a standard Cauchy cdf—see Section 5.3.4.

23. **VCHI2**   Calculates the inverse of the chi-square cdf—see Section 5.3.11. This routine can also be used to generate Erlang and gamma variates—see Section 5.3.9.

24. **VCHISQ**   Calculates the inverse of the chi-square cdf—see Section 5.3.11 (faster but less accurate than VCHI2). This routine can also be used to generate Erlang and gamma variates—see Section 5.3.9.

25. **VF**   Calculates the inverse of the F cdf—see Section 5.3.12.

26. **VNHOMO**   Generates a deviate from a nonhomogeneous Poisson distribution—see Section 5.3.18.

27. VNORM   Calculates the inverse of the standard normal cdf using a rational approximation.
28. VSTUD   Calculates the inverse of the *t* cdf—see Section 5.3.13.

Subroutines

29. BINSRH   Performs a binary search in a tabulated cdf.
30. FNORM   Calculates areas under a standard normal curve, and also the normal loss function.
31. TBINOM   Tabulates the binomial cdf.
32. TTPOSN   Tabulates the cdf of a truncated Poisson.

```
      FUNCTION ALOGAM( Y)
C
C          THIS PROCEDURE EVALUATES THE NATURAL LOGARITHM OF GAMMA(Y)
C          FOR ALL Y>0, ACCURATE TO APPROXIMATELY MACHINE PRECISION
C          OR 10 DECIMAL DIGITS, WHICHEVER IS LESS.
C          STIRLING'S FORMULA IS USED FOR THE CENTRAL POLYNOMIAL PART
C          OF THE PROCEDURE.
C          BASED ON ALG. 291 IN CACM, VOL. 9, NO. 9 SEPT. 66, P. 384,
C          BY M. C. PIKE AND I. D. HILL.
C
      X = Y
      IF ( X .GE. 7.0) GO TO 20
      F = 1.0
      Z = X - 1.0
      GO TO 7
    5 X = Z
      F = F* Z
    7 Z = Z + 1.
      IF ( Z .LT. 7) GO TO 5
      X = X + 1.
      F = - ALOG( F)
      GO TO 30
   20 F = 0.
   30 Z = ( 1.0 / X) ** 2.0
      ALOGAM = F + ( X - 0.5) * ALOG( X) - X + 0.918938533204673 +
     *         (((-0.000595238095238 * Z + 0.000793650793651) * Z -
     *         0.002777777777778) * Z + 0.083333333333333) / X
      RETURN
      END
```

Figure L.1. The function ALOGAM.

```
      FUNCTION BETACH( IY, A, B, CON)
      REAL CON( 3)
C
C  GENERATE A BETA DEVIATE
C  REF.:  CHENG, R., GENERATING BETA VARIATES WITH NONINTEGRAL
C  SHAPE PARAMETERS,  CACM, VOL. 21,  APRIL, 1978
C
C  INPUTS:
C    A, B = THE TWO PARAMETERS OF THE STANDARD BETA,
C    CON( ) = A REAL WORK VECTOR OF LENGTH AT LEAST 3.
C    AT THE FIRST CALL CON( 1) SHOULD BE 0.
C    IT IS SET TO A FUNCTION OF A & B FOR USE IN SUBSEQUENT CALLS
C    IY = A RANDOM NUMBER SEED.
C
C  AUXILIARY ROUTINE:
C    UNIF
C
      IF ( CON( 1) .GT. 0.) GO TO 200
      CON( 1) = AMIN1( A, B)
      IF ( CON( 1) .GT. 1.) GO TO 100
      CON( 1) = 1./ CON( 1)
      GO TO 150
  100 CON( 1) = SQRT(( A + B - 2.)/( 2.* A* B- A- B))
  150 CON( 2)= A + B
      CON ( 3) = A + 1./ CON( 1)
C
C  SUBSEQUENT ENTRIES ARE HERE
C
  200 U1 = UNIF( IY)
      U2 = UNIF( IY)
      V = CON( 1)* ALOG( U1/(1.- U1))
      W = A* EXP( V)
      IF (( CON( 2))*ALOG(( CON( 2))/(B+W)) + (CON( 3))
     1 *V - 1.3862944 .LT. ALOG( U1*U1*U2)) GO TO 200
      BETACH = W/( B + W)
      RETURN
      END
```

Figure L.2. The function BETACH.

```
      FUNCTION BETAFX( IX, NA, NB, WORK)
      REAL WORK( NA)
C
C  GENERATE A BETA DEVIATE
C  REF.: FOX, B.(1963), GENERATION OF RANDOM SAMPLES FROM THE
C  BETA AND F DISTRIBUTIONS, TECHNOMETRICS, VOL. 5, 269-270.
C
C  INPUTS:
C    IX = A RANDOM NUMBER SEED,
C    NA, NB = PARAMETERS OF THE BETA DISTRIBUTION,
C    MUST BE INTEGER
C    WORK( ) = A WORK VECTOR OF LENGTH AT LEAST NA + NB.
C
C  OUTPUTS:
C    BETAFX = A DEVIATE FROM THE  RELEVANT DISTRIBUTION
C
C  AUXILIARY ROUTINE;
C    UNIF
C
C  SETUP WORK VECTOR FOR SORTING
C
      DO 100 I =  1, NA
      WORK( I) = 2.
  100 CONTINUE
      NAB = NA + NB - 1
C
C  GENERATE NAB UNIFORMS AND FIND THE NA-TH SMALLEST
C
      DO 200 I = 1, NAB
      FY = UNIF( IX)
      IF ( FY .GE. WORK( 1)) GO TO 200
      IF ( NA .LE. 1) GO TO 170
      DO 160 K = 2, NA
      IF ( FY .LT.  WORK( K)) GO TO 150
      WORK( K - 1) = FY
      GO TO 200
  150 WORK( K - 1) = WORK( K)
  160 CONTINUE
  170 WORK( NA) = FY
  200 CONTINUE
C
      BETAFX = WORK( 1)
      RETURN
      END
```

Figure L.3. The function BETAFX.

```
      FUNCTION BOXNRM( IY, U1)
C
C RETURN A UNIT NORMAL(OR GAUSSIAN) RANDOM VARIABLE
C USING THE BOX-MULLER METHOD.
C
C INPUT:
C    IY = RANDOM NUMBER SEED
C    U1 = A WORK SCALER WHICH SHOULD BE > 254 ON FIRST CALL.
C       IT STORES DATA FOR SUBSEQUENT CALLS
C AUXILIARY ROUTINE:
C    UNIF
C
      IF ( U1 .GE. 254.) GO TO 100
C
C A DEVIATE IS LEFT FROM LAST TIME
C
      BOXNRM =  U1
C
C INDICATE THAT NONE LEFT. NOTE THAT PROBABILITY THAT
C A DEVIATE >  254 IS EXTREMELY SMALL.
C
      U1 = 255.
      RETURN
C
C TIME TO GENERATE A PAIR
C
  100 U1 = UNIF( IY)
      U1 = SQRT(-2.* ALOG( U1))
      U2 = 6.2831852* UNIF( IY)
C
C USE ONE...
C
      BOXNRM = U1* COS( U2)
C
C SAVE THE OTHER FOR THE NEXT PASS
C
      U1 = U1* SIN( U2)
      RETURN
      END
```

Figure L.4. The function BOXNRM.

```
      FUNCTION D2( Z)
C
C  EVALUATE ( EXP( X ) - 1)/X
C
      DOUBLE PRECISION P1, P2, Q1, Q2, Q3, PV, ZZ
C ON COMPILERS WITHOUT DOUBLE PRECISION THE ABOVE MUST BE
C REPLACED BY REAL P1, ETC., AND THE DATA INITIALIZATION
C MUST USE E RATHER THAN D FORMAT.
C
      DATA P1, P2, Q1, Q2, Q3/.84006685253648 3239D3,
     1 .200011141589964569D2,
     2 .168013370507926648D4,
     3 .18013370407390023D3,
     4 1.D0/
C
C  THE APPROXIMATION 1801 FROM HART ET. AL.(1968, P. 213)
C     COMPUTER APPROXIMATIONS, NEW YORK, JOHN WILEY.
C
      IF ( ABS( Z) .GT. 0.1) GO TO 100
      ZZ = Z * Z
      PV = P1 + ZZ* P2
      D2 = 2.D0* PV/( Q1+ ZZ*( Q2+ ZZ* Q3) - Z* PV)
      RETURN
  100 D2 = ( EXP( Z) - 1.) / Z
      RETURN
      END
```

Figure L.5. The function D2.

```
      FUNCTION GAMAIN( X, P, G, IFAULT)
C
C     ALGORITHM AS 32 J.R.STATIST.SOC. C. (1970) VOL.19 NO.3
C
C     COMPUTES INCOMPLETE GAMMA RATIO FOR POSITIVE VALUES OF
C     ARGUMENTS X AND P.  G MUST BE SUPPLIED AND SHOULD BE EQUAL TO
C     LN(GAMMA(P)).
C
C     IFAULT = 1 IF P.LE.0 ELSE 2 IF X.LT.0 ELSE 0.
C     USES SERIES EXPANSION IF P.GT.X OR X.LE.1, OTHERWISE A
C     CONTINUED FRACTION APPROXIMATION.
C
      DIMENSION PN(6)
C
C     DEFINE ACCURACY AND INITIALIZE
C
      DATA ACU/1.E-8/, OFLO/1.E30/
      GIN=0.0
      IFAULT=0
```

(continued)

```
C
C          TEST FOR ADMISSIBILITY OF ARGUMENTS
C
      IF(P.LE.0.0) IFAULT=1
      IF(X.LT.0.0) IFAULT=2
      IF(IFAULT.GT.0.OR.X.EQ.0.0) GO TO 50
      FACTOR=EXP(P*ALOG(X)-X-G)
      IF(X.GT.1.0.AND.X.GE.P) GO TO 30
C
C          CALCULATION BY SERIES EXPANSION
C
      GIN=1.0
      TERM=1.0
      RN=P
   20 RN=RN+1.0
      TERM=TERM*X/RN
      GIN=GIN+TERM
      IF(TERM.GT.ACU) GO TO 20
      GIN=GIN*FACTOR/P
      GO TO 50
C
C          CALCULATION BY CONTINUED FRACTION
C
   30 A=1.0-P
      B=A+X+1.0
      TERM=0.0
      PN(1)=1.0
      PN(2)=X
      PN(3)=X+1.0
      PN(4)=X*B
      GIN=PN(3)/PN(4)
   32 A=A+1.0
      B=B+2.0
      TERM=TERM+1.0
      AN=A*TERM
      DO 33 I=1,2
   33 PN(I+4)=B*PN(I+2)-AN*PN(I)
      IF(PN(6).EQ.0.0) GO TO 35
      RN=PN(5)/PN(6)
      DIF=ABS(GIN-RN)
      IF(DIF.GT.ACU) GO TO 34
      IF(DIF.LE.ACU*RN) GO TO 42
   34 GIN=RN
   35 DO 36 I=1,4
   36 PN(I)=PN(I+2)
      IF(ABS(PN(5)).LT.OFLO) GO TO 32
      DO 41 I=1,4
   41 PN(I)=PN(I)/OFLO
      GO TO 32
   42 GIN=1.0-FACTOR*GIN
C
   50 GAMAIN=GIN
      RETURN
      END
```

Figure L.6. The function GAMAIN.

```
      FUNCTION IALIAS( PHI, N, A, R)
      INTEGER A( N)
      REAL R( N)
C
C  GENERATE A DISCRETE R.V. BY THE ALIAS METHOD
C
C  INPUTS:
C    PHI = NUMBER BETWEEN 0 AND 1
C    N = RANGE OF R. V., I.E., 1, 2, ..., N
C    A( I) = ALIAS OF I
C    R( I) = ALIASING PROBABILITY
C
C  REF.: KRONMAL AND PETERSON, THE AMERICAN STATISTICIAN,
C      VOL 33, 1979.
C
      V = PHI *N
      JALIAS = V + 1.
      IF( JALIAS - V .GT. R( JALIAS)) JALIAS = A( JALIAS)
      IALIAS = JALIAS
      RETURN
      END
C
      SUBROUTINE ALSETP( N, P, A, R, L)
      INTEGER A( N), L( N)
      REAL P( N), R( N)
C
C  SETUP TABLES FOR USING THE ALIAS METHOD OF GENERATING
C   DISCRETE RANDOM VARIABLES USING FUNCTION IALIAS.
C
C  INPUTS:  N = RANGE OF THE RANDOM VARIABLE, I.E. 1, 2,..., N
C    P( I) = PROB( R.V. = I)
C    L(  ) A WORK VECTOR OF SIZE N
C
C  OUTPUTS:  A( I) = ALIAS OF I
C    R( I) ALIASING PROBABILITY OF I
C
      LOW = 0
      MID = N+ 1
C
C  PUT AT THE BOTTOM, OUTCOMES FOR WHICH THE UNIFORM DISTRIBUTION
C  ASSIGNS TOO MUCH PROBABILITY.  PUT THE REST AT THE TOP.
C
      DO 100 I = 1, N
      A( I) = I
      R( I) = N*P( I)
      IF ( R( I) .GE. 1.) GO TO 50
C
C  TOO MUCH PROBABILITY ASSIGNED TO I
C
      LOW = LOW + 1
      L(LOW) = I
      GO TO 100
C
C  TOO LITTLE PROBABILITY ASSIGNED TO I
C
   50 MID = MID - 1
```

(continued)

```
        L( MID) = I
100 CONTINUE
C
C  NOW GIVE THE OUTCOMES AT THE BOTTOM(WITH TOO MUCH PROBABILITY)
C  AN ALIAS AT THE TOP.
C
        N1 = N - 1
        DO 200 I = 1, N1
        K = L( MID)
        J = L( I)
        A( J)   = K
        R( K) = R( K) + R( J) - 1.
        IF ( R( K) .GE. 1.) GO TO 200
        MID = MID + 1
200 CONTINUE
        RETURN
        END
```

Figure L.7. The function IALIAS and its initialization subroutine ALSETP.

```fortran
      FUNCTION IVDISC( U, N, M, CDF, LUCKY)
      REAL CDF( N)
      INTEGER LUCKY( M)
C
C   INVERT A DISCRETE CDF USING AN INDEX VECTOR TO SPEED THINGS UP
C   REF. AHRENS AND KOHRT,  COMPUTING, VOL 26, P19, 1981.
C
C   INPUTS:
C     U = A NUMBER 0 <= U <= 1.
C     N = SIZE OF CDF; = RANGE OF THE RANDOM VARIABLE; = 1,2 ...,N
C     M = SIZE OF LUCKY
C     CDF = THE CDF VECTOR.
C     LUCKY = THE POINTER VECTOR, SETUP BY VDSETP
C
C   OUTPUT:
C     IVDISC = F INVERSE OF U
C
C   TAKE CARE OF CASE U = 1.
C
      IF ( U .LT. CDF( N)) GO TO 50
      IVDISC = N
      RETURN
C
   50 NDX = U* M + 1.
      JVDISC = LUCKY( NDX)
C
C   IS U IN A GRID INTERVAL FOR WHICH THERE IS A JVDISC
C   SUCH THAT CDF( JVDISC - 1) < INTERVAL <= CDF( JVDISC) ?
C
      IF ( JVDISC .GT. 0) GO TO 300
C   HARD LUCK, WE MUST SEARCH THE CDF FOR SMALLEST JVDISC
C   SUCH THAT CDF( JVDISC) >= U.
C
      JVDISC = - JVDISC
      GO TO 200
  100 JVDISC = JVDISC + 1
  200 IF ( CDF( JVDISC) .LT. U) GO TO 100
  300 IVDISC = JVDISC
      RETURN
      END
C
      SUBROUTINE VDSETP( N, M, CDF, LUCKY)
      REAL CDF( N)
      INTEGER LUCKY( M)
C
C   SETUP POINTER VECTOR LUCKY FOR INDEXED INVERSION
C   OF DISCRETE CDF BY THE FUNCTION IVDISC.
C
C   INPUTS:
C     N = NO. POINTS IN CDF
C     M = NO. LOCATIONS ALLOCATED FOR POINTERS
C     CDF = THE CDF VECTOR
C
C   OUTPUTS:
C     LUCKY = THE POINTER VECTOR; > 0 IMPLIES THIS POINTER
```

(continued)

```
C       GRID INTERVAL FALLS COMPLETELY WITHIN SOME CDF INTERVAL;
C       < 0 IMPLIES THE GRID INTERVAL IS SPLIT BY A CDF POINT.
C
C DATE 10 SEPT 1986
C
C   TAKE CARE OF DEGENERATE CASE OF N = 1
C
      DO 25 I = 1, M
      LUCKY( I)= 1
   25 CONTINUE
      IF ( N .EQ. 1) RETURN
C
C INITIALIZE FOR GENERAL CASE
C
      NOLD = M
      I = N
      DO 300 II = 2, N
      I = I - 1
C
C FIND THE GRID INTERVAL IN WHICH CDF( I) FALLS
C
      NDX = CDF( I)* M +1
      IF ( NDX .GT. M) GO TO 300
      LUCKY( NDX) = - I
C
C ANY GRID INTERVALS FALLING STRICTLY BETWEEN THIS
C CDF POINT AND THE THE NEXT HIGHER, MAP TO THE NEXT HIGHER
C
      ND1 = NDX +1
      IF ( ND1 .GT. NOLD) GO TO 200
      DO 100 K = ND1, NOLD
      LUCKY( K) = I + 1
  100 CONTINUE
  200 NOLD = NDX - 1
  300 CONTINUE
      RETURN
C
      END
```

Figure L.8. The function IVDISC and its initialization subroutine VDSETP.

```
      FUNCTION IVUNIP( IX, L, P)
      IMPLICIT INTEGER(A-Z)
C
C  SOME COMPILERS REQUIRE THE ABOVE TO BE INTEGER*4
C
C  PORTABLE CODE TO GENERATE AN INTEGER UNIFORM OVER [ 1, L],
C  USING INVERSION.
C
C  INPUTS:
C     IX = A RANDOM INTEGER, 0 =< IX < P
C     L = AN INTEGER, 0 < L < P,
C     P = AN INTEGER, 1 < P < 2**31 , E.G., MODULUS OF RANDOM NUMBER
C            GENERATOR.
C
C  OUTPUTS:
C     IVUNIP = INTEGER PART OF( IX * L / P) + 1,
C              EFFECTIVELY, A RANDOM INTEGER UNIFORM ON [ 1, L].
C     PROCEDURE IS VERY SLOW FOR P << 2**31.
C
C  REFERENCE: PROF. A. PERKO, UNIVERSITY OF JYVASKYLA, FINLAND
C     DATA M/2147483647/, P15/32768/, P16/65536/
C
C  DATE 30 SEPT 1986
C
C  CALCULATE IX* L IN DOUBLE PRECISION, I.E., FIND A, B  SO:
C     IX* L  =  A*(2**31) + B
C
      X = L/ P16
      Y = L - P16* X
      U = IX/ P15
      V = IX - P15* U
C
C  NOW CALCULATE:
C     A*(2**31) + B  =  Y * V  +  V * X * P16  +  U * P15 * R * 2
C     + U * P15 * S  +  U * P15 * X* P16.
C
      YV = Y * V
      YV1 = YV/ P15
      YV2 = YV - P15* YV1
      R = Y/ 2
      S = Y - 2* R
      UR = U * R
      UR1 = UR/ P15
      UR2 = UR - P15* UR1
C
C  EXPLOIT THE POSSIBILITY THAT X = 0:
C
      IF ( X .GT. 0) GO TO 100
      VX1 = 0
      VX2 = 0
      GO TO 200
  100 VX = V* X
      VX1 =   VX/ P15
      VX2 = VX - P15* VX1
      VX1 = VX1 + U * X
  200 AB = YV1 + 2* ( UR2 + VX2) + S* U
```

(continued)

```
            A2 = AB/ P16
            B = P15 *( AB - P16* A2) + YV2
            A = A2 + UR1 + VX1
C
C  NOW THAT WE HAVE A AND B
C    NOTE THAT ( A * 2**31 + B)/ P =
C    ( A * ( P + IE) + B)/ P = A + ( A * IE + B ) / P, SO
C    COMPUTE K = INTEGER PART OF ( ( A * IE + B )/ P)
C    NOTE THAT K = SMALLEST INTEGER SATISFYING:
C    A * IE + B - ( K + 1) * P < 0
C
            IE = M - P + 1
            MOE = M / IE
            INA = A
            K = 0
            KUM = B - P
      300 IF ( KUM .LT. 0) GO TO  400
            KUM = KUM - P
            K = K + 1
            GO TO  300
C
C  ADD IN SOME MORE IE'S UNTIL A OF THEM HAVE BEEN ADDED
C
      400 IF ( INA .EQ. 0) GO TO  500
            IEFIT = MIN( MOE, INA)
            INA = INA - IEFIT
            KUM = KUM + IE * IEFIT
            GO TO  300
C
      500 IVUNIP = A + K + 1
            RETURN
            END
```

Figure L.9. The function IVUNIP.

```
      FUNCTION LUKBIN( PHI, N, R, T, M, NK)
      REAL R( N)
      INTEGER NK( M)
C
C  THE BUCKET-BINARY SEARCH METHOD OF FOX
C  REF. OPERATIONS RESEARCH, VOL 25, 1978
C   FOR GENERATING(QUICKLY) DISCRETE RANDOM VARIABLES
C
C  INPUTS:
C    PHI = A PROBABILITY, 0 < PHI < 1
C    N =  NO. POSSIBLE OUTCOMES FOR THE R. V.
C    R = A VECTOR SET BY SETLKB
C    T = A SCALER THRESHOLD SET BY SETLKB
C    M = NUMBER OF CELLS ALLOWED FOR THE POINTER VECTOR NK( ).
C    NK = A POINTER VECTOR SET BY SETLKB
C
C  OUTPUTS:
C    LUKBIN = RANDOM VARIABLE VALUE
C
      IF ( PHI .GE. T) GO TO 50
C
C  HERE IS THE QUICK BUCKET METHOD
C
      I = M * PHI
      LUKBIN = NK( I + 1)
      RETURN
C
C  WE MUST RESORT TO BINARY SEARCH
C
   50 IF (PHI .LE. R( 1)) GO TO 500
      MAXI = N
      MINI = 2
  100 K = ( MAXI + MINI)/ 2
      IF ( PHI .GT. R(K)) GO TO 300
      IF ( PHI .GT. R( K - 1)) GO TO 500
      MAXI = K - 1
      GO TO 100
  300 MINI = K + 1
      GO TO 100
  500 LUKBIN = K
      RETURN
      END
C
      SUBROUTINE SETLKB( M, N, CDF, T, NK, R)
      REAL CDF( N), R( N)
      INTEGER NK( M)
C
C  SETUP FOR THE LUKBIN METHOD.  SEE FUNCTION LUKBIN
C
C  INPUTS:
C    M = NUMBER OF CELLS ALLOWED FOR THE POINTER VECTOR NK( ).
C    M > 0.  LARGER M LEADS TO GREATER SPEED.
C    N =  NO. POSSIBLE OUTCOMES FOR THE R. V.
C    CDF( ) = CDF OF THE R. V.
C
C  OUTPUTS:  T = A THRESHOLD TO BE USED BY LUKBIN
```

(continued)

```
C     NK( ) = A POINTER VECTOR WHICH SAVES TIME
C     R() = A THRESHOLD PROBABILITY VECTOR.
C
      EXCESS = 0.
      P0 = 0.
      NLIM = 0
      NUSED = 0
      DO 400 I = 1, N
      PI = CDF( I) - P0
      LOT = M* PI
      P0 = CDF( I)
      IF ( LOT .LE. 0) GO TO 200
      NUSED = NLIM + 1
      NLIM = NLIM + LOT
      DO 100 J = NUSED, NLIM
      NK( J) = I
 100  CONTINUE
 200  EXCESS = EXCESS + PI - FLOAT(LOT)/FLOAT( M)
      R( I) = EXCESS
 400  CONTINUE
      T  = FLOAT( NLIM)/ FLOAT( M)
      DO 500 I = 1, N
      R( I) = R( I) + T
 500  CONTINUE
      RETURN
      END
```

Figure L.10. The function LUKBIN and its initialization subroutine SETLKB.

```
      FUNCTION REMPIR( U, N, K, X, THETA, IFAULT)
      DIMENSION X(N)
C
C  INVERSION OF AN EMPIRICAL DISTRIBUTION BUT WITH AN
C  EXPONENTIAL APPROXIMATION TO THE K LARGEST POINTS.
C
C  INPUTS:
C    U = PROBABILITY BEING INVERTED
C    N = NO. OF OBSERVATIONS
C    K = NO. OBS. TO BE USED TO GET EXPONENTIAL RIGHT TAIL
C    X = ORDERED VECTOR OF OBSERVATIONS
C    THETA = MEAN OF THE EXPONENTIAL TAIL,  < 0 ON FIRST CALL
C
C  OUTPUTS:
C    REMPIR = EMPIRICAL F INVERSE OF U
C    THETA = MEAN OF THE EXPONENTIAL TAIL
C    IFAULT = 0 IF NO ERROR IN INPUT DATA, 3 OTHERWISE
C
      IF ( THETA .GT. 0.) GO TO 100
C
C  FIRST CALL; CHECK FOR ERRORS...
C
      IF (( N .LE. 0) .OR. ( K .GT. N) .OR. ( K .LT. 0)) GO TO 9000
      X0 = 0.
C
C  E.G.,  IF ORDERED
C
      DO 20 I = 1, N
      IF ( X(I) .LT. X0) GO TO 9000
      X0 = X(I)
   20 CONTINUE
C
C  INPUT IS OK
C
      IFAULT = 0
      IF ( K .GT. 0) GO TO 45
C
C  IT IS A PURE EMPIRICAL DISTRIBUTION, FLAG THETA
C
      THETA = .000001
      GO TO 100
C
C  CALCULATE MEAN OF EXPONENTIAL TAIL
C  TAKE CARE OF SPECIAL CASE OF PURE EXPONENTIAL
C
   45 NK = N - K + 1
C
C  IN PURE EXPONENTIAL CASE, THETA = 0.
C
      THETA = 0.
      IF ( K .LT. N) THETA = - X( N- K)*( K - .5)
   48 DO 50 I = NK, N
      THETA = THETA + X(I)
   50 CONTINUE
      THETA = THETA/ K
C
```

(continued)

```
C  HERE IS THE CODE FOR SUBSEQUENT CALLS
C
  100 IF ( U .LE. 1. - FLOAT(K)/FLOAT(N)) GO TO 200
C
C  PULL IT BY(FROM) THE TAIL
C
      XNK = 0.
      IF ( K .LT. N) XNK = X(N - K)
      REMPIR = XNK - THETA*ALOG(N*(1.- U)/ K)
      RETURN
C
C  PULL IT FROM THE EMPIRICAL PART
C
  200 V = N* U
      I = V
      X0 = 0.
      IF ( I .GT. 0) X0 = X(I)
      REMPIR = X0 + ( V - I)*( X(I+1) - X0)
      RETURN
C
 9000 IFAULT = 3
      RETURN
      END
```

Figure L.11. The function REMPIR.

```
      FUNCTION RGKM3( ALP, WORK, K, IX )
      REAL WORK(6)
C
C  GENERATE A GAMMA VARIATE WITH PARAMETER ALP
C
C  INPUTS:
C    ALP = DISTRIBUTION PARAMETER, ALP .GE. 1.
C    WORK() = VECTOR OF WORK CELLS; ON FIRST CALL WORK(1) = -1.
C      WORK() MUST BE PRESERVED BY CALLER BETWEEN CALLS.
C    K = WORK CELL. K MUST BE PRESERVED BY CALLER BETWEEN CALLS.
C    IX = RANDOM NUMBER SEED
C
C  OUTPUTS:
C    RGKM3 = A GAMMA VARIATE
C
C  REFERENCES: CHENG, R.C.  AND G.M. FEAST(1979), SOME SIMPLE GAMMA
C    VARIATE GENERATORS, APPLIED STAT. VOL 28, PP. 290-295.
C      TADIKAMALLA, P.R. AND M.E. JOHNSON(1981), A COMPLETE GUIDE TO
C    GAMMA VARIATE GENERATION, AMER. J. OF MATH. AND MAN. SCI.,
C    VOL. 1, PP. 213-236.
C
      IF ( WORK(1) .EQ. ALP) GO TO 100
C
C  INITIALIZATION IF THIS IS FIRST CALL
C
      WORK(1) = ALP
      K = 1
      IF ( ALP .GT. 2.5) K = 2
      WORK(2) = ALP - 1.
      WORK(3) = ( ALP - 1./( 6.* ALP )) / WORK(2)
      WORK(4) = 2. / WORK(2)
      WORK(5) = WORK(4) + 2.
      GO TO ( 1, 11) , K
C
C  CODE FOR SUBSEQUENT CALLS STARTS HERE
C
    1 U1 = UNIF( IX)
      U2 = UNIF( IX)
      GO TO 20
   11 WORK(6) = SQRT( ALP )
   15 U1 = UNIF( IX)
      U = UNIF( IX)
      U2 = U1 + ( 1. - 1.86* U ) / WORK(6)
      IF (( U2 .LE. 0.) .OR. ( U2 .GE. 1. ) ) GO TO 15
   20 W = WORK(3) * U1 / U2
      IF (( WORK(4)* U2 - WORK(5)+ W + 1. /  W ) .LE. 0.) GO TO 200
      IF (( WORK(4)* ALOG( U2)- ALOG( W)+ W - 1.) .LT. 0.) GO TO 200
  100 GO TO ( 1, 15) , K
  200 RGKM3 = WORK(2) * W
      RETURN
      END
```

Figure L.12. The function RGKM3.

```
      FUNCTION RGS( ALP, IX)
C
C GENERATE A GAMMA VARIATE WITH PARAMETER ALP
C
C INPUTS:
C   ALP = DISTRIBUTION PARAMETER; 0.0 < ALP <= 1.0.
C   IX = RANDOM NUMBER SEED
C
C OUTPUTS:
C   RGS = A GAMMA VARIATE
C
C REFERENCES: AHRENS, J.H.   AND U. DIETER(1972), COMPUTER METHODS FOR
C SAMPLING FROM GAMMA, BETA, POISSON AND BINOMIAL DISTRIBUTIONS,
C COMPUTING, VOL. 12, PP. 223-246
C TADIKAMALLA, P.R. AND M.E. JOHNSON(1981), A COMPLETE GUIDE TO
C GAMMA VARIATE GENERATION, AMER. J. OF MATH. AND MAN. SCI.,
C VOL. 1, PP. 213-236.
C
C DATE 10 SEPT 1986
C
    1 U1 = UNIF( IX)
      B = ( 2.718281828 + ALP) / 2.718281828
      P = B * U1
      U2 = UNIF( IX)
      IF ( P .GT. 1.) GO TO 3
    2 X = EXP( ALOG( P) / ALP)
      IF ( U2 .GT. EXP( - X)) GO TO 1
      RGS = X
      RETURN
    3 X = - ALOG(( B - P) / ALP)
      IF ( ALOG( U2) .GT. ( ALP - 1.)* ALOG( X)) GO TO 1
      RGS = X
      RETURN
      END
```

Figure L.13. The function RGS.

```
      FUNCTION RSTAB( ALPHA, BPRIME, U, W)
C  GENERATE A STABLE PARETIAN VARIATE
C
C  INPUTS:
C     ALPHA = CHARACTERISTIC EXPONENT, 0<= ALPHA <= 2.
C     BPRIME = SKEWNESS IN REVISED PARAMETERIZATION, SEE REF.
C            = 0 GIVES A SYMMETRIC DISTRIBUTION.
C     U = UNIFORM VARIATE ON (0, 1.)
C     W = EXPONENTAIL DISTRIBUTED VARIATE WITH MEAN 1.
C
C  REFERENCE: CHAMBERS, J. M., C. L. MALLOWS AND B. W. STUCK(1976),
C     "A METHOD FOR SIMULATING STABLE RANDOM VARIABLES", JASA, VOL. 71,
C     NO. 354, PP. 340-344
C
C  AUXILIARY ROUTINES:
C     D2, TAN, TAN2
C
      DOUBLE PRECISION DA, DB
C
C  ON COMPILERS WITHOUT DOUBLE PRECISION THE ABOVE MUST BE
C  REPLACED BY REAL DA, DB
C
      DATA PIBY2/1.57079633E0/
      DATA THR1/0.99/
      EPS = 1. - ALPHA
C
C  COMPUTE SOME TANGENTS
C
      PHIBY2 = PIBY2*( U - .5)
      A = PHIBY2* TAN2( PHIBY2)
      BB = TAN2( EPS* PHIBY2)
      B = EPS* PHIBY2* BB
      IF ( EPS .GT. -.99) TAU = BPRIME/( TAN2( EPS* PIBY2)* PIBY2)
      IF ( EPS .LE. -.99) TAU = BPRIME*PIBY2* EPS*(1. - EPS) *
     1 TAN2(( 1. - EPS)* PIBY2)
C
C  COMPUTE SOME NECESSARY SUBEXPRESSIONS
C  IF PHI NEAR PI BY 2, USE DOUBLE PRECISION.
C
      IF ( A .GT. THR1) GO TO 50
C
C  SINGLE PRECISION
C
      A2 = A**2
      A2P = 1. + A2
      A2 = 1. - A2
      B2 = B**2
      B2P = 1. + B2
      B2 = 1. - B2
      GO TO 100
C
C  DOUBLE PRECISION
C
   50 DA = A
      DA = DA**2
      DB = B
```

(continued)

```
          DB = DB**2
          A2 = 1.D0 - DA
          A2P = 1.D0 + DA
          B2 = 1.D0 - DB
          B2P = 1.D0 + DB
C
C   COMPUTE COEFFICIENT
C
  100 Z = A2P*( B2 + 2.* PHIBY2* BB * TAU)/( W* A2 * B2P)
C
C   COMPUTE THE EXPONENTIAL-TYPE EXPRESSION
C
          ALOGZ = ALOG( Z)
          D = D2( EPS* ALOGZ/(1. - EPS))*( ALOGZ/(1.- EPS))
C
C   COMPUTE STABLE
C
          RSTAB = ( 1. + EPS* D) * 2.*(( A - B)*(1. + A* B) - PHIBY2*
        1 TAU* BB*( B* A2 - 2. * A))/( A2 * B2P) + TAU* D
          RETURN
          END
```

Figure L.14. The function RSTAB.

```
          FUNCTION SUMNRM( IX)
C
C   SUM 12 UNIFORMS AND SUBTRACT 6 TO APPROXIMATE A NORMAL
C   ***NOT RECOMMENDED!***
C
C   INPUT:
C     IX = A RANDOM NUMBER SEED
C   AUXILIARY ROUTINE:
C     UNIF
C
          SUM = - 6.
          DO 100 I = 1, 12
          SUM = SUM + UNIF( IX)
  100 CONTINUE
          SUMNRM = SUM
          RETURN
          END
```

Figure L.15. The function SUMNRM.

```
      FUNCTION TAN( XARG)
C
C  EVALUATE THE TANGENT OF XARG
C  THE APPROXIMATION 4283 FROM HART ET. AL.( 1968, P. 251)
C     COMPUTER APPROXIMATIONS, NEW YORK, JOHN WILEY.
C
      LOGICAL NEG, INV
      DATA P0, P1, P2, Q0, Q1, Q2
     1 /.129221035E+3,-.887662377E+1,.528644456E-1,
     2 .164529332E+3,-.451320561E+2,1./
      DATA PIBY4/.785398163E0/,PIBY2/1.57079633E0/
      DATA PI /3.14159265E0/
C
C  DATE 13 SEPT 1986
C
      NEG = .FALSE.
      INV = .FALSE.
      X = XARG
      NEG = X .LT. 0.
      X = ABS( X)
C
C  PERFORM RANGE REDUCTION IF NECESSARY
C
      IF ( X .LE. PIBY4) GO TO 50
      X = AMOD( X, PI)
      IF ( X .LE. PIBY2) GO TO 30
      NEG = .NOT. NEG
      X = PI - X
   30 IF ( X .LE. PIBY4) GO TO 50
      INV = .TRUE.
      X = PIBY2 - X
   50 X = X/ PIBY4
C
C  CONVERT TO RANGE OF RATIONAL
C
      XX = X * X
      TANT = X*( P0+ XX*( P1+ XX* P2))/( Q0 + XX*( Q1+ XX* Q2))
      IF ( NEG) TANT = - TANT
      IF ( INV) TANT = 1./ TANT
      TAN = TANT
      RETURN
      END
```

Figure L.16. The function TAN.

```
        FUNCTION TAN2( XARG)
C
C   COMPUTE TAN( XARG)/ XARG
C   DEFINED ONLY FOR ABS( XARG) .LE. PI BY 4
C   FOR OTHER ARGUMENTS RETURNS TAN( X)/X COMPUTED DIRECTLY.
C   THE APPROXIMATION 4283 FROM HART ET. AL(1968, P. 251)
C      COMPUTER APPROXIMATIONS, NEW YORK, JOHN WILEY.
C
        DATA P0, P1, P2, Q0, Q1, Q2
       1 /.129221035E+3,-.887662377E+1,.528644456E-1,
       2  .164529332E+3,-.451320561E+2,1.0/
        DATA PIBY4/.785398163E0/
C
C   DATE 13 SEPT 1986
C
        X = ABS( XARG)
        IF ( X .GT. PIBY4) GO TO 200
        X = X/ PIBY4
C
C   CONVERT TO RANGE OF RATIONAL
C
        XX = X* X
        TAN2 =( P0+ XX*( P1 + XX* P2))/( PIBY4*( Q0+ XX*( Q1+ XX* Q2)))
        RETURN
   200 TAN2 = TAN( XARG)/ XARG
        RETURN
        END
```

Figure L.17. The function TAN2.

```
      FUNCTION TRPNRM( IX)
C
C  GENERATE UNIT NORMAL DEVIATE BY COMPOSITION METHOD
C  OF AHRENS AND DIETER
C  THE   AREA UNDER THE NORMAL CURVE IS DIVIDED INTO 5
C    DIFFERENT AREAS.
C
C  INPUT:
C    IX = A RANDOM NUMBER SEED
C
C  AUXILIARY ROUTINE:
C    UNIF
C
C  DATE 30 SEPT 1986
C
      U = UNIF( IX)
      UO = UNIF( IX)
 120  IF ( U .GE. .919544) GO TO 160
C
C  AREA A,  THE TRAPEZOID IN THE MIDDLE
C
      TRPNRM = 2.40376*( UO+ U*.825339)-2.11403
      RETURN
 160  IF ( U .LT. .965487) GO TO 210
C
C  AREA B
C
 180  TRPTMP = SQRT(4.46911-2* ALOG( UNIF( IX)))
 190  IF ( TRPTMP* UNIF( IX) .GT. 2.11403) GO TO 180
 200  GO TO 340
 210  IF ( U .LT. .949991) GO TO 260
C
C  AREA C
C
 230  TRPTMP = 1.8404+ UNIF( IX)*.273629
 240  IF (.398942* EXP(- TRPTMP* TRPTMP/2)-.443299 +
     1 TRPTMP*.209694 .LT. UNIF( IX)*  4.27026E-02) GO TO 230
 250  GO TO 340
 260  IF ( U .LT. .925852) GO TO 310
C
C  AREA D
C
 280  TRPTMP = .28973+ UNIF( IX)*1.55067
 290  IF (.398942* EXP(- TRPTMP* TRPTMP/2)-.443299 +
     1 TRPTMP*.209694 .LT. UNIF( IX)* 1.59745E-02) GO TO 280
 300  GO TO 340
 310  TRPTMP = UNIF( IX)*.28973
C
C  AREA E
C
 330  IF (.398942* EXP( - TRPTMP* TRPTMP/2)-.382545
     1 .LT. UNIF( IX)*1.63977E-02)   GO TO 310
 340  IF ( UO .GT. .5) GO TO 370
 360  TRPTMP = - TRPTMP
 370  TRPNRM = TRPTMP
      RETURN
      END
```

Figure L.18. The function TRPNRM.

```
      FUNCTION UNIF( IX)
C  PORTABLE RANDOM NUMBER GENERATOR USING THE RECURSION:
C     IX = 16807 * IX MOD (2**(31) - 1)
C  USING ONLY 32 BITS, INCLUDING SIGN.
C  SOME COMPILERS REQUIRE THE DECLARATION:
C       INTEGER*4 IX, K1
C
C  INPUTS:
C    IX = INTEGER GREATER THAN 0 AND LESS THAN 2147483647
C
C  OUTPUTS:
C    IX = NEW PSEUDORANDOM VALUE,
C    UNIF = A UNIFOM FRACTION BETWEEN 0 AND 1.
C
      K1 = IX/127773
      IX = 16807*( IX - K1*127773) - K1 * 2836
      IF ( IX .LT. 0) IX = IX + 2147483647
      UNIF = IX*4.656612875E-10
      RETURN
      END
```

Figure L.19. The function UNIF.

```
      FUNCTION UNIFL( JX)
C        INTEGER*4 JX
      DIMENSION JX(3)
C
C INPUTS:
C   JX(2), JX(3) = 2 RANDOM INTEGERS WITH:
C                     0 < JX(2) < 2147483563
C                     0 < JX(3) < 2147483399
C
C OUTPUTS:
C   JX(1) = PSEUDO RANDOM INTEGER, 0 < JX(1) < 2147483563
C           JX(1) SHOULD BE A HIGH QUALITY RANDOM INTEGER.
C           PERIOD OF JX(1) IS ABOUT 2.30584 E 18.
C   JX(2) = PSEUDO RANDOM INTEGER, 0 < JX(2) < 2147483563
C   JX(3) = PSEUDO RANDOM INTEGER, 0 < JX(3) < 2147483399
C   UNIFL = PSEUDO RANDOM REAL, 0. < UNIFL < 1.
C
C REF.: L'ECUYER, P.(1986),"EFFICIENT AND PORTABLE COMBINED PSEUDO
C   RANDOM NUMBER GENERATORS", UNIV. LAVAL, P. Q., CANADA.
C
C FOR ANY GIVEN SEED FOR STREAM 1, I.E., JX(2), THE FOLLOWING
C   SEEDS FOR STREAM 2, I.E., JX(3), WILL GIVE NONOVERLAPPING
C   SEQUENCES OF LENGTH 2147483562 FOR THE OUTPUT STREAM, JX(1):
C   1, 1408222472,  905932980,   1185805228,   1749585844,  563802022
C   1147463446,  1209938334,  155409276,  617052445,
C   ETC., I.E., EVERY 164TH TERM FROM STREAM JX(3).
C
C        INTEGER*4 K
C SOME COMPILERS MAY REQUIRE THE INTEGER*4 DECLARATIONS TURNED ON
C
C DATE 22 SEPT 1986
C
C GET NEXT TERM IN THE FIRST STREAM = 40014* JX(2) MOD 2147483563
C
      K = JX( 2)/ 53668
      JX( 2) = 40014 * ( JX( 2) - K * 53668) - K * 12211
      IF ( JX( 2) .LT. 0) JX( 2) = JX( 2) + 2147483563
C
C GET NEXT TERM IN THE SECOND STREAM = 40692* JX(3) MOD 2147483399
      K = JX( 3)/ 52774
      JX( 3) = 40692 * ( JX( 3) - K * 52774) - K * 3791
      IF ( JX( 3) .LT. 0) JX( 3) = JX( 3) + 2147483399
C
C SET JX(1) = (( JX(3) + 2147483562 - JX(2)) MOD 2147483562) + 1
C
      K = JX( 3) - JX( 2)
      IF ( K .LE. 0) K = K + 2147483562
C
C PUT THE COMBINATION BACK INTO JX( 1)
      JX( 1) = K
C
C PUT IT ON THE INTERVAL (0,1)
      UNIFL = K * 4.656613 E -10
      RETURN
      END
```

Figure L.20. The function UNIFL.

```
      FUNCTION VBETA( PHI, A, B)
C
C  INVERSE OF THE BETA CDF
C
C  INPUTS:
C    PHI = PROBABILITY TO BE INVERTED, 0 < PHI < 1.
C    A, B = THE 2 SHAPE PARAMETERS OF THE BETA
C
C  OUTPUTS:
C    VBETA = INVERSE OF THE BETA CDF
C
C  AUXILIARY ROUTINE:
C    VNORM
C
C  REF.: ABRAMOWITZ, M. AND STEGUN, I., HANDBOOK OF MATHEMATICAL
C    FUNCTIONS WITH FORMULAS, GRAPHS, AND MATHEMATICAL TABLES,
C    NAT. BUR. STANDARDS, GOVT. PRINTING OFF.
C
C  ACCURACY: ABOUT 1 DECIMAL DIGIT
C    EXCEPT IF A OR B =1., IN WHICH CASE ACCURACY= MACHINE PRECISION.
C
      DATA DWARF /.1E-9/
C
C DATE 30 SEPT 1986
C
      IF ( PHI .GT. DWARF) GO TO 100
      VBETA = 0.
      RETURN
  100 IF ( PHI .LT. (1. - DWARF)) GO TO 150
      VBETA = 1.
      RETURN
  150 IF ( B .NE. 1.) GO TO 200
      VBETA = PHI**(1./A)
      RETURN
  200 IF ( A .NE. 1.) GO TO 300
      VBETA = 1. - (1. - PHI)**(1./B)
      RETURN
  300 YP = - VNORM( PHI, IFAULT)
      GL = ( YP* YP - 3.)/ 6.
      AD = 1./( 2.* A - 1.)
      BD = 1./( 2.* B - 1.)
      H = 2./( AD + BD)
      W =( YP* SQRT( H+ GL)/ H)-
     1( BD- AD)*( GL+.83333333-.66666666/H)
      VBETA = A/( A + B* EXP(2.* W))
      RETURN
      END
```

Figure L.21. The function VBETA.

```
      FUNCTION VCAUCH( PHI)
C
C   INVERSE OF STANDARD CAUCHY DISTRIBUTION CDF.
C
C   MACHINE EPSILON
      DATA DWARF/ .1E-30/
C
      ARG = 3.1415926*( 1. - PHI)
      SINARG = AMAX1( DWARF, SIN( ARG))
      VCAUCH = COS( ARG)/ SINARG
      RETURN
      END
```

Figure L.22. The function VCAUCH.

```
      FUNCTION VCHI2( PHI, V, G, IFAULT)
C
C   EVALUATE THE PERCENTAGE POINTS OF THE CHI-SQUARED
C   PROBABILITY DISTRIBUTION FUNCTION.
C   SLOW BUT ACCURATE INVERSION OF CHI-SQUARED CDF.
C
C   BASED ON ALGORITHM AS 91 APPL. STATIST. (1975) VOL.24, NO.3
C
C   INPUTS:
C     PHI = PROBABILITY TO BE INVERTED.
C     PHI SHOULD LIE IN THE RANGE 0.000002 TO 0.999998,
C     V = DEGREES OF FREEDOM, AND MUST BE POSITIVE,
C     G MUST BE SUPPLIED AND SHOULD BE EQUAL TO LN(GAMMA( V/2.0))
C     E.G. USING FUNCTION ALOGAM.
C
C   OUTPUT:
C     VCHI2 = INVERSE OF  CHI-SQUARE CDF OF PHI
C     IFAULT = 4 IF INPUT ERROR, ELSE 0
C
C   AUXILIARY ROUTINES:
C     VNORM, GAMAIN
C
C   DESIRED ACCURACY AND LN( 2.):
C
      DATA E, AA/0.5E-5, 0.6931471805E0/
C
C   DATE 10 SEPT 1986
C
C   CHECK FOR UNUSUAL INPUT
C
      IF( V .LE. 0.) GO TO 9000
      IF ( PHI .LE. .000002 .OR. PHI .GE. .999998 ) GO TO 9000
      IFAULT = 0
      XX = 0.5 * V
      C = XX - 1.0
```

(continued)

```
C
C             STARTING APPROXIMATION FOR SMALL CHI-SQUARED
C
      IF ( V .GE. -1.24 * ALOG( PHI)) GOTO 1
      CH = ( PHI * XX * EXP( G + XX * AA)) ** (1.0 / XX)
      IF (CH - E) 6, 4, 4
C
C             STARTING APPROXIMATION FOR V LESS THAN OR EQUAL TO 0.32
C
    1 IF ( V .GT. 0.32) GOTO 3
      CH = 0.4
      A = ALOG(1.0 - PHI)
    2 Q = CH
      P1 = 1.0 + CH * (4.67 + CH)
      P2 = CH * (6.73 + CH * (6.66 + CH))
      T = -0.5 + (4.67 + 2.0 * CH) / P1 -
     *    (6.73 + CH * (13.32 + 3.0 * CH)) / P2
      CH = CH - (1.0 - EXP(A + G + 0.5 * CH + C * AA) * P2 / P1) / T
      IF (ABS(Q / CH - 1.0) - 0.01) 4, 4, 2
C
C  GET THE CORRESPONDING NORMAL DEVIATE:
C
    3 X = VNORM( PHI, IFAULT)
C
C             STARTING APPROXIMATION USING WILSON AND HILFERTY ESTIMATE
C
      P1 = 0.222222 / V
      CH = V * (X * SQRT(P1) + 1.0 - P1) ** 3
C
C             STARTING APPROXIMATION FOR P TENDING TO 1
C
      IF (CH .GT. 2.2 * V + 6.0)
     *    CH = -2.0 * (ALOG(1.0 - PHI) - C * ALOG(0.5 * CH) + G)
C
C             CALL TO ALGORITHM AS 32 AND CALCULATION OF SEVEN TERM
C             TAYLOR SERIES
C
    4 Q = CH
      P1 = 0.5 * CH
      P2 = PHI - GAMAIN(P1,XX,G,IF1)
      IF (IF1 .EQ. 0) GOTO 5
      IFAULT = 4
      RETURN
    5 T = P2 * EXP(XX * AA + G + P1 - C * ALOG(CH))
      B = T / CH
      A = 0.5 * T - B * C
      S1 = (210.0+A*(140.0+A*(105.0+A*(84.0+A*(70.0+60.0*A))))) / 420.0
      S2 = (420.0+A*(735.0+A*(966.0+A*(1141.0+1278.0*A)))) / 2520.0
      S3 = (210.0 + A * (462.0 + A * (707.0 + 932.0 * A))) / 2520.0
      S4 =(252.0+A*(672.0+1182.0*A)+C*(294.0+A*(889.0+1740.0*A)))/5040.0
      S5 = (84.0 + 264.0 * A + C * (175.0 + 606.0 * A)) / 2520.0
      S6 = (120.0 + C * (346.0 + 127.0 * C)) / 5040.0
      CH = CH+T*(1.0+0.5*T*S1-B*C*(S1-B*(S2-B*(S3-B*(S4-B*(S5-B*S6))))))
      IF (ABS(Q / CH - 1.0) .GT. E) GOTO 4
C
    6 VCHI2 = CH
      RETURN
 9000 IFAULT = 4
C
C  TRY TO GIVE SENSIBLE RESPONSE
C
      VCHI2 = 0.
      IF ( PHI .GT. .999998) VCHI2 = V + 4.*SQRT(2.*V)
      RETURN
      END
```

Figure L.23. The function VCHI2.

```fortran
      FUNCTION VCHISQ( PHI, DF)
C
C INVERSE OF THE CHI-SQUARE C.D.F.
C TO GET INVERSE CDF OF K-ERLANG SIMPLY USE:
C .5* VCHISQ( PHI, K/2.);
C WARNING: THE METHOD IS ONLY ACCURATE TO ABOUT 2 PLACES FOR
C DF < 5 BUT NOT EQUAL TO 1. OR 2.
C
C INPUTS:
C   PHI = PROBABILITY
C   DF= DEGREES OF FREEDOM, NOTE, THIS IS REAL
C
C OUTPUTS:
C   VCHISQ = F INVERSE OF PHI.
C
C REF.:  ABRAMOWITZ AND STEGUN
C
C   SQRT(.5)
C
      DATA SQP5 /.7071067811E0/
      DATA DWARF /.1E-15/
      IF ( DF .NE. 1.) GO TO 100
C
C WHEN DF=1 WE CAN COMPUTE IT QUITE ACCURATELY
C BASED ON FACT IT IS A SQUARED STANDARD NORMAL
C
      Z = VNORM( 1. -(1. - PHI)/2., IFAULT)
      VCHISQ = Z* Z
      RETURN
  100 IF ( DF .NE. 2.) GO TO 200
C
C WHEN DF = 2 IT CORRESPONDS TO EXPONENTIAL WITH MEAN 2.
C
      ARG = 1. - PHI
      IF ( ARG .LT. DWARF) ARG = DWARF
      VCHISQ = - ALOG( ARG)* 2.
      RETURN
  200 IF ( PHI .GT. DWARF) GO TO 300
      VCHISQ = 0.
      RETURN
  300 Z = VNORM( PHI, IFAULT)
      SQDF = SQRT(DF)
      ZSQ = Z* Z
      CH =-(((3753.*ZSQ+4353.)*ZSQ-289517.)*ZSQ-289717.)*
     1 Z*SQP5/9185400.
      CH=CH/SQDF+(((12.*ZSQ-243.)*ZSQ-923.)*ZSQ+1472.)/25515.
      CH=CH/SQDF+((9.*ZSQ+256.)*ZSQ-433.)*Z*SQP5/4860.
      CH=CH/SQDF-((6.*ZSQ+14.)*ZSQ-32.)/405.
      CH=CH/SQDF+(ZSQ-7.)*Z*SQP5/9.
      CH=CH/SQDF+2.*(ZSQ-1.)/3.
      CH=CH/SQDF+Z/SQP5
      VCHISQ = DF*( CH/SQDF + 1.)
      RETURN
      END
```

Figure L.24. The function VCHISQ.

```
      FUNCTION VF( PHI, DFN, DFD, W)
      DIMENSION W(4)
C
C  INVERSE OF THE F CDF
C  REF.: ABRAMOWITZ AND STEGUN
C
C  INPUTS:
C    PHI = PROBABILITY TO BE INVERTED.  0. < PHI < 1.
C    DFN= DEGREES OF FREEDOM OF THE NUMERATOR
C    DFD = DEGREES OF FREEDOM OF THE DENOMINATOR
C       OF THE F DISTRIBUTION.
C    W = WORK VECTOR OF LENGTH 4. W(1) MUST BE SET TO -1 ON ALL
C        INITIAL CALLS TO VF WHENEVER DFN = 1 OR DFD = 1. AS LONG AS
C        DFN AND DFD DO NOT CHANGE, W SHOULD NOT BE ALTERED. IF,
C        HOWEVER, DFN OR DFD SHOULD CHANGE, W(1) MUST BE RESET TO -1.
C
C  OUTPUTS:
C    VF = INVERSE OF F CDF EVALUATED AT PHI.
C    W  = WORK VECTOR OF LENGTH 4. W RETAINS INFORMATION BETWEEN
C        SUCCESSIVE CALLS TO VF WHENEVER DFN OR DFD EQUAL 1.
C
C  ACCURACY: ABOUT 1 DECIMAL DIGIT, EXCEPT
C  IF DFN OR DFD =1. IN WHICH CASE ACCURACY = ACCURACY OF VSTUD.
C  ELSE ACCURACY IS THAT OF VBETA.
C
      DATA DWARF /.1E-15/
C
C  AUXILIARY ROUTINES:
C    VBETA, VSTUD
C
      IF ( DFN .NE. 1.) GO TO 200
      FV = VSTUD(( 1. + PHI)/2., DFD, W, IFAULT)
      VF = FV* FV
      RETURN
C
  200 IF ( DFD .NE. 1.) GO TO 300
      FV = VSTUD(( 2. - PHI)/2., DFN, W, IFAULT)
      IF ( FV .LT. DWARF) FV = DWARF
      VF = 1./(FV* FV)
      RETURN
C
  300 FV = 1. - VBETA( PHI, DFN/2., DFD/2.)
      IF ( FV .LT. DWARF) FV = DWARF
      VF = (DFD/DFN)*( 1. - FV)/( FV)
      RETURN
      END
```

Figure L.25. The function VF.

```
        FUNCTION VNHOMO( U, N, RATE, T, R, K, IFAULT)
        REAL RATE( N), T( N)
C
C   INVERSE OF THE DYNAMIC EXPONENTIAL,
C   A.K.A. NONHOMOGENOUS POISSON
C
C   INPUTS:
C     U = PROBABILITY TO BE INVERTED
C     N = NO. INTERVALS IN A CYCLE
C     RATE(K) = ARRIVAL RATE IN INTERVAL K OF CYCLE
C     T(K) = END POINT OF INTERVAL K, SO T(N) = CYCLE LENGTH
C     R = CURRENT TIME MOD CYCLE LENGTH, E.G. 0 AT TIME 0
C     K = CURRENT INTERVAL OF CYCLE, E.G. 1 AT TIME 0
C     NOTE: R AND K WILL BE UPDATED FOR THE NEXT CALL
C
C   OUTPUTS:
C     VNHOMO = INTERARRIVAL TIME
C     R = TIME OF NEXT ARRIVAL MOD CYCLE LENGTH
C     K = INTERVAL IN WHICH NEXT ARRIVAL WILL OCCUR
C     IFAULT = 5 IF ERROR IN INPUT, ELSE 0
C
        IFAULT = 0
        X = 0.
C
C   GENERATE INTERARRIVAL TIME WITH MEAN 1
C
        E = - ALOG(1. - U)
C
C   SCALED TO CURRENT RATE DOES IT FALL IN CURRENT INTERVAL?
C
  100 STEP = T(K) - R
        IF ( STEP .LT. 0.) GO TO 900
        RNOW = RATE( K)
        IF ( RNOW .GT. 0.) GO TO 110
        IF ( RNOW .EQ. 0.) GO TO 120
        GO TO 900
C
C   INTERARRIVAL TIME IF AT CURRENT RATE
C
  110 U = E/ RNOW
        IF ( U .LE. STEP) GO TO 300
C
C   NO, JUMP TO END OF THIS INTERVAL
C
  120 X = X + STEP
        E = E - STEP* RNOW
        IF ( K .EQ. N) GO TO 200
        R = T(K)
        K = K+ 1
        GO TO 100
  200 R = 0.
        K = 1
        GO TO 100
C
C   YES, WE STAYED IN INTERVAL K
C
  300 VNHOMO = X + U
        R = R + U
        RETURN
  900 IFAULT = 5
        RETURN
        END
```

Figure L.26. The function VNHOMO.

```
      FUNCTION VNORM( PHI, IFAULT)
C
C  VNORM RETURNS THE INVERSE OF THE CDF OF THE NORMAL   DISTRIBUTION.
C    IT USES A RATIONAL APPROXIMATION WHICH SEEMS TO HAVE A RELATIVE
C    ACCURACY OF ABOUT 5 DECIMAL PLACES.
C    REF.: KENNEDY AND GENTLE, STATISTICAL COMPUTING, DEKKER, 1980.
C
C  INPUTS:
C    PHI = PROBABILITY, 0 <= PHI <= 1.
C
C  OUTPUTS:
C    F INVERSE OF PHI, I.E., A VALUE SUCH THAT
C    PROB( X <= VNORM) = PHI.
C    IFAULT = 6 IF PHI OUT OF RANGE, ELSE 0
C
      DATA PLIM /1.0E-18/
      DATA P0/-0.322232431088E0/, P1 / -1.0/, P2 / -0.342242088547E0/
      DATA P3 / -0.0204231210245E0/,    P4/-0.453642210148E-4/
      DATA Q0/0.099348462606E0/, Q1/0.588581570495E0/
      DATA Q2/0.531103462366E0/, Q3/0.10353775285E0/
      DATA Q4/0.38560700634E-2/
C
      IFAULT = 0
      P = PHI
      IF (P.GT.0.5) P = 1. - P
      IF ( P .GE. PLIM) GO TO 100
C
C  THIS IS AS FAR OUT IN THE TAILS AS WE GO
C
      VTEMP =  8.
C  CHECK FOR INPUT ERROR
      IF ( P .LT. 0.) GO TO 9000
      GO TO 200
  100 Y = SQRT(-ALOG(P*P))
      VTEMP = Y + ((((Y*P4 + P3)*Y + P2)*Y + P1)*Y + P0)/
     1        ((((Y*Q4 + Q3)*Y + Q2)*Y + Q1)*Y + Q0)
  200 IF ( PHI .LT. 0.5) VTEMP = - VTEMP
      VNORM = VTEMP
      RETURN
 9000 IFAULT = 6
      RETURN
      END
```

Figure L.27. The function VNORM.

```
      FUNCTION VSTUD( PHI, DF, W, IFAULT)
      DIMENSION W(4)
C
C  GENERATE THE INVERSE CDF OF THE STUDENT T
C
C  INPUTS:
C     PHI = PROBABILITY , 0 <= PHI <= 1
C     DF = DEGREES OF FREEDOM
C     W = WORK VECTOR OF LENGTH 4 WHICH DEPENDS ONLY ON DF.
C        VSTUD WILL RECALCULATE,  E.G., ON FIRST CALL, IF
C        W(1) < 0. ON INPUT.
C
C  OUTPUTS:
C     VSTUD = F INVERSE OF PHI, I.E., VALUE SUCH THAT
C        PROB( X <= VSTUD) = PHI.
C     W = WORK VECTOR OF LENGTH 4. W RETAINS INFORMATION BETWEEN CALLS
C           WHEN DF DOES NOT CHANGE.
C     IFAULT = 7 IF INPUT ERROR, ELSE 0.
C
C  ACCURACY IN DECIMAL DIGITS OF ABOUT:
C     5 IF DF >= 8.,
C     MACHINE PRECISION IF DF = 1. OR 2.,
C     3 OTHERWISE.
C
C  REFERENCE: G. W. HILL, COMMUNICATIONS OF THE ACM, VOL. 13, NO. 10,
C                OCT. 1970.
C
C  MACHINE EPSILON
      DATA DWARF/ .1E-30/
C
C  PI OVER 2
      DATA PIOVR2 /1.5707963268E0/
C
C  DATE 30 SEPT 1986
C
C  CHECK FOR INPUT ERROR
C
      IFAULT = 0
      IF (   PHI .LT. 0. .OR. PHI .GT. 1.) GO TO 9000
      IF ( DF .LE. 0.) GO TO 9000
C
C  IF IT IS A CAUCHY USE SPECIAL METHOD
C
      IF ( DF .EQ. 1.) GO TO 300
C
C  THERE IS ALSO A SPECIAL, EXACT METHOD IF DF = 2
C
      IF ( DF .EQ. 2.) GO TO 400
C
C  CHECK TO SEE IF WE'RE NOT TOO FAR OUT IN THE TAILS.
C
      IF (PHI.GT.DWARF .AND. (1. - PHI).GT.DWARF) GO TO 205
         T = 10.E30
         GO TO 290
C
C  GENERAL CASE
C
C  FIRST, CONVERT THE CUMULATIVE PROBABILITY TO THE TWO-TAILED
C  EQUIVALENT USED BY THIS METHOD.
C
  205 PHIX2 = 2. * PHI
      P2TAIL = AMIN1(PHIX2, 2.0 - PHIX2)
C
C  BEGIN THE APROXIMATION
C
C  IF W(1) >= 0 THEN SKIP THE COMPUTATIONS THAT DEPEND SOLELY ON DF.
C     THE VALUES FOR THESE VARIABLES WERE STORED IN W DURING A PREVIOUS
C     CALL.
```

(continued)

```
C
      IF ( W(1) .GT. 0.) GO TO 250
         W(2) = 1.0/(DF - 0.5)
         W(1) = 48.0/(W(2) * W(2))
         W(3) = ((20700.0*W(2)/W(1)-98.)*W(2)-16.)*W(2) + 96.36
         W(4)=((94.5/(W(1)+W(3))-3.)/W(1) + 1.)*SQRT(W(2)*PIOVR2)*DF
  250 X = W(4) * P2TAIL
      C = W(3)
      Y = X ** (2./DF)
      IF (Y.LE.(.05 + W(2))) GO TO 270
C
C         ASYMPTOTIC INVERSE EXPANSION ABOUT NORMAL
C
          X = VNORM( P2TAIL * 0.5, IFAULT)
          Y = X * X
          IF (DF.LT.5) C = C + 0.3 * (DF - 4.5) * (X + .6)
          C = (((.05*W(4)*X-5.)*X-7.)*X-2.)*X+W(1)+C
          Y = ((((((.4*Y+6.3)*Y+36.)*Y+94.5)/C-Y-3.)/W(1)+1.)*X
          Y = W(2) * Y * Y
          IF (Y.LE.0.002) GO TO 260
              Y = EXP(Y) - 1.
              GO TO 280
  260         Y = .5 * Y * Y + Y
              GO TO 280
C
  270     Y2 = (DF + 6.)/(DF + Y) - 0.089 * W(4) - .822
          Y2 = Y2 * (DF + 2.) * 3.
          Y2 = (1./Y2 + .5/(DF + 4.)) * Y - 1.
          Y2 = Y2 * (DF + 1.)/(DF + 2.)
          Y  = Y2 + 1./Y
C
  280 T = SQRT(DF * Y)
  290 IF (PHI.LT.0.5) T = - T
      VSTUD = T
C
      RETURN
C
C   INVERSE OF STANDARD CAUCHY DISTRIBUTION CDF.
C  RELATIVE ACCURACY = MACHINE PRECISION, E.G. 5 DIGITS
C
  300 ARG =  3.1415926*( 1. - PHI)
      SINARG = AMAX1( DWARF, SIN( ARG))
      VSTUD = COS( ARG)/ SINARG
      RETURN
C
C  SPECIAL CASE OF DF = 2.
C   RELATIVE ACCURACY = MACHINE PRECISION, E.G., 5 DIGITS
C
  400 IF  ( PHI .GT. .5) GO TO 440
      T = 2.* PHI
      GO TO 450
  440 T = 2.*( 1. - PHI)
  450 IF ( T .LE. DWARF) T = DWARF
      VTEMP = SQRT((2./(T*( 2. - T))) - 2.)
      IF ( PHI .LE. .5) VTEMP = - VTEMP
      VSTUD = VTEMP
      RETURN
 9000 IFAULT = 7
      RETURN
      END
```

Figure L.28. The function VSTUD.

```
      SUBROUTINE BINSRH( U, K, N, CDF)
      REAL CDF(N)
C
C  FIND SMALLEST K SUCH THAT CDF( K) >= U BY BINARY SEARCH
C
C  INPUTS:
C    U = PROBABILITY
C    CDF = A CDF FOR A VARIABLE DEFINED ON (1, 2,..., N)
C    N = DIMENSION OF CDF
C
      K = 1
      IF ( U .LE. CDF(1)) GO TO 300
      MAXI = N
      MINI = 2
C
  100 K = ( MAXI + MINI)/ 2
      IF ( U .GT. CDF(K)) GO TO 200
      IF ( U .GT. CDF( K - 1)) GO TO 300
      MAXI = K - 1
      GO TO 100
  200 MINI = K + 1
      GO TO 100
  300 RETURN
      END
```

Figure L.29. The subroutine BINSRH.

```
      SUBROUTINE FNORM( X , PHI, UNL)
C
C EVALUATE THE STANDARD NORMAL CDF AND THE UNIT LINEAR LOSS
C
C INPUT:
C   X = A REAL NUMBER
V
C OUTPUTS:
C   PHI = AREA UNDER UNIT NORMAL CURVE TO THE LEFT OF X.
C   UNL = E(MAX(0,X-Z)) WHERE Z IS UNIT NORMAL,
C   I.E., UNL = UNIT NORMAL LOSS FUNCTION =
C   INTEGRAL FROM MINUS INFINITY TO X OF ( X - Z) F( Z) DZ, WHERE
C   F(Z) IS THE UNIT NORMAL DENSITY.
C
C   WARNING: RESULTS ARE ACCURATE ONLY TO ABOUT 5 PLACES
C
      Z = X
      IF ( Z .LT.  0.) Z = -Z
      T = 1./(1. + .2316419 * Z)
      PHI = T*(.31938153 + T*(-.356563782 + T*(1.781477937 +
     1  T*( -1.821255978 + T*1.330274429))))
      E2 = 0.
C
C  6 S. D. OUT GETS TREATED AS INFINITY
C
      IF ( Z .LE. 6.) E2 = EXP(-Z*Z/2.)*.3989422803
      PHI = 1. - E2* PHI
      IF ( X .LT. 0.) PHI = 1. - PHI
      UNL = X* PHI + E2
      RETURN
      END
```

Figure L.30. The subroutine FNORM.

```
      SUBROUTINE TBINOM( N, P, CDF, IFAULT)
      REAL CDF( N)
C
C  TABULATE THE BINOMIAL CDF
C
C  INPUTS:
C     N = NO. OF POSSIBLE OUTCOMES
C     (SHIFTED BY 1 SO THEY ARE 1,2...N)
C     P = SECOND PARAMETER OF BINOMIAL, 0 <= P <= 1.
C     E.G., A BINOMIAL WITH 10 TRIALS AND A PROBABILITY OF
C     SUCCESS OF .2 AT EACH TRIAL HAS 11 POSSIBLE
C     OUTCOMES,  THUS TBINOM IS CALLED WITH N = 11 AND P = .2
C
C  OUTPUTS:
C     CDF( I) = PROB( X <= I-1)
C        = PROBABILITY OF LESS THAN I SUCCESSES
C     IFAULT = 8 IF INPUT ERROR, ELSE 0
C
C  SOMEWHAT CIRCUITOUS METHOD IS USED TO AVOID UNDERFLOW
C  AND OVERFLOW WHICH MIGHT OTHERWISE OCCUR FOR N > 120.
C
C  TOLERANCE FOR SMALL VALUES
      DATA TOLR /1.E-10/
C
C  DATE 13 SEPT 1986
C
C  CHECK FOR BAD INPUT
C
      IF (( N .LE. 0).OR.( P .LT. 0.).OR.( P .GT. 1.)) GO TO 9000
      IFAULT = 0
C
C  WATCH OUT FOR BOUNDARY CASES OF P
C
      IF ( P .GT. 0.) GO TO 50
      DO 30 J = 1,N
      CDF( J) = 1.
   30 CONTINUE
      RETURN
   50 IF ( P .LT. 1.) GO TO 70
      DO 60 J = 1, N
      CDF(J) = 0.
   60 CONTINUE
      CDF( N) = 1.
      RETURN
C
C  TYPICAL VALUES OF P START HERE
C
   70    QOP = ( 1. - P)/ P
      POQ = P/( 1. - P)
      R = 1.
C
C  SET K APPROXIMATELY = MEAN
C
      K = N* P + 1.
      IF ( K .GT. N) K = N
      CDF(K) = 1.
      K1 = K-1
      TOT = 1.
      IF ( K1 .LE. 0) GO TO 150
      J = K
```

(continued)

```
C
C   CALCULATE P.M.F. PROBABILITIES RELATIVE TO P(K).
C   FIRST BELOW K
C
      DO 100 I = 1, K1
      J = J - 1
C
C   RELATIVELY SMALL PROBABILITIES GET SET TO 0.
C
      IF ( R .LE. TOLR) R = 0.
      R = R* QOP *J/( N - J)
      CDF( J) = R
      TOT = TOT + R
  100 CONTINUE
C
C   NOW ABOVE K
C
  150 IF ( K .EQ. N) GO TO 300
      R = 1.
      K1 = K + 1
      DO 200 J = K1 , N
C
C   RELATIVELY SMALL PROBABILITIES GET SET TO 0.
C
      IF ( R .LE. TOLR) R = 0.
      R = R * POQ*( N - J + 1)/( J -1)
      TOT = TOT + R
      CDF( J ) = R
  200 CONTINUE
C
C   NOW RESCALE AND CONVERT P.M.F. TO C.D.F.
C
  300 RUN = 0.
      DO 400 J = 1, N
      RUN = RUN + CDF( J)/ TOT
      CDF( J) = RUN
  400 CONTINUE
      RETURN
 9000 IFAULT = 8
      RETURN
      END
```

Figure L.31. The subroutine TBINOM.

```
      SUBROUTINE TTPOSN( N, NSTART, P, CDF, IFAULT)
      REAL CDF( N)
C
C TABULATE THE TRUNCATED POISSON CDF.
C
C INPUTS:
C   N = NUMBER OF VALUES TO TABULATE
C   NSTART = STARTING VALUE
C   P = MEAN OF THE DISTRIBUTION
C
C OUTPUTS:
C   CDF( I) = THE CDF OF THE POISON DISTRIBUTION
C     SHIFTED SO THAT CDF( I) = PROB( X <= I + NSTART - 1).
C   IFAULT = 9 IF INPUT ERROR, ELSE 0
C
C TOLERANCE FOR SMALL VALUES
      DATA TOLR/.1E-18/
C
C CHECK FOR BAD INPUT
C
      IFAULT = 0
      IF (( N .LE. 0) .OR. ( P .LT. 0.)
     1    .OR. ( NSTART .LT. 0)) GO TO 9000
C
C WATCH OUT FOR BOUNDARY CASES OF P
C
      IF ( P .GT. 0.) GO TO 50
      DO 30 J = 1, N
      CDF( J) = 1.
   30 CONTINUE
      RETURN
C
C TYPICAL VALUES OF P START HERE
C
   50 R = 1.
      TOT = 1.
      K = P + 1.
      NLIM = N + NSTART
      IF ( K .GT. NLIM) K = NLIM
      IF ( K .LE. NSTART) K = NSTART + 1
      CDF( K - NSTART) = 1.
      K1 = K - 1 - NSTART
      IF ( K1 .LE. 0) GO TO 150
      J = K
C
C CALCULATE P.M.F. PROBABILITIES RELATIVE TO P(K)
C FIRST GO DOWN FROM K
C
      DO 100 I = 1, K1
C
C RELATIVELY SMALL PROBABILITIES GET SET TO 0.
C
      IF ( R .LE. TOLR) R = 0.
      J = J - 1
      R = R* J/ P
      CDF( J - NSTART) = R
      TOT = TOT + R
  100 CONTINUE
```

(continued)

```
C
C   NOW GO UP
C
  150 K1 = K1 + 2
      IF ( K1 .GT. N) GO TO 300
      R = 1.
      JJ = K
      DO 200 J = K1 , N
C
C  RELATIVELY SMALL PROBABILITIES GET SET TO 0.
C
      IF ( R .LE. TOLR) R = 0.
      R = R* P/ JJ
      TOT = TOT + R
      CDF( J ) = R
      JJ = JJ + 1
  200 CONTINUE
C
C  NOW RESCALE AND CONVERT TO C.D.F.
C
  300 RUN = 0.
      DO 400 J = 1, N
      RUN = RUN + CDF( J)/ TOT
      CDF( J) = RUN
  400 CONTINUE
      RETURN
C
 9000 IFAULT = 9
      RETURN
      END
```

Figure L.32. The subroutine TTPOSN.

Examples of Simulation Programming

This appendix contains most of the programs and examples referred to in Chapters 7 and 8. A number of comments are in order.

First and foremost, these programs are illustrations: they are not intended to be copied slavishly. The Fortran programs, for example, are written in the current ANSI standard Fortran [ANSI (1978)], using the full language. They will not run on machines still using Fortran IV, nor on compilers which accept only the official ANSI subset. Simula programs are presented with keywords underlined and in lower case, since we believe this is easier to read than the all-upper-case representation imposed by some machines. All the programs presented have been tested in one form or another, though not necessarily with exactly the text given in this appendix.

While writing this book we used a variety of machines. Most of the numerical results in Chapter 8 were obtained using a CDC Cyber 173; almost all of them were checked on a XDS Sigma 6. The examples in Simula and GPSS were also run on a Cyber 173. The examples in Simscript were run on an IBM 4341. Some of the GPSS programs were also checked on this machine.

The reader should be aware that if he *does* run our programs on a different machine, he may get different answers despite using the same portable random number generator. Such functions as ERLANG and NXTARR use floating-point arithmetic; NXTARR in particular cumulates floating-point values. A very minor difference in floating-point representations on two different machines may be just enough to throw two supposedly identical simulations out of synchronization: once this happens, all further resemblance between their outputs may be lost.

```
      PROGRAM BOMBER
      COMMON /TIMCOM/TIME
      INTEGER INRNGE, BOMB, READY, BURST, ENDSIM
      INTEGER SEED, EVENT, I, N, GOUT, PDOWN
      REAL TIME, DIST, TOFFLT, NORMAL, UNIF, R
      LOGICAL PLANOK(5), GUNOK(3), GUNRDY(3), TRACE
      PARAMETER (INRNGE = 1, BOMB = 2, READY = 3,
     X           BURST = 4, ENDSIM = 5)
      EXTERNAL PUT, GET, INIT, UNIF, NORMAL, DIST, TOFFLT
C
C  EXAMPLE 7.1.1 FROM SECTION 7.1.4
C
C  TIME IS MEASURED IN SECONDS.
C  TIME 0 IS WHEN THE FIRST AIRCRAFT
C  COMES INTO RANGE.
C
C  CHANGE THE FOLLOWING TO SUPPRESS TRACE
C
      TRACE = .TRUE.
C
C  INITIALIZE - ALL PLANES ARE OUT OF RANGE, ALL GUNS ARE
C  INTACT AND READY TO FIRE.
C  NO PLANES HAVE BEEN SHOT DOWN.
C
      CALL INIT
      DO 1 I=1,5
    1 PLANOK(I) = .FALSE.
      PDOWN = 0
      DO 2 I=1,3
      GUNOK(I) = .TRUE.
    2 GUNRDY(I) = .TRUE.
      GOUT = 0
      SEED = 12345
C
C  R IS THE RANGE FOR OPENING FIRE.
C  THE SCHEDULE INITIALLY CONTAINS JUST TWO EVENTS: PLANE 1
C  COMES INTO RANGE AND THE END OF THE SIMULATION.
C
      R = 3000.0
      CALL PUT(INRNGE, (3900.0-R)/300.0, 1)
      CALL PUT(ENDSIM, 50.0, 0)
   99 CALL GET(EVENT, N)
      IF (EVENT .EQ. INRNGE) GO TO 100
      IF (EVENT .EQ. BOMB)   GO TO 200
      IF (EVENT .EQ. READY)  GO TO 300
      IF (EVENT .EQ. BURST)  GO TO 400
      IF (EVENT .EQ. ENDSIM) GO TO 500
C
C  THIS SECTION SIMULATES AN AIRCRAFT COMING INTO RANGE
C
C  IF IT IS NOT THE LAST AIRCRAFT, SCHEDULE ITS SUCCESSOR.
C  IF THERE IS A GUN READY TO FIRE, IT WILL DO SO.
C  WHEN A GUN FIRES, WE MUST SCHEDULE THE BURST AND ALSO
C  THE TIME WHEN THE GUN IS RELOADED.
C  FINALLY, SCHEDULE AN ARRIVAL AT THE BOMBING POINT.
```

(continued)

```
C
  100 IF (TRACE) PRINT 910,TIME,N
  910 FORMAT (F6.2,'   AIRCRAFT',I4,' IN RANGE')
      IF (N .LT. 5) CALL PUT(INRNGE, TIME+2.0, N+1)
      PLANOK(N) = .TRUE.
      DO 101 I=1,3
      IF (.NOT. GUNOK(I) .OR. .NOT. GUNRDY(I)) GO TO 101
      IF (TRACE) PRINT 911,TIME,I,N
  911 FORMAT (F6.2,'   GUN',I4,' FIRES AT AIRCRAFT',I4)
      CALL PUT(BURST, TIME+TOFFLT(N), N)
      CALL PUT(READY, TIME+NORMAL(5.0, 0.5, SEED), I)
      GUNRDY(I) = .FALSE.
  101 CONTINUE
      CALL PUT(BOMB, TIME+R/300.0, N)
      GO TO 99
C
C  THIS SECTION SIMULATES AN AIRCRAFT DROPPING ITS BOMBS
C
C  IF THE PLANE HAS BEEN DESTROYED, THE EVENT IS IGNORED.
C  OTHERWISE, WE DECIDE FOR EACH GUN IN TURN IF IT IS
C  HIT OR NOT.
C
  200 IF (PLANOK(N)) THEN
         IF (TRACE) PRINT 920,TIME,N
  920    FORMAT (F6.2,'   AIRCRAFT',I4,' DROPS ITS BOMBS')
         DO 201 I=1,3
         IF ( (UNIF(SEED) .LE. 0.2) .AND. GUNOK(I) ) THEN
            GUNOK(I) = .FALSE.
            GOUT = GOUT + 1
            IF (TRACE) PRINT 921,TIME,I
  921       FORMAT (F6.2,'   GUN',I4,' DESTROYED')
         ENDIF
  201    CONTINUE
      ENDIF
      GO TO 99
C
C  THIS SECTION SIMULATES THE END OF LOADING A GUN
C
C  IF THE GUN HAS BEEN DESTROYED, THE EVENT IS IGNORED.
C  OTHERWISE, LOOK FOR A TARGET: WE TRY FIRST THE NEAREST
C  ONCOMING PLANE, THEN THE NEAREST RECEDING PLANE.  IF
C  A TARGET IS FOUND, SCHEDULE THE BURST AND THE END OF
C  RELOADING; OTHERWISE, MARK THE GUN AS READY.
C
  300 IF (GUNOK(N)) THEN
         DO 301 I=1,5
         IF (PLANOK(I) .AND. DIST(I) .LE. 0) GO TO 303
  301    CONTINUE
         DO 302 I = 5, 1, -1
         IF (PLANOK(I) .AND. DIST(I) .LE. 2100.0) GO TO 303
  302    CONTINUE
         IF (TRACE) PRINT 930,TIME,N
  930    FORMAT (F6.2,'   GUN',I4,' READY - NO TARGET')
         GUNRDY(N) = .TRUE.
         GO TO 99
  303    IF (TRACE) PRINT 931,TIME,N,I
  931    FORMAT (F6.2,'   GUN',I4,' FIRES AT AIRCRAFT',I4)
         CALL PUT(BURST, TIME+TOFFLT(I), I)
         CALL PUT(READY, TIME+NORMAL(5.0, 0.5, SEED), N)
      ENDIF
      GO TO 99
```

(continued)

```
C
C   THIS SECTION SIMULATES A SHELL BURST
C
   400 IF (PLANOK(N)) THEN
          IF (TRACE) PRINT 940,TIME,N
   940    FORMAT (F6.2,'   SHOT BURSTS BY AIRCRAFT',I4)
          IF (UNIF(SEED).GT.0.3-ABS(DIST(N))/10000.0) THEN
             IF (TRACE) PRINT 941,TIME,N
   941       FORMAT (F6.2,'   AIRCRAFT',I4,' DESTROYED')
             PLANOK(N) = .FALSE.
             PDOWN = PDOWN + 1
          ELSE
             IF (TRACE) PRINT 942,TIME
   942       FORMAT (F6.2,'   NO DAMAGE')
          ENDIF
       ELSE
          IF (TRACE) PRINT 943,TIME,N
   943    FORMAT (F6.2,'   SHOT AIMED AT AIRCRAFT',I4,
      X         ' BURSTS IN THE AIR')
       ENDIF
       GO TO 99
C
C   THIS IS THE END OF THE SIMULATION
C
   500 PRINT 950,GOUT,PDOWN
   950 FORMAT (///,I5,' GUNS DESTROYED',/,I5,
      X         ' PLANES SHOT DOWN')
       STOP
       END
```

Figure X.7.1. A simulation in Fortran.

```
          PROGRAM SIMNAI
          COMMON /TIMCOM/ TIME
          INTEGER ARRVAL,ENDSRV,OPNDOR,SHTDOR
          INTEGER ARRSD,SRVSD,TELSD,RNGSD
          INTEGER QUEUE,TLATWK,TLFREE,NCUST,NSERV
          INTEGER NREPLS,REPL,EVENT,I
          REAL    TIME,NXTARR,ERLANG,UNIF,X
          LOGICAL TRACE
          PARAMETER (ARRVAL = 1, ENDSRV = 2, OPNDOR = 3, SHTDOR = 4)
          EXTERNAL NXTARR, UNIF, ERLANG, INIT, PUT, GET, SETARR
C
C   EXAMPLE 1.4.1 - SEE SECTION 7.1.6
C
C   THE TIME UNIT IS 1 MINUTE.
C   TIME 0 CORRESPONDS TO 9 A.M.
C
C   CHANGE THE FOLLOWING TO SUPPRESS TRACE
C
          TRACE = .TRUE.
C
C   SET RANDOM NUMBER SEEDS SPACED 200000 VALUES APART
C
          ARRSD = 1234567890
          SRVSD = 1933985544
          TELSD = 2050954260
          RNGSD =  918807827
C
C   SET NUMBER OF RUNS TO CARRY OUT
C
          NREPLS = 200
          DO 98 REPL = 1,NREPLS
C
C   INITIALIZE FOR A NEW RUN
C
C   SET NUMBER OF TELLERS AT WORK.
C   INITIALLY NO TELLERS ARE FREE SINCE THE DOOR IS NOT OPEN
C
          X = UNIF(TELSD)
          TLATWK = 3
          IF (X.LT.0.20) TLATWK = 2
          IF (X.LT.0.05) TLATWK = 1
          TLFREE = 0
          IF (TRACE) PRINT 950,TLATWK
      950 FORMAT (I3,' TELLERS AT WORK')
C
C   SET QUEUE EMPTY, NO CUSTOMERS OR SERVICES SO FAR
C
          QUEUE = 0
          NCUST = 0
          NSERV = 0
C
C   SET TIME 0, SCHEDULE EMPTY, ARRIVAL DISTBN INITIALIZED
C
          CALL INIT
          X = SETARR(0)
C
C   INITIALIZE SCHEDULE WITH FIRST ARRIVAL, OPENING OF DOOR,
C   AND CLOSING OF DOOR ( = END OF 1 RUN)
C
          CALL PUT (ARRVAL, NXTARR(ARRSD))
          CALL PUT (OPNDOR, 60.0)
          CALL PUT (SHTDOR, 360.0)
```

(continued)

```
C
C     *****************
C
C     START OF MAIN LOOP
C
C     *****************
C
   99 CALL GET (EVENT)
      IF (EVENT .EQ. ARRVAL) GO TO 100
      IF (EVENT .EQ. ENDSRV) GO TO 200
      IF (EVENT .EQ. OPNDOR) GO TO 300
      IF (EVENT .EQ. SHTDOR) GO TO 400
C
C     THIS SECTION SIMULATES THE ARRIVAL OF A CUSTOMER
C
  100 NCUST = NCUST + 1
      IF (TRACE) PRINT 951,TIME,NCUST
  951 FORMAT (F10.2,' ARRIVAL OF CUSTOMER',I6)
C
C     AT EACH ARRIVAL, WE MUST SCHEDULE THE NEXT
C
      CALL PUT (ARRVAL, NXTARR(ARRSD))
C
C     IF THERE IS A TELLER FREE, WE START SERVICE
C
      IF (TLFREE.GT.0) THEN
        TLFREE = TLFREE - 1
        NSERV = NSERV + 1
        IF (TRACE) PRINT 952,TIME,NSERV
  952   FORMAT (F10.2,' START OF SERVICE',I6)
        CALL PUT (ENDSRV, TIME + ERLANG(2.0,2,SRVSD))
C
C     OTHERWISE THE CUSTOMER MAY OR MAY NOT STAY,
C     DEPENDING ON THE QUEUE LENGTH
C
      ELSE
        IF (UNIF(RNGSD).GT.(QUEUE-5)/5.0) THEN
          QUEUE = QUEUE + 1
          IF (TRACE) PRINT 953,TIME,QUEUE
  953     FORMAT (F10.2,' QUEUE IS NOW',I6)
        ELSE
          IF (TRACE) PRINT 954,TIME
  954     FORMAT (F10.2,' CUSTOMER DOES NOT JOIN QUEUE')
        ENDIF
      ENDIF
      GO TO 99
C
C     IN THIS SECTION WE SIMULATE THE END OF A PERIOD OF SERVICE
C
C     IF THE QUEUE IS NOT EMPTY, THE FIRST CUSTOMER WILL START
C     SERVICE AND WE MUST SCHEDULE THE CORRESPONDING END
C     SERVICE. OTHERWISE THE TELLER IS NOW FREE.
C
  200 IF (QUEUE.NE.0) THEN
        NSERV = NSERV + 1
        QUEUE = QUEUE - 1
        IF (TRACE) PRINT 955,TIME,NSERV
  955   FORMAT (F10.2,' START OF SERVICE',I6)
        IF (TRACE) PRINT 956,TIME,QUEUE
  956   FORMAT (F10.2,' QUEUE IS NOW',I6)
        CALL PUT (ENDSRV, TIME + ERLANG(2.0,2,SRVSD))
```

(continued)

```
          ELSE
            TLFREE = TLFREE + 1
            IF (TRACE) PRINT 957,TIME,TLFREE
     957    FORMAT (F10.2,' NUMBER OF TELLERS FREE',I6)
          ENDIF
          GO TO 99
C
C   IN THIS SECTION WE SIMULATE THE OPENING OF THE DOOR
C
C   ALL THE TELLERS BECOME AVAILABLE. IF THERE ARE CUSTOMERS
C   IN THE QUEUE, SOME OF THEM CAN START SERVICE AND WE MUST
C   SCHEDULE THE CORRESPONDING END SERVICE EVENTS.
C
     300  TLFREE = TLATWK
          IF (TRACE) PRINT 958,TIME
     958  FORMAT (F10.2,' THE DOOR OPENS')
          DO 301 I = 1,TLATWK
          IF (QUEUE.NE.0) THEN
            NSERV = NSERV + 1
            QUEUE = QUEUE -1
            TLFREE = TLFREE - 1
            IF (TRACE) PRINT 959,TIME,NSERV
     959    FORMAT (F10.2,' START OF SERVICE',I6)
            IF (TRACE) PRINT 960,TIME,QUEUE
     960    FORMAT (F10.2,' NUMBER IN QUEUE IS NOW',I6)
            CALL PUT (ENDSRV, TIME + ERLANG(2.0,2,SRVSD))
          ENDIF
     301  CONTINUE
          IF (TRACE) PRINT 961,TIME,TLFREE
     961  FORMAT (F10.2,' NUMBER OF TELLERS FREE',I6)
          GO TO 99
C
C   IN THIS SECTION WE SIMULATE THE CLOSING OF THE DOOR
C
C   SINCE WE KNOW ALL CUSTOMERS ALREADY IN THE QUEUE WILL
C   EVENTUALLY RECEIVE SERVICE, WE CAN SIMPLY COUNT THEM
C   AND STOP SIMULATING AT THIS POINT.
C
     400  CONTINUE
          IF (TRACE) PRINT 962,TIME,QUEUE
     962  FORMAT (F10.2,' DOOR CLOSES : NUMBER IN QUEUE',I6)
          NSERV = NSERV + QUEUE
          PRINT 990, TLATWK, NCUST, NSERV
     990  FORMAT (I6,' TELLERS,',I6,' CUSTOMERS,',I6,' WERE SERVED')
          WRITE (1,991) TLATWK,NCUST,NSERV
     991  FORMAT (3I6)
C
C   THIS IS THE END OF ONE RUN
C
      98  CONTINUE
          STOP
          END
```

Figure X.7.2. The standard example in Fortran.

```
      REAL FUNCTION ERLANG(MEAN,SHAPE,SEED)
      REAL MEAN,UNIF,X
      INTEGER SHAPE,SEED,I
      EXTERNAL UNIF
      X = 1.0
      DO 1 I = 1,SHAPE
    1 X = X * UNIF(SEED)
      ERLANG = -MEAN * ALOG(X) / SHAPE
      RETURN
      END
```

Figure X.7.3. The function ERLANG.

```
      REAL FUNCTION NXTARR(SEED)
      REAL X(4),Y(4),MEAN(4),CUMUL,UNIF
      INTEGER SEG,SEED
      EXTERNAL UNIF
      SAVE CUMUL, SEG, X, Y, MEAN
C
C  THIS SUBROUTINE IMPLEMENTS A PROCEDURE
C  FOR NONSTATIONARY PROCESSES: SEE 5.3.18, 4.9.
C  CUMUL GIVES THE SUCCESSIVE VALUES OF Y, WHILE SEG TELLS US
C  WHICH SEGMENT OF THE PIECEWISE-LINEAR INTEGRATED RATE
C  FUNCTION WE ARE CURRENTLY ON.
C
C  THE TABLE BELOW DEFINES THE PIECEWISE-LINEAR FUNCTION.
C  THE DUMMY VALUE OF Y(4) SERVES TO STOP US OVER-RUNNING
C  THE TABLE.
C
      DATA X    / 45.0,120.0,300.0,  0.0/
      DATA Y    /  0.0, 37.5,217.5, 1E10/
      DATA MEAN /  2.0,  1.0,  2.0,  0.0/
      CUMUL = CUMUL - ALOG(UNIF(SEED))
C
C  FIRST WE MUST CHECK IF WE ARE STILL IN THE SAME SEGMENT
C
    1 IF (CUMUL.LT.Y(SEG+1)) GO TO 2
      SEG = SEG + 1
      GO TO 1
C
C  NOW WE KNOW THE SEGMENT, WE CAN CALCULATE THE INVERSE
C
    2 NXTARR = X(SEG) + (CUMUL - Y(SEG)) * MEAN(SEG)
      RETURN
C
C  THIS ENTRY POINT INITIALIZES CUMUL AND SEG
C
      ENTRY SETARR
      CUMUL = 0.0
      SEG = 1
      SETARR = 0.0
      RETURN
      END
```

Figure X.7.4. The function NXTARR.

```
''EXAMPLE OF SIMSCRIPT MODELING

''EXAMPLE 7.2.1 OF SECTION 7.2.3

PREAMBLE

    EVENT NOTICES INCLUDE CUSTOMER.ARRIVAL, NEW.HAMBURGER,
        AND STOP.SIMULATION
    EVERY END.SERVICE HAS A PATRON

    TEMPORARY ENTITIES
        EVERY ORDER HAS A SIZE
            AND A TIME.OF.ARRIVAL
            AND MAY BELONG TO THE QUEUE
        DEFINE TIME.OF.ARRIVAL AS A REAL VARIABLE
        DEFINE SIZE AS AN INTEGER VARIABLE

    THE SYSTEM HAS A SERVER.STATUS
        AND A TOTAL.CUSTOMERS
        AND A LOST.CUSTOMERS
        AND A STOCK
        AND A SERVICE.TIME
        AND A STREAM.NO
        AND AN ORDER.SIZE RANDOM STEP VARIABLE
        AND OWNS THE QUEUE

    DEFINE SERVICE.TIME AS A REAL VARIABLE
    DEFINE TOTAL.CUSTOMERS, LOST.CUSTOMERS, STOCK,
        SERVER.STATUS AND STREAM.NO AS INTEGER VARIABLES
    DEFINE ORDER.SIZE AS AN INTEGER,STREAM 7 VARIABLE
    DEFINE QUEUE AS A FIFO SET

    ACCUMULATE UTILIZATION AS THE MEAN OF SERVER.STATUS
    TALLY MEAN.SERVICE.TIME AS THE MEAN OF SERVICE.TIME

    DEFINE .BUSY TO MEAN 1
    DEFINE .IDLE TO MEAN 0
    DEFINE SECONDS TO MEAN UNITS
    DEFINE MINUTES TO MEAN *60 UNITS
    DEFINE HOURS TO MEAN *60 MINUTES
END ''PREAMBLE


MAIN
    LET STREAM.NO = 1
    READ ORDER.SIZE
    SCHEDULE A STOP.SIMULATION IN 4 HOURS
    SCHEDULE A NEW.HAMBURGER IN
        UNIFORM.F(30.0,50.0,STREAM.NO) SECONDS
    SCHEDULE A CUSTOMER.ARRIVAL IN
        EXPONENTIAL.F(2.5,STREAM.NO) MINUTES
    START SIMULATION
END ''MAIN


EVENT CUSTOMER.ARRIVAL
    ADD 1 TO TOTAL.CUSTOMERS
    SCHEDULE A CUSTOMER.ARRIVAL IN
        EXPONENTIAL.F(2.5,STREAM.NO) MINUTES
    IF N.QUEUE = 3,
        ADD 1 TO LOST.CUSTOMERS
```

(continued)

```
         ELSE
            CREATE AN ORDER
            LET SIZE(ORDER) = ORDER.SIZE
            LET TIME.OF.ARRIVAL = TIME.V
            FILE ORDER IN QUEUE
            CALL START.SERVICE
         ALWAYS
         RETURN
   END ''EVENT CUSTOMER.ARRIVAL


   ROUTINE TO START.SERVICE
         IF SERVER.STATUS = .IDLE
            AND QUEUE IS NOT EMPTY
            AND SIZE(F.QUEUE) < STOCK
            REMOVE FIRST ORDER FROM QUEUE
            SUBTRACT SIZE(ORDER) FROM STOCK
            LET SERVER.STATUS = .BUSY
            SCHEDULE AN END.SERVICE GIVING ORDER IN
               45.0 + 25.0*SIZE(ORDER) SECONDS
         ALWAYS
         RETURN
   END ''ROUTINE TO START.SERVICE


   EVENT NEW.HAMBURGER
         ADD 1 TO STOCK
            SCHEDULE A NEW.HAMBURGER IN
            UNIFORM.F(30.0,50.0,STREAM.NO) SECONDS
         CALL START.SERVICE
         RETURN
   END ''EVENT NEW.HAMBURGER


   EVENT END.SERVICE GIVEN ORDER
         DEFINE ORDER AS AN INTEGER VARIABLE
         LET SERVICE.TIME = TIME.V - TIME.OF.ARRIVAL(ORDER)
         DESTROY THIS ORDER
         LET SERVER.STATUS = .IDLE
         CALL START.SERVICE
         RETURN
   END ''EVENT END.SERVICE


   EVENT STOP.SIMULATION
         PRINT 3 LINES WITH TOTAL.CUSTOMERS, LOST.CUSTOMERS,
            MEAN.SERVICE.TIME AND UTILIZATION THUS
*** CUSTOMERS ARRIVED.    ***WERE LOST.
MEAN SERVICE TIME WAS    *** SECONDS.
THE SERVER UTILIZATION WAS *.**
         STOP
   END ''STOP.SIMULATION
```

Figure X.7.5. A simulation in Simscript.

```
''  THE STANDARD EXAMPLE IN SIMSCRIPT II.5

''  THE TIME UNIT IS 1 MINUTE, WITH TIME 0 AT 9.00 A.M.

PREAMBLE

    EVENT NOTICES INCLUDE ARRIVAL, DEPARTURE, OPENING,
                        AND CLOSING

    TEMPORARY ENTITIES
        EVERY CUSTOMER HAS A TIME.OF.ARRIVAL AND MAY BELONG
                        TO THE QUEUE

    THE SYSTEM OWNS THE QUEUE

    DEFINE TEL.FREE, TEL.AT.WORK, NUM.CUSTOMERS,
            NUM.SERVICES, AND NUM.BALKS AS INTEGER VARIABLES
    DEFINE WAITING.TIME AS A REAL VARIABLE
    DEFINE NEXT.ARRIVAL AS A REAL FUNCTION

    DEFINE ARRIVAL.SEED TO MEAN 1
    DEFINE SERVICE.SEED TO MEAN 2
    DEFINE TELLER.SEED  TO MEAN 3
    DEFINE BALK.SEED    TO MEAN 4

    ACCUMULATE AVG.QUEUE AS THE AVERAGE OF N.QUEUE
    TALLY AVG.WAIT  AS THE AVERAGE OF WAITING.TIME

END

MAIN

    DEFINE X AS A REAL VARIABLE

    LET X = UNIFORM.F(0.0, 1.0, TELLER.SEED.)
    LET TEL.AT.WORK = 3    '' NOW FIX UP NUMBER OF TELLERS
    IF X < 0.2 LET TEL.AT.WORK = 2
    ALWAYS IF X < 0.05 LET TEL.AT.WORK = 1

    ALWAYS

    LET TEL.FREE = 0
    LET NUM.CUSTOMERS = 0
    LET NUM.SERVICES  = 0
    LET NUM.BALKS     = 0

    SCHEDULE AN OPENING AT 60.0
    SCHEDULE A CLOSING  AT 360.0
    SCHEDULE AN ARRIVAL AT NEXT.ARRIVAL
    START SIMULATION

    ''CONTROL WILL RETURN WHEN ALL CUSTOMERS HAVE BEEN SERVED

    PRINT 3 LINES WITH TEL.AT.WORK,
        NUM.CUSTOMERS, NUM.SERVICES,
        NUM.BALKS, AVG.QUEUE, AND AVG.WAIT THUS
**  TELLERS AT WORK
*** CUSTOMERS, *** SERVICES, *** BALKS
AVERAGE QUEUE *.**, AVERAGE WAIT *.** MINUTES

END
```

(continued)

```
EVENT ARRIVAL
    ADD 1 TO NUM.CUSTOMERS
    SCHEDULE AN ARRIVAL AT NEXT.ARRIVAL
    IF (N.QUEUE - 5)/5.0 > UNIFORM.F(0.0, 1.0, BALK.SEED)
        ADD 1 TO NUM.BALKS
        RETURN
    ALWAYS
    IF TEL.FREE = 0
        CREATE A CUSTOMER
        LET TIME.OF.ARRIVAL(CUSTOMER) = TIME.V
        FILE CUSTOMER IN QUEUE
    ELSE
        LET WAITING.TIME = 0.0
        SCHEDULE A DEPARTURE IN ERLANG.F(2.0, 2, SERVICE.SEED)
                                    UNITS
        ADD 1 TO NUM.SERVICES
        SUBTRACT 1 FROM TEL.FREE
    ALWAYS
    RETURN
END

EVENT DEPARTURE
    IF QUEUE IS EMPTY
        ADD 1 TO TEL.FREE
    ELSE
        REMOVE FIRST CUSTOMER FROM QUEUE
        LET WAITING.TIME = TIME.V - TIME.OF.ARRIVAL(CUSTOMER)
        DESTROY CUSTOMER
        SCHEDULE A DEPARTURE IN ERLANG.F(2.0, 2, SERVICE.SEED)
                                    UNITS
        ADD 1 TO NUM.SERVICES
    ALWAYS
    RETURN
END

EVENT OPENING
    DEFINE I AS AN INTEGER VARIABLE
    RESET TOTALS OF N.QUEUE
    FOR I = 1 TO TEL.AT.WORK DO
        SCHEDULE A DEPARTURE NOW
    LOOP
    '' FREEING ALL THE TELLERS HAS EXACTLY THE SAME EFFECT
    '' AS TEL.AT.WORK DEPARTURES
END

EVENT CLOSING
    CANCEL THE ARRIVAL
    RETURN
END

ROUTINE NEXT.ARRIVAL

    DEFINE X, Y, AND M AS REAL, SAVED, 1-DIMENSIONAL ARRAYS
    DEFINE CUMUL AS A REAL, SAVED VARIABLE
    DEFINE SEG AS AN INTEGER, SAVED VARIABLE

    IF SEG = 0 ''THIS IS THE FIRST CALL OF THE ROUTINE
        LET SEG = 1
        LET CUMUL = 0.0
        RESERVE X(*), Y(*), AND M(*) AS 4
        LET X(1) = 45.0
        LET X(2) = 120.0
```

(continued)

```
        LET X(3) = 300.0
        LET X(4) = 0.0
        LET Y(1) = 0.0
        LET Y(2) = 37.5
        LET Y(3) = 217.5
        LET Y(4) = 99999.9
        LET M(1) = 2.0
        LET M(2) = 1.0
        LET M(3) = 2.0
        LET M(4) = 0.0
    ALWAYS
    ADD EXPONENTIAL.F(1.0, ARRIVAL.SEED) TO CUMUL
    'SEARCH' IF CUMUL >= Y(SEG+1)
        ADD 1 TO SEG
        GO TO SEARCH
    ALWAYS
    RETURN WITH X(SEG) + (CUMUL - Y(SEG)) * M(SEG)

END
```

Figure X.7.6. The standard example in Simscript.

```
          SIMULATE

* EXAMPLE 7.3.1 OF SECTION 7.3.4
* 1 SIMULATED TIME UNIT REPRESENTS 100 MSEC.

EXPO      FUNCTION      RN8,C24
0.0,0/0.1,0.104/0.2,0.222/0.3,0.355/0.4,0.509/0.5,0.69
0.6,0.915/0.7,1.2/0.75,1.38/0.8,1.6/0.84,1.83/0.88,2.12
0.9,2.3/0.92,2.52/0.94,2.81/0.95,2.99/0.96,3.2/0.97,3.5
0.98,3.9/0.99,4.6/0.995,5.3/0.998,6.2/0.999,7.0/0.9997,8.0

CLASS     FUNCTION      RN7,D3
0.6,1/0.9,2/1.0,3

TIME      FUNCTION      PF1,L3
1,1/2,5/3,20

CORE      FUNCTION      PF1,L3
1,16/2,32/3,64

MEMRY     STORAGE       256
ALL       EQU           4,Q
CLS1      QTABLE        1,10,10,15
CLS2      QTABLE        2,10,10,15
CLS3      QTABLE        3,50,50,10
TOTAL     QTABLE        ALL,20,20,30

          GENERATE      600              TIMER - DECREMENTS
          TERMINATE     1                COUNTER EVERY MINUTE

          GENERATE      ,,,20,,3PF       CREATE 20 TRANSACTIONS
                                         WITH 3 FULL-WORD
                                         PARAMETERS EACH.
CYCLE     ADVANCE       100,FN$EXPO      TYPE A COMMAND.
          ASSIGN        1,FN$CLASS,PF
          ASSIGN        2,FN$TIME,PF
          ASSIGN        3,FN$CORE,PF
          QUEUE         PF1              COMMAND TRANSMITTED.
          QUEUE         ALL
          ENTER         MEMRY,PF3        CONTEND FOR CORE.
RROB      SEIZE         CPU
          ADVANCE       1                USE 1 QUANTUM OF CPU.
          RELEASE       CPU
          ASSIGN        2-,1,PF          IF REQUIRED TIME IS NOW
          TEST G        PF2,0,DONE       ZERO, GO TO DONE;
          BUFFER                         OTHERWISE GO TO BACK OF
          TRANSFER      ,RROB            CPU QUEUE AND WAIT TURN.
DONE      LEAVE         MEMRY,PF3        RELEASE CORE.
          DEPART        PF1              REPLY READY : MEASURE
          DEPART        ALL              RESPONSE TIME.
          TRANSFER      ,CYCLE

          START         1,NP             INITIALIZE WITHOUT
                                         PRINTING.
          RESET                          REINITIALIZE STATISTICS.
          START         10               SIMULATE 10 MINUTES.
          END
```

Figure X.7.7. A simulation in GPSS.

```
        SIMULATE
*
* THE RUNNING EXAMPLE - SEE SECTION 1.4
*
* THE TIME UNIT IS 1 SECOND, WITH TIME 0 AT 9.45 A.M.
*      THUS TIME    900 = 10.00 A.M.
*                  4500 = 11.00 A.M.
*                 15300 =  2.00 P.M.
*                 18900 =  3.00 P.M.
*
* THE FOLLOWING ENTITIES ARE USED :
*
* FACILITIES - TELL1, TELL2, TELL3 : THE THREE TELLERS
* QUEUES     - BANK  : KEEPS TRACK OF HOW MANY PEOPLE ARE
*                      IN THE SYSTEM
*            - WLINE : KEEPS TRACK OF HOW MANY PEOPLE ARE
*                      IN THE WAITING LINE
* SWITCHES   - DOOR  : SET WHEN THE DOOR OF THE BANK IS OPEN
*
* NOW DEFINE SOME NECESSARY FUNCTIONS :
*
* THE EXPONENTIAL DISTRIBUTION WITH MEAN 1
* THIS IS THE STANDARD APPROXIMATION TAKEN FROM THE
* IBM MANUAL
*
EXPON  FUNCTION  RN1,C24
0,0/.1,.104/.2,.222/.3,.355/.4,.509/.5,.69/.6,.915
.7,1.2/.75,1.38/.8,1.6/.84,1.83/.88,2.12/.9,2.3
.92,2.52/.94,2.81/.95,2.99/.96,3.2/.97,3.5
.98,3.9/.99,4.6/.995,5.3/.998,6.2/.999,7/.9997,8
*
* MEAN INTERARRIVAL TIME OF CUSTOMERS AS A FUNCTION
* OF THE TIME OF DAY
*
MEAN   FUNCTION  C1,D3
4500,120/15300,60/18900,120
*
* NUMBER OF TELLERS AT WORK
*
TELLS  FUNCTION  RN1,D3
.8,3/.95,2/1,1
*
* PROBABILITY OF BALKING ( TIMES 1000 ) AS A FUNCTION
* OF WAITING-LINE LENGTH
*
BALK   FUNCTION  Q$WLINE,D6
5,0/6,200/7,400/8,600/9,800/10,1000
*
* THE ERLANG DISTRIBUTION WITH MEAN AND SHAPE 2 :
* THIS IS CALCULATED SIMPLY AS THE SUM OF TWO EXPONENTIALS
*
ERLAN  FVARIABLE  60*(FN$EXPON+FN$EXPON)
*
* THE FOLLOWING SECTION GIVES THE CODE FOR A CUSTOMER.
* AS DISCUSSED IN THE TEXT, THE ARRIVAL DISTRIBUTION IS
* FUDGED AT THE POINTS WHERE THE ARRIVAL RATE CHANGES.
* A 'CUSTOMER' TRANSACTION HAS ONE HALF-WORD PARAMETER
* USED TO HOLD THE NUMBER OF THE TELLER HE IS USING.
*
```

(continued)

```
          GENERATE    FN$MEAN,FN$EXPON,,,,1PH   GENERATE A NEW
*                                               CUSTOMER
          ADVANCE     0           TO CLEAR THE 'GENERATE' BLOCK
          GATE LS     DOOR        IF THE DOOR IS CLOSED, WAIT
          QUEUE       BANK        ENTER THE SYSTEM
          TRANSFER    .FN$BALK,,BALK    BALK IF QUEUE TOO LONG
          QUEUE       WLINE       ENTER WAITING LINE
          TRANSFER    ALL,GET1,GET3,3   TRY ALL THREE TELLERS
GET1      SEIZE       TELL1
          ASSIGN      1,1,PH      GET TELLER 1
          TRANSFER    ,SERV
GET2      SEIZE       TELL2
          ASSIGN      1,2,PH      GET TELLER 2
          TRANSFER    ,SERV
GET3      SEIZE       TELL3
          ASSIGN      1,3,PH      GET TELLER 3
SERV      DEPART      WLINE       LEAVE WAITING LINE
          ADVANCE     V$ERLAN     RECEIVE SERVICE
          RELEASE     PH1         TELLER IS FREED
          DEPART      BANK        LEAVE THE SYSTEM
          TERMINATE
*
BALK      DEPART      BANK        AFTER BALKING, LEAVE THE SYSTEM
          TERMINATE
*
* THIS SECTION HANDLES THE SETUP OF TELLERS, OPENS
* AND CLOSES THE DOOR OF THE BANK, AND CONTROLS
* THE ENDING OF THE SIMULATION
*
          GENERATE    1,,,1,127,1PH    GENERATE ONE TOP-PRIORITY
*                                      TRANSACTION
          FUNAVAIL    TELL1-TELL3      AT 9.45 ALL TELLERS ARE
*                                      UNAVAILABLE
          ADVANCE     899         WAIT UNTIL 10.00
          LOGIC S     DOOR        OPEN THE DOOR
          ASSIGN      1,FN$TELLS,PH
          FAVAIL      TELL1
          TEST G      PH1,1,*+4   MAKE 1 TO 3 TELLERS
          FAVAIL      TELL2       AVAILABLE
          TEST G      PH1,2,*+2
          FAVAIL      TELL3
          ADVANCE     18000       WAIT UNTIL 3.00
          LOGIC R     DOOR        CLOSE THE DOOR
          TEST E      Q$BANK,0    WAIT UNTIL CUSTOMERS LEAVE
          TERMINATE   1           END RUN
*
*** CONTROL CARDS ***
*
          START       1
          END
```

Figure X.7.8. The standard example in GPSS.

```
simulation begin

comment Example 7.4.1 of section 7.4.3;

process class passenger;
     begin
     Boolean taxpaid, repeat;
     real timein;
     taxpaid := draw (0.75, seeda);
     timein := time;
     activate new passenger delay negexp (1, seedb);
     if not freeclerks.empty then begin
                 activate freeclerks.first after current;
                 freeclerks.first.out end;
        wait (checkinq)
        end passenger;



process class clerk;
     begin
clerkloop: inspect checkinq.first
     when passenger do begin
                 checkinq.first.out;
                 if repeat then hold (1) else hold (normal
                         (3,1,seedc));
                 if taxpaid then begin
                             totpass := totpass + 1;
                             totwait := totwait + time - time in end
                 else begin
                             if not freetaxmen.empty then begin
                                 activate freetaxmen.first after current;
                                 freetaxmen.first.out end;
                         this passenger.into (taxq) end end
            otherwise wait (free clerks);
            goto clerkloop
            end clerk;



process class taxman;
     begin
taxloop: inspect taxq.first
     when passenger do begin
                 taxq.first.out; hold (negexp (2,seedd));
                 taxpaid := repeat := true;
                 if not freeclerks.empty then begin
                             activate freeclerks.first after current;
                             freeclerks.first.out end;
                     this passenger.into (checkinq) end
            otherwise wait (freetaxmen);
            goto taxloop
            end taxman;
            integer seeda, seedb, seedc, seedd, i;
            integer nclerks, ntaxmen, totpass;
            real totwait;
            ref (head) checkinq, taxq, freeclerks, freetaxmen;
```
                                                                              (continued)

```
seeda := 12345;    {and so on for the other seeds}
nclerks := inint; ntaxmen := inint;    {read these values}
checkinq :- new head; freetaxmen :- new head;
for i := 1  step 1 until nclerks do activate new clerk;
for i := 1  step 1 until ntaxmen do activate new taxman;
activate new passenger delay negexp (1, seedb);
hold (120);
outtext ('number of clerks'); outint (nclerks, 6); outimage;
outtext ('number of taxmen'); outint (ntaxmen, 6); outimage;
outtext ('mean wait'); outfix (totwait/totpass, 2,6); outimage

end simulation
```

Figure X.7.9. A simulation in Simula.

```
begin external class Demos;

comment Example 7.3.1 of section 7.3.4 recoded in Demos;

Demos begin
      ref (negexp) thinktime;
      ref (randint) aux ;
      ref (res) core, cpu ;
      ref (histogram) overall ;
      ref (histogram) array response [1:3] ;
      integer array time needed, core needed [1:3] ;
      integer i ;

      comment The unit of time is 1 msec ;

      entity class terminal ;
             begin
             integer class, time left, i ;
             real time began ;
             hold (thinktime.sample) ;
             i := aux.sample ;
             class := if i < 60 then 1 else
                          if i < 90 then 2 else 3 ;
             time began := time ;
             time left := time needed [class] ;
             core.acquire (core needed [class]) ;
             while time left > 0 do
                     begin cpu.acquire (1) ;
                           hold (100) ;
                           time left := time left -100 ;
                           cpu.release (1)
                     end ;
             core.release (core needed [class]) ;
             overall.update (time-time began) ;
             response [class].update (time-time began) ;
             repeat
             end terminal ;
             thinktime :- new negexp ("THINK TIME", 0.0001) ;
             aux    :- new randint ("AUXILIARY", 0,99) ;
             core   :- new res ("CORE", 256) ;
             cpu    :- new res ("CPU", 1) ;
             overall :- new histogram ("OVERALL", 0.0, 60000.0, 30) ;
             response [1] :-
                new histogram ("CLASS 1", 0.0, 15000.0, 15) ;
             response [2] :-
                new histogram ("CLASS 2", 0.0, 15000.0, 15) ;
             response [3] :-
                new histogram ("CLASS 3", 0.0, 60000.0, 30) ;
             time needed [1] := 100 ;
             time needed [2] := 500 ;
             time needed [3] := 2000 ;
             core needed [1] := 16 ;
             core needed [2] := 32 ;
             core needed [3] := 64 ;

             for i := 1 step 1 until 20 do
                     new terminal ("TERMINAL").schedule (0.0) ;

             hold (60000.0) ;
             reset ;
             hold (20.0 * 60000.0)
             end
      end
```

Figure X.7.10. A simulation using Demos.

```
begin
external class Demos;
Demos begin

comment    -    the running example - see section 1.4.

The time unit is one minute with time 0 at 9.00 a.m.;

integer n;
real x;
ref (count) arrivals, services, balks;
ref (erlang) servicetime;
ref (res) tellers;
ref (arrdist) nextarrival;
ref (balkdist) balk;

rdist class arrdist;
begin
real cumul; integer seg;
array x, y, mean [1 : 4];
ref (negexp) aux;

    real procedure sample;
    begin
    cumul := cumul + aux . sample;
    while cumul >= y[seg+1] do seg := seg + 1;
    sample := x[seg] + (cumul - y[seg]) * mean[seg] - time
    end;

x[1] :=  45.0; y[1] :=   0.0; mean[1] := 2.0;
x[2] := 120.0; y[2] :=  37.5; mean[2] := 1.0;
x[3] := 300.0; y[3] := 217.5; mean[3] := 2.0;
x[4] :=   0.0; y[4] := 1.0&9; mean[4] := 0.0;

aux :- new negexp ("arr aux", 1.0);
cumul := 0.0;
seg := 1
end *** definition of arrdist *** ;

bdist class balkdist (pr);
ref (res) pr;
begin

    Boolean procedure sample;
    sample := (pr.length - 5) / 5 > zyqsample;

    comment - zyqsample is the (0,1) uniform generator
             supplied by Demos;

end *** definition of balkdist *** ;
entity class customer;
begin
arrivals.update (1);
new customer ("customer") . schedule (nextarrival . sample);
if not balk . sample then
    begin
    tellers . acquire (1);
    services . update (1);
    hold (servicetime . sample);
    tellers . release (1)
    end
```

(continued)

```
else
    balks . update (1)
end *** definition of customer *** ;

trace;
    arrivals     :- new count ("arrivals");
    services     :- new count ("services");
    balks        :- new count ("balks");
    servicetime  :- new erlang ("servicetime", 2.0, 2);
    nextarrival  :- new arrdist ("interval");

    comment - draw number of tellers;
    x := new uniform ("aux unif", 0.0, 1.0) . sample;
    n := if x < 0.05 then 1
            else if x < 0.2 then 2
                else 3;

    comment - set up required tellers, though they are not
              yet available;
    tellers :- new res ("tellers", n);
    balk    :- new balkdist ("balks", tellers);
    tellers . acquire (n);

    comment - schedule first customer;
    new customer ("customer") . schedule (nextarrival . sample);

    comment - open door and make tellers available at 10.00 a.m.;
    hold (60.0);
    tellers . release (n);

    comment - wait until 3.00 p.m. and wrap up;
    hold (300.0);
    services . update (tellers . length)

end
end;
```

Figure X.7.11. The standard example in Demos.

```
      REAL FUNCTION ERLANG (MEAN, SHAPE, SEED)
      INTEGER I,J,K,SEED
      REAL X(25),FX(25),U,UNIF
      EXTERNAL UNIF
      DATA X /.0,.2,.4,.6,.8,1.0,1.2,1.4,1.6,1.8,2.0,
     X        2.2,2.4,2.6,2.8,3.0,3.2,3.4,3.6,3.8,4.0,
     X        5.0,6.0,7.0,7.79998/
      DATA FX/.0,.01752,.06155,.12190,.19121,
     X        .26424,.33737,.40817,.47507,.53716,
     X        .59399,.64543,.69156,.73262,.76892,
     X        .80085,.82880,.85316,.87431,.89262,
     X        .90842,.95957,.98265,.99270,1.0/
C
C     THIS FUNCTION GENERATES AN ERLANG WITH MEAN 2.0
C     AND SHAPE 2 USING LINEAR INTERPOLATION IN A TABLE
C     OF THE DISTRIBUTION
C
C     THE PARAMETERS MEAN AND SHAPE ARE NOT USED, BUT
C     REMAIN FOR COMPATIBILITY WITH THE MAIN PROGRAM.
C
      U = UNIF(SEED)
      I = 1
      J = 25
C
C     BINARY SEARCH FOR U IN TABLE
C
    1 K = (I + J)/2
      IF (FX(K).LE.U) GO TO 2
      J = K - 1
      GO TO 1
    2 IF (FX(K+1).GT.U) GO TO 3
      I = K + 1
      GO TO 1
    3 ERLANG = X(K) + (U - FX(K)) * (X(K+1) - X(K)) /
     X                    (FX(K+1) - FX(K))
      RETURN
      END
```

Figure X.8.1. A table-driven Erlang generator.

```
C
C    INITIALIZE FOR A NEW RUN
C
        IF (EVEN) GO TO 96
C
C ON AN ODD-NUMBERED RUN, WE START A NEW SEGMENT
C OF RANDOM NUMBERS AND SET AVGSRV = 2.0
C
        DO 95 I = 1,100
        X = UNIF(ARRSD)
        X = UNIF(SRVSD)
   95 X = UNIF(RNGSD)
        HLDARR = ARRSD
        HLDSRV = SRVSD
        HLDRNG = RNGSD
        AVGSRV = 2.0
        GO TO 97
C
C ON AN EVEN-NUMBERED RUN, WE USE THE SAME SEEDS
C AS BEFORE FOR ALL GENERATORS, BUT SET AVGSRV = 1.8
C
   96 ARRSD = HLDARR
        SRVSD = HLDSRV
        RNGSD = HLDRNG
        AVGSERV = 1.8
C
   97 EVEN = .NOT. EVEN
C
```

Figure X.8.2. Additional code when using common random numbers.

```
          PROGRAM STAREG
          INTEGER NTELL,NARR,NSERV,N,NM,I
          REAL X(200),Y(200),Z
          REAL ARR,SERV,SLOPE
          REAL SIGX,SIGY,SIGXY,SIGX2,SIGY2,SIGZ,SIGZ2
C
C   REGRESSION OF NSERV = Y AGAINST NARR = X
C   WE SUBTRACT OFF THE KNOWN MEAN 247.5 OF X
C   TO MAKE LIFE MORE SIMPLE
C
C   ESTIMATION IS BY SPLITTING
C
          N = 0
          SIGX = 0.0
          SIGY = 0.0
          SIGXY = 0.0
          SIGX2 = 0.0
          SIGY2 = 0.0
C
C   GET THE DATA AND CALCULATE THE APPROPRIATE SUMS
C
        1 READ (*,900, END = 2) NTELL,NARR,NSERV
      900 FORMAT (3I6)
C
C   NTELL IS IN FACT CONSTANT = 2
C
          ARR = NARR - 247.5
          SERV = NSERV
          N = N + 1
          SIGX = SIGX + ARR
          SIGY = SIGY + SERV
          SIGXY = SIGXY + ARR * SERV
          SIGX2 = SIGX2 + ARR * ARR
          SIGY2 = SIGY2 + SERV * SERV
          X(N) = ARR
          Y(N) = SERV
          GO TO 1
C
C   START BY PRINTING THE NAIVE ESTIMATE AS A CHECK
C
        2 PRINT 901,SIGY/N,(SIGY2 - SIGY*SIGY/N) /
      X        (N * (N-1))
      901 FORMAT (' NAIVE EST',F10.4,' EST VAR',F10.4)
C
C   NOW DO REGRESSION WITH SPLITTING
C   WE WILL CALL THE OBSERVATIONS Z
C
          NM = N - 1
          SIGZ = 0.0
          SIGZ2 = 0.0
          DO 3 I = 1,N
C
C   SUBTRACT OFF CURRENT X AND Y
C
          SIGX = SIGX - X(I)
          SIGY = SIGY - Y(I)
          SIGXY = SIGXY - X(I)*Y(I)
          SIGX2 = SIGX2 - X(I)*X(I)
          SIGY2 = SIGY2 - Y(I)*Y(I)
C
C   CALCULATE THE SLOPE USING THE REMAINING VALUES
```

(continued)

```
C
      SLOPE = (NM*SIGXY - SIGX*SIGY) / (NM*SIGX2 - SIGX*SIGX)
C
C  AND HENCE ONE OBSERVATION
C
      Z = Y(I) - SLOPE * X(I)
      SIGZ = SIGZ + Z
      SIGZ2 = SIGZ2 + Z * Z
C
C  PUT BACK THE CURRENT X AND Y
C
      SIGX = SIGX + X(I)
      SIGY = SIGY + Y(I)
      SIGXY = SIGXY + X(I)*Y(I)
      SIGX2 = SIGX2 + X(I)*X(I)
      SIGY2 = SIGY2 + Y(I)*Y(I)
    3 CONTINUE
C
C  WE SHOULD NOW HAVE A BETTER ESTIMATE
C
      PRINT 902,SIGZ/N,(SIGZ2 - SIGZ*SIGZ/N) /
     X      (N * (N-1))
  902 FORMAT (' EST WITH SPLITTING',F10.4,' EST VAR',
     X      F10.4,' --- BIASED --- ')
      STOP
      END
```

Figure X.8.3. Variance reduction by splitting.

```
      PROGRAM STASPL
      INTEGER NTELL,NARR,NSERV,N,NM,I
      REAL X(200),Y(200),ZN,Z
      REAL ARR,SERV,SLOPE
      REAL SIGX,SIGY,SIGXY,SIGX2,SIGY2,SIGZ,SIGZ2
C
C  REGRESSION OF NSERV = Y AGAINST NARR = X
C  WE SUBTRACT OFF THE KNOWN MEAN 247.5 OF X
C  TO MAKE LIFE MORE SIMPLE
C
C  ESTIMATION IS BY JACKKNIFING
C
      N = 0
      SIGX = 0.0
      SIGY = 0.0
      SIGXY = 0.0
      SIGX2 = 0.0
      SIGY2 = 0.0
C
C  GET THE DATA AND CALCULATE THE APPROPRIATE SUMS
C
    1 READ (*,900, END = 2) NTELL,NARR,NSERV
  900 FORMAT (3I6)
C
C     NTELL IS IN FACT CONSTANT = 2
C
      ARR = NARR - 247.5
      SERV = NSERV
      N = N + 1
      SIGX = SIGX + ARR
      SIGY = SIGY + SERV
      SIGXY = SIGXY + ARR * SERV
      SIGX2 = SIGX2 + ARR * ARR
      SIGY2 = SIGY2 + SERV * SERV
      X(N) = ARR
      Y(N) = SERV
      GO TO 1
C
C  START BY PRINTING THE NAIVE ESTIMATE AS A CHECK
C
    2 PRINT 901,SIGY/N,(SIGY2 - SIGY*SIGY/N) /
    X       (N * (N-1))
  901 FORMAT (' NAIVE EST',F10.4,' EST VAR',F10.4)
C
C  FIRST CALCULATE ZN USING ALL N VALUES
C
      ZN = (SIGY*SIGX2 - SIGX*SIGXY) / (N*SIGX2 - SIGX*SIGX)
      PRINT 903,ZN
  903 FORMAT (' ESTIMATE USING ALL N VALUES',F10.4)
C
C  NOW CALCULATE THE PSEUDOVALUES
C
      NM = N - 1
      SIGZ = 0.0
      SIGZ2 = 0.0
      DO 3 I = 1,N
C
C  SUBTRACT OFF CURRENT X AND Y
C
      SIGX = SIGX - X(I)
      SIGY = SIGY - Y(I)
```

(continued)

```
      SIGXY = SIGXY - X(I)*Y(I)
      SIGX2 = SIGX2 - X(I)*X(I)
      SIGY2 = SIGY2 - Y(I)*Y(I)
C
C  CALCULATE AN ESTIMATE USING ONLY N-1 VALUES
C
      Z = (SIGY*SIGX2 - SIGX*SIGXY) / (NM*SIGX2 - SIGX*SIGX)
C
C  AND HENCE CALCULATE THE NEXT PSEUDOVALUE
C
      Z = N * ZN - NM * Z
      SIGZ = SIGZ + Z
      SIGZ2 = SIGZ2 + Z * Z
C
C  PUT BACK THE CURRENT X AND Y
C
      SIGX = SIGX + X(I)
      SIGY = SIGY + Y(I)
      SIGXY = SIGXY + X(I)*Y(I)
      SIGX2 = SIGX2 + X(I)*X(I)
      SIGY2 = SIGY2 + Y(I)*Y(I)
    3 CONTINUE
C
C  WE SHOULD NOW HAVE A BETTER ESTIMATE
C
      PRINT 902,SIGZ/N,(SIGZ2 - SIGZ*SIGZ/N) /
     X     (N * (N-1))
  902 FORMAT (' EST WITH JACKKNIFING',F10.4,' EST VAR',
     X     F10.4,' --- BIASED --- ')
      STOP
      END
```

Figure X.8.4. Variance reduction by jackknifing.

```
      REAL FUNCTION B(SRVSD)
      REAL X,UNIF,ERLANG,BC,BT,BLIM,BMULT
      INTEGER SRVSD
      EXTERNAL UNIF, ERLANG
      COMMON /BPARAM/ BC,BT,BLIM,BMULT
C
      IF (UNIF(SRVSD).GT.BLIM) THEN
C
C  USE TRUNCATED ERLANG
C
    1     B = ERLANG(2.0,2,SRVSD)
          IF (B.GT.BT) GO TO 1
          BMULT = BMULT * BC
      ELSE
C
C  USE TRUNCATED EXPONENTIAL
C
          B = BT - ALOG(UNIF(SRVSD))
          BMULT = BMULT * BC * B / BT
      ENDIF
      RETURN
      END
```

Figure X.8.5. Importance sampling—the function B.

References

Abramowitz, M., and Stegun, I. A. (1964). *Handbook of Mathematical Functions with Formulas, Graphs, and Mathematical Tables.* Dover: New York.

Adelson, R. M. (1966). Compound Poisson distributions. *Oper. Res. Quart.*, **17**, 73–75.

Ahrens, J. H., and Dieter, U. (1972a). Computer methods for sampling from the exponential and normal distributions. *Commun. ACM*, **15**, 873–882.

Ahrens, J. H., and Dieter, U. (1972b). Computer methods for sampling from gamma, beta, Poisson and binomial distributions. *Computing*, **12**, 223–246.

Ahrens, J. H., and Dieter, U. (1980). Sampling from binomial and Poisson distributions: a method with bounded computation times. *Computing*, **25**, 193–208.

Ahrens, J. H., and Dieter, U. (1982a). Generating gamma variates by a modified rejection technique. *Commun. ACM*, **25**, 47–54.

Ahrens, J. H., and Dieter, U. (1982b). Computer generation of Poisson deviates from modified normal distributions. *ACM Trans. Math. Software*, **8**, 163–179.

Ahrens, J. H., and Kohrt, K. D. (1981). Computer methods for efficient sampling from largely arbitrary statistical distributions. *Computing*, **26**, 19–31.

Albin, S. L. (1982). On the Poisson approximations for superposition arrival processes in queues. *Manage. Sci.*, **28**, 126–137.

Anderson, T. W., and Darling, D. A. (1954). A test of goodness of fit. *JASA*, **49**, 765–769.

Andrews, D. F., Bickel, P. J., Hampel, F. R., Huber, P. J., Rogers, W. H., and Tukey, J. W. (1972). *Robust Estimates of Location.* Princeton University Press: Princeton, N.J.

ANSI (1978). *American National Standard Programming Language FORTRAN.* American National Standards Institute: New York.

Arvidsen, N. I., and Johnsson, T. (1982). Variance reduction through negative correlation, a simulation study. *J. Stat. Comput. Simul.*, **15**, 119–127.

Asmussen, S. (1985). Conjugate processes and the simulation of ruin problems. *Stochastic Process. Appl.*, **20**, 213–229.

Avriel, M. (1976). *Nonlinear Programming: Analysis and Methods.* Prentice-Hall: Englewood Cliffs, N.J.

Babad, J. M. (1975). The IBM GPSS random number generator. Technical Report, University of Chicago, Graduate School of Business.

Barlow, R. E., and Proschan, F. (1965). *Mathematical Theory of Reliability.* Wiley: New York.

Barr, D. R., and Slezak, N. L. (1972). A comparison of multivariate normal generators. *Commun. ACM*, **15**, 1048–1049.

Bazaraa, M. S., and Shetty, C. M. (1979). *Nonlinear Programming: Theory and Algorithms.* Wiley: New York.

Bennett, C. H. (1979). On random and hard-to-describe numbers. IBM Watson Research Center Report RC 7483 (No. 32272), Yorktown Heights, N.Y.

Bentley, J. L., and Saxe, J. B. (1980). Generating sorted lists of random numbers. *ACM Trans. Math. Software*, **6**, 359–364.

Bergström, H. (1952). On some expansions of stable distributions. *Ark. Mat. II*, **18**, 375–378.

Berman, M. B. (1970). Generating random variates from gamma distributions with non-integer shape parameters. Report R-641-PR, The RAND Corporation, Santa Monica, Calif.

Best, D. J., and Roberts, D. E. (1975). Algorithm AS 91: the percentage points of the χ^2 distribution. *Appl. Stat.*, **24**, 385–388.

Bhattacharjee, G. P. (1970). Algorithm AS 32—The incomplete gamma integral. *Appl. Stat.*, **19**, 285–287.

Billingsley, P. (1968). *Convergence of Probability Measures.* Wiley: New York.

Billingsley, P. (1986). *Probability and Measure.* Wiley: New York.

Birtwhistle, G. M. (1979a). *Demos Reference Manual.* Dep. of Computer Science, University of Bradford: Bradford.

Birtwhistle, G. M. (1979b). *Demos—Discrete Event Modelling on Simula.* Macmillan: London.

Birtwhistle, G. M., Dahl, O.-J., Myhrhaug, B., and Nygaard, K. (1973). *Simula Begin.* Studentlitteratur: Oslo.

Bobillier, P. A., Kahan, B. C., and Probst, A. R. (1976). *Simulation with GPSS and GPSS V.* Prentice-Hall: Englewood Cliffs, N.J.

Boender, C. G. E., Rinnooy Kan, A. H. G., Timmer, G. T., and Stougie, L. (1982). A stochastic method for global optimization. *Math. Program.*, **22**, 125–140.

Boneh, A. (1983). Preduce—a probabilistic algorithm identifying redundancy by a random feasible point generation. (Lecture Notes in Mathematics, #206, Chap. 10). *Redundancy in Mathematical Programming* (M. H. Karman, V. Lofti, and J. Telgen, eds.). Springer-Verlag: New York.

Box, G. E. P., Hunter, W. G., and Hunter, J. S. (1978). *Statistics for Experimenters.* Wiley: New York.

Box, G. E. P., and Muller, M. E. (1958). A note on the generation of random normal deviates. *Ann. Math. Stat.*, **29**, 610–611.

Bratley, P., and Fox, B. L. (1986). Implementing Sobol's quasirandom sequence generator. Technical Report. Université de Montréal.

Braun, J. E. (editor) (1983). *Simscript II.5 Reference Handbook*, 2nd edn., CACI Inc.: Los Angeles.

Bright, H. S., and Enison, R. L. (1979). Quasi-random number sequences from a long-period TLP generator with remarks on application to cryptography. *Comput. Surv.*, **11**, 357–370.

Brillinger, D. R. (1981). *Time Series: Data Analysis and Theory.* Holden-Day: San Francisco.

Brinch Hansen, P. (1973). *Operating System Principles.* Prentice-Hall: Englewood Cliffs, N.J.

Brinch Hansen, P. (1977). *The Architecture of Concurrent Programs.* Prentice-Hall: Englewood Cliffs, N.J.

Brown, C. A., and Purdom, P. W. (1981). An average time analysis of backtracking. *SIAM J. Comput.*, **10**, 583–593.

Brown, M., and Solomon, H. (1979). On combining pseudorandom number generators. *Ann. Stat.*, **1**, 691–695.

Brown, M., Solomon, H., and Stephens, M. A. (1981). Monte Carlo simulation of the renewal process. *J. Appl. Probab.*, **18**, 426–434.

Brumelle, S. L. (1971). Some inequalities for parallel service queues. *Oper. Res.*, **19**, 402–413.

Callaert, H., and Janssen, P. (1981). The convergence rate of fixed-width confidence intervals for the mean. *Sankyā*, **43**, 211–219.

Carson, J. S. (1978). Variance reduction techniques for simulated queuing processes. Report No. 78–8, Dept. of Industrial Engineering, University of Wisconsin, Madison.

Carson, J. S. (1980). Linear combinations of indirect estimators for variance reduction in regenerative simulations. Technical Report, Dept. of Industrial and Systems Engineering, Georgia Institute of Technology, Atlanta.

Carson, J. S., and Law, A. M. (1980). Conservation equations and variance reduction in queuing simulations. *Oper. Res.*, **28**, 535–546.

Carter, G., and Ignall, E. J. (1975). Virtual measures: a variance reduction technique for simulation. *Manage. Sci.*, **21**, 607–617.

CDC (1972). *MIMIC Digital Simulation Language Reference Manual*. Publication No. 44610400. Control Data Corporation: Sunnyvale, Calif.

Chambers, J. M., Mallows, C. L., and Stuck, B. W. (1976). A method for simulating stable random variables. *JASA*, **71**, 340–344.

Chen, H. C., and Asau, Y. (1974). On generating random variates from an empirical distribution. *AIEE Trans.*, **6**, 163–166.

Cheng, R. C. H. (1978). Generating beta variates with nonintegral shape parameters. *Commun. ACM*, **21**, 317–322.

Cheng, R. C. H. (1981). The use of antithetic control variates in computer simulation. *Proceedings 1981 Winter Simulation Conference*, pp. 313–318.

Cheng, R. C. H. (1982). The use of antithetic variates in computer simulations. *J. Opl. Res. Soc.*, **33**, 229–237.

Cheng, R. C. H., and Feast, G. M. (1979). Some simple gamma variate generators. *Appl. Stat.*, **28**, 290–295.

Chouinard, A., and McDonald, D. (1982). A rank test for non-homogeneous Poisson processes. Manuscript, Dept. of Mathematics, University of Ottawa.

Chow, Y. S., and Robbins, H. (1965). On the asymptotic theory of fixed-width sequential confidence intervals for the mean. *Ann. Math. Stat.*, **36**, 457–462.

Chu, Y. (1969). *Digital Simulation of Continuous Systems*. McGraw-Hill: New York.

Çinlar, E. (1972). Superposition of point processes. In *Stochastic Point Processes: Statistical Analysis, Theory, and Applications*, (P. A. W. Lewis, editor), pp. 549–606. Wiley: New York.

Çinlar, E. (1975). *Introduction to Stochastic Processes*. Prentice-Hall: Englewood Cliffs, N.J.

Clark, G. M. (1981). Use of Polya distributions in approximate solutions to nonstationary $M/M/s$ queues. *Commun. ACM*, **24**, 206–217.

Clark, R. L. (1973). A linguistic contribution to GOTO-less programming. *Datamation*, December, 62–63.

Clementson, A. T. (1973). *ECSL User's Manual*. Department of Engineering Production, University of Birmingham: Birmingham.

Cochran, W. G. (1977). *Sampling Techniques*. Wiley: New York.

Conte, S. D., and de Boor, C. (1980). *Elementary Numerical Analysis*. McGraw-Hill: New York.

Conway, R. W. (1963). Some tactical problems in digital simulation. *Manage. Sci.*, **10**, 47–61.

Cooper, R. B. (1981). *Introduction to Queueing Theory*, 2nd edn. North Holland: New York.

Coveyou, R. R., and McPherson, R. D. (1967). Fourier analysis of uniform random number generators. *J. ACM*, **14**, 100–119.

Cox, D. R., and Smith, W. L. (1961). *Queues*. Wiley: New York.

Crane, M. A., and Lemoine, A. J. (1977). *An Introduction to the Regenerative Method for Simulation Analysis*. (Lecture Notes in Control and Information Sciences.) Springer-Verlag: New York, Heidelberg, Berlin.

Dahl, O.-J., Myhrhaug, B., and Nygaard, K. (1970). *Common Base Language*. Publication S-22. Norwegian Computing Center: Oslo.

Dahl, O.-J., and Nygaard, K. (1967). *Simula: A Language for Programming and Description of Discrete Event Systems. Introduction and user's manual*, 5th edn. Norwegian Computing Center: Oslo.

Dannenbring, D. G. (1977). Procedures for estimating optimal solution values for large combinatorial problems. *Manage. Sci.*, **23**, 1273–1283.

Davey, D. (1982). Une contribution aux méthodes de construction des langages de simulation discrets. Ph.D. Thesis, Département d'informatique et de recherche opérationnelle, Université de Montréal.

Davey, D., and Vaucher, J. G. (1980). Self-optimizing partitioned sequencing sets for discrete event simulation. *INFOR*, **18**, 41–61.

Deák, I. (1980). Three digit accurate multiple normal probabilities. *Numer. Math.*, **35**, 369–380.

De Millo, R. A., Lipton, R. J., and Perlis, A. J. (1979). Social processes and proofs of theorems and programs. *Commun. ACM*, **22**, 271–280.

Denardo, E. V., and Fox, B. L. (1979). Shortest-route methods: 1, reaching, pruning, and buckets. *Oper. Res.* **27**, 161–186.

Devroye, L. (1979). Inequalities for the completion times of stochastic PERT networks. *Math. Oper. Res.*, **4**, 441–447.

Devroye, L. (1980). The computer generation of binomial random variables. Technical Report, School of Computer Science, McGill University, Montreal.

Devroye, L. (1981a). The computer generation of Poisson random variables. *Computing*, **26**, 197–207.

Devroye, L. (1981c). Personal communication and unpublished course notes.

Devroye, L. (1982). A note on approximations in random variate generation. *J. Stat. Comput. Simul.*, **14**, 149–158.

Devroye, L. (1984a). Random variate generation for unimodal and monotone densities. *Computing*, **32**, 43–68.

Devroye, L. (1984b). A simple algorithm for generating random variates with a log-concave density. *Computing*, **33**, 247–257.

Devroye, L. (1986). *Non-uniform Random Variate Generation*. Springer-Verlag: New York.

Dieter, U. (1971). Pseudo-random numbers: the exact distribution of pairs. *Math. Comput.*, **25**, 855–883.

Dieter, U. (1975). How to calculate shortest vectors in a lattice. *Math. Comput.*, **29**, 827–833.

Dieter, U. (1982). Personal communication.

Dieter, U., and Ahrens, J. (1971). An exact determination of serial correlations of pseudo-random numbers. *Numer. Math.*, **17**, 101–123.

Disney, R. L. (1980). A tutorial on Markov renewal theory, semi-regenerative processes, and their applications. Technical Report, Virginia Polytechnic Institute and State University, Blacksburg, Va.

Downham, D. Y., and Roberts, F. D. K. (1967). Multiplicative congruential pseudo-random number generators. *Comput. J.*, **10**, 74–77.

Draper, N. R., and Smith, H. (1981). *Applied Regression Analysis*. Wiley: New York.

Duersch, R., and Schruben, L. (1986). An interactive run length control for simulation on PC's. *Proceedings Winter Simulation Conference*, Washington, D.C.

Duket, S. D., and Pritsker, A. A. B. (1978). Examination of simulation output using spectral methods. *Math. Comput. Simul.*, **20**, 53–60.

Dussault, J.-P. (1980). Réduction du biais des estimateurs de la variance dans une expérience avec variables de contrôle. Publication No. 389, Département d'informatique et de recherche opérationnelle, Université de Montréal.

Edwards, P. (1967). *The Encyclopedia of Philosophy*. Macmillan: New York.

Efron, B. (1982). *The Jackknife, the Bootstrap and other Resampling Plans*. (CBMS-NSF Series.) SIAM: Philadelphia.

Efron, B., and Stein, C. (1981). The jackknife estimate of variance. *Ann. Stat.*, **9**, 586–596.

Epstein, B. (1960). Testing for the validity of the assumption that the underlying distribution of life is exponential. *Technometrics*, **2**, 83–101.

Fama, E., and Roll, R. (1968). Some properties of symmetric stable distributions. *J. Amer. Stat. Assoc.*, **63**, 817–836.

Fama, E., and Roll, R. (1971). Parameter estimates for symmetric stable distributions. *J. Amer. Stat. Assoc.*, **66**, 331–338.

Fan, C. T., Muller, M. E., and Rezucha, I. (1962). Development of sampling plans by using sequential (item by item) selection techniques and digital computers. *JASA*, **57**, 387–402.

Federgruen, A., and Schweitzer, P. J. (1978). A survey of asymptotic value-iteration for undiscounted Markovian decision problems. Working Paper No. 7833, Graduate School of Management, University of Rochester.

Feller, W. (1971). *An Introduction to Probability Theory and Its Applications*, Vol. 2, 2nd edn. Wiley: New York.

Fishman, G. S. (1977). Achieving specific accuracy in simulation output analysis. *Commun. ACM*, **20**, 310–315.

Fishman, G. S. (1978). *Principles of Discrete Event Simulation*. Wiley: New York.

Fishman, G. S., and Moore, L. R. (1986). An exhaustive analysis of multiplicative congruential random number generators with modulus $2^{31} - 1$. *SIAM J. Sci. Statist. Comput.*, **7**, 24–45.

Ford, L. R., Jr., and Fulkerson, D. R. (1962). *Flows in Networks*. Princeton University Press: Princeton, N.J.

Forsythe, G. E. (1972). Von Neumann's comparison method for random sampling from the normal and other distributions. *Math. Comput.*, **26**, 817–826.

Fox, B. L. (1963). Generation of random samples from the beta and F distributions. *Technometrics*, **5**, 269–270.

Fox, B. L. (1968). Letter to the editor. *Technometrics*, **10**, 883–884.

Fox, B. L. (1978a). Estimation and simulation (communication to the editor). *Manage. Sci.* **24**, 860–861.

Fox, B. L. (1978b). Data structures and computer science techniques in operations research. *Oper. Res.*, **26**, 686–717.

Fox, B. L. (1981). Fitting 'standard' distributions to data is necessarily good: dogma or myth? *Proceedings 1981 Winter Simulation Conference*, pp. 305–307.

Fox, B. L. (1986). Implementation and relative efficiency of quasirandom sequence generators. *ACM Trans. Math. Software* (to appear).

Fox, B. L., and Glynn, P. W. (1986a). Discrete-time conversion for simulating semi-Markov processes. *Oper. Res. Lett.*, **5**, 191–196.

Fox, B. L., and Glynn, P. W. (1986b). Estimating time averages via randomly-spaced observations. *SIAM J. Appl. Math.* (to appear).

Fox, B. L., and Glynn, P. W. (1986c). Conditional confidence intervals. Manuscript. University of Wisconsin.

Fox, B. L., and Glynn, P. W. (1986d). Numerical methods for transient Markov chains. Manuscript. Université de Montréal.

Fox, B. L., and Glynn, P. W. (1986e). Computing Poisson probabilities. Technical Report. Université de Montréal.

Franta, W., R. (1975). A note on random variate generators and antithetic sampling. *INFOR*, **13**, 112–117.

Franta, W. R. (1977). *The Process View of Simulation*. North Holland: New York.

Fulkerson, D. R. (1962). Expected critical path lengths in PERT networks. *Oper. Res.*, **10**, 808–817.

Gafarian, A. V., Ancker, C. J., Jr., and Morisaku, T. (1978). Evaluation of commonly used rules for detecting 'steady state' in computer simulation. *Nav. Res. Logistics Quart.*, **25**, 511–529.

Galambos, J. (1978). *The Asymptotic Theory of Extreme Order Statistics*. Wiley: New York.

Garey, M. R., and Johnson, D. S. (1979). *Computers and Intractability*. Freeman: San Francisco.

Gavish, B., and Merchant, D. K. (1978). Binary level testing of pseudo-random number generators. Technical Report, University of Rochester.

Gerontidis, I., and Smith, R. L. (1982). Monte Carlo generation of order statistics from general distributions. *Appl. Stat.* **31**, 238–243.

Glynn, P. W. (1982a). Some new results in regenerative process theory. Technical Report 16, Dept. of Operations Research, Stanford University, Stanford, Calif.

Glynn, P. W. (1982b). Regenerative aspects of the steady-state simulation problem for Markov chains. Technical Report 17, Dept. of Operations Research, Stanford University, Stanford, Calif.

Glynn, P. W. (1982c). Regenerative simulation of Harris recurrent Markov chains. Technical Report 18, Dept. of Operations Research, Stanford University, Stanford, Calif.

Glynn, P. W. (1982d). Coverage error for confidence intervals in simulation output analysis. Technical Report 15, Dept. of Operations Research, Stanford University, Stanford, Calif.

Glynn, P. W. (1982e). Asymptotic theory for nonparametric confidence intervals. Technical Report 19, Dept. of Operations Research, Stanford University, Stanford, Calif.

Glynn, P. W. (1986a). Optimization of stochastic systems. *Proceedings Winter Simulation Conference*, Washington, D.C.

Glynn, P. W. (1986b). Sensitivity analysis for stationary probabilities of a Markov chain. *Proceedings Winter Simulation Conference*, Washington, D.C.

Glynn, P. W., and Iglehart, D. L. (1985). Large-sample theory for standardized time series: an overview. *Proceedings Winter Simulation Conference*.

Glynn, P. W., and Iglehart, D. L. (1986). Consistent variance estimation for steady-state simulation.

Glynn, P. W., and Iglehart, D. L. (1988). The theory of standardized time series. *Math. Oper. Res.* (to appear).

Glynn, P. W., and Whitt, W. (1986a). Extensions of the queueing relations $L = \lambda W$ and $H = \lambda G$.

Glynn, P. W., and Whitt, W. (1986b). A central-limit-theorem-version of $L = \lambda W$. *Queueing Systems*, **2**, 191–215.

Glynn, P. W., and Whitt, W. (1986c). The efficiency of simulation estimators. Technical report. AT & T Bell Laboratories, Holmdel, New Jersey.

Glynn, P. W., and Whitt, W. (1987). Indirect estimation via $L = \lambda W$. *Operations Research* (to appear).

Gorenstein, S. (1967). Testing a random number generator. *Commun. ACM*, **10**, 111–118.

Granovsky, B. L. (1981). Optimal formulae of the conditional Monte Carlo. *SIAM J. Alg. Discrete Math.*, **2**, 289–294.

Grassmann, W. K. (1981). The optimal estimation of the expected number in a $M/D/\infty$ queueing system. *Oper. Res.*, **29**, 1208–1211.

Gray, H. L., and Schucany, W. R. (1972). *The Generalized Jackknife Statistic*. Dekker: New York.

Greenberger, M. (1961). Notes on a new pseudo-random number generator. *J. ACM*, **8**, 163–167.

Guibas, L. J., and Odlyzko, A. M. (1980). Long repetitive patterns in random sequences. *Z. Wahrsch.*, **53**, 241–262.

Gumbel, E. J. (1958). *Statistics of Extremes*. Columbia University Press: New York.

Gunther, F. L., and Wolff, R. W. (1980). The almost regenerative method for stochastic system simulations. *Oper. Res.*, **28**, 375–386.

Halfin, S. (1982). Linear estimators for a class of stationary queueing processes. *Oper. Res.*, **30**, 515–529.

Halton, J. (1970). A retrospective and prospective survey of the Monte Carlo method. *SIAM Rev.*, **12**, 1–63.

Hammersley, J. M., and Handscomb, D. C. (1964). *Monte Carlo Methods*. Methuen: London.

Hannan, E. J. (1970). *Multiple Time Series*. Wiley: New York.

Harris, C. M. (1976). A note on testing for exponentiality. *Nav. Res. Logistics Quart.*, **23**, 169–175.

Harrison, J. M., and Lemoine, A. J. (1977). Limit theorems for periodic queues. *J. Appl. Probab.*, **14**, 566–576.

Hart, J. F. *et al.* (1968). *Computer Approximations*. SIAM Series in Applied Mathematics, Wiley: New York.

✓Heath, D., and Sanchez, P. (1986). On the adequacy of pseudorandom number generators. *Oper. Res. Lett.*, **5**, 3–6.

Heildelberger, P. (1986). Limitations of infinitesimal perturbation analysis. Technical Report. IBM T. J. Watson Research Center. Yorktown Heights, N.Y.

Heidelberger, P., and Iglehart, D. L. (1979). Comparing stochastic systems using regenerative simulations with common random numbers. *Adv. Appl. Probab.*, **11**, 804–819.

Heidelberger, P., and Lewis, P. A. W. (1981). Regression-adjusted estimates for regenerative simulations, with graphics. *Commun. ACM*, **24**, 260–273.

Heidelberger, P., and Welch, P. D. (1981a). A spectral method for simulation confidence interval generation and run length control. *Commun. ACM*, **24**, 233–245.

Heidelberger, P., and Welch, P. D. (1981b). Adaptive spectral methods for simulation output analysis. *IBM J. Res. Devel.*, **25**, 860–876.

Heidelberger, P., and Welch, P. D. (1982). Simulation run length control in the presence of an initial transient. IBM Technical Report.

Heidelberger, P., and Welch, P. (1983). Simulation run length control in the presence of an initial transient. *Oper. Res.* **31**, 1109–1144.

Henriksen, J. O. (1979). An interactive debugging facility for GPSS. *ACM SIGSIM Simuletter*, **10**, 60–67.

⎧ Heyman, D. P., and Sobel, M. J. (1982). *Stochastic Models in Operations Research*,
⎨   Vol. 1, McGraw-Hill: New York.
⎩ Heyman, D. P., and Sobel, M. J. (1984). *Stochastic Models in Operations Research*, Vol. 2, McGraw-Hill: New York.

Heyman, D. P., and Stidham, Jr., S., (1980). The relation between customer and time averages in queues. *Oper. Res.*, **28**, 983–994.

Hill, G. W. (1970). Algorithm 395: Student's *t*-distribution. *Commun. ACM*, **13**, 617–619.

Hinderer, K. (1978). Approximate solutions of finite-state dynamic programs. In *Dynamic Programming and Its Applications* (M. L. Puterman, editor), pp. 289–317, Academic Press: New York.

Hoare, C. A. R. (1981). The emperor's old clothes. *Commun. ACM*, **24**, 75–83.

Hobson, E. W. (1957). *The Theory of Functions of a Real Variable*, Vol. 1. Dover: New York (reprint of the 1927 edition).

Hordijk, A., Iglehart, D. L., and Schassberger, R. (1976). Discrete-time methods of simulating continuous-time Markov chains. *Adv. Appl. Probab.*, **8**, 772–788.

IBM (1971). *General Purpose Simulation System V User's Manual*, 2nd edn. Publication SH20-0851-1. IBM: New York.

IBM (1972a). *SIMPL/1 Program Reference Manual*. Publication SH19-5060-0. IBM: New York.

IBM (1972b). *SIMPL/1 Operation Guide*. Publication SH19-5038-0. IBM: New York.

Iglehart, D. L. (1978). The regenerative method for simulation analysis. In *Current Trends in Programming Methodology—Software Modeling* (K. M. Chandy and R. T. Yeh, editors). Prentice-Hall: Englewood Cliffs, N.J.

Iglehart, D. L., and Lewis, P. A. W. (1979). Regenerative simulation with internal controls. *J. ACM*, **26**, 271–282.

Iglehart, D. L., and Shedler, G. S. (1983). Statistical efficiency of regenerative simulation methods for networks of queues. *Adv. Appl. Prob.* **15**, 183–197.

Iglehart, D. L., and Stone, M. L. (1982). Regenerative simulation for estimating extreme values. Technical Report 20, Dept. of Operations Research, Stanford University, Stanford, Calif.

Ignall, E. J., Kolesar, P., and Walker, W. E. (1978). Using simulation to develop and validate analytic models: some case studies. *Oper. Res.*, **26**, 237–253.

IMSL (1980). *IMSL Library*, Vol. I, 8th edn. Distributed by International Mathematical and Statistical Libraries, Inc., Houston, Tex.

Jensen, K., and Wirth, N. (1976). *PASCAL User Manual and Report*, 2nd edn., 2nd printing. Springer-Verlag: New York, Heidelberg, Berlin.

Jöhnk, M. D. (1964). Erzeugung von Betaverteilten und Gammaverteilten Zufallszahlen. *Metrika*, **8**, 5–15.

Kachitvichyanukul, V., and Schmeiser, B. (1985). Computer generation of hypergeometric random variates. *J. Statist. Comput. Simul.*, **22**, 127–145.

Karlin, S., and Taylor, H. M. (1975). *A First Course in Stochastic Processes*, 2nd edn. Academic Press: New York.

Katzan, H., Jr., (1971). *APL User's Guide*. Van Nostrand Reinhold: New York.

Kelton, W. D. (1980). The startup problem in discrete-event simulation. Report 80-1, Dept. of Industrial Engineering, University of Wisconsin, Madison.

Kelton, W. D., and Law, A. M. (1984). An analytical evaluation of alternative strategies in steady-state simulation. *Oper. Res.*, **32**, 169–184.

Kennedy, Jr., W. J., and Gentle, J. E. (1980). *Statistical Computing*. Dekker: New York.

Kershenbaum, A., and Van Slyke, R. (1972). Computing minimum spanning trees efficiently. *Proceedings of the 25th ACM National Conference*, pp. 518–527.

Khintchine, A. Y. (1969). *Mathematical Methods in the Theory of Queueing*. Hafner: New York.

Kinderman, A. J., and Ramage, J. G. (1976). Computer generation of normal random variables. *JASA*, **71**, 893–896.

Kleijnen, J. P. C. (1974). *Statistical Techniques in Simulation*, Part 1. Dekker: New York.

Kleijnen, J. P. C. (1975). *Statistical Techniques in Simulation*, Part 2. Dekker: New York.

Kleijnen, J. P. C. (1978). Communication to the editor (reply to Fox (1978a) and Schruben (1978)). *Manage. Sci.*, **24**, 1772–1774.

Kleinrock, L. (1975). *Queueing Systems*, Vol. 1. Wiley: New York.

Kleinrock, L. (1976). *Queueing Systems*, Vol. 2. Wiley: New York.

Knuth, D. E. (1973). *The Art of Computer Programming*, Vol. 1, 2nd edn. Addison-Wesley: Reading, Mass.

Knuth, D. E. (1975). Estimating the efficiency of backtrack programs. *Math. Comp.*, **29**, 121–136.

Knuth, D. E. (1981). *The Art of Computer Programming*, Vol. 2, 2nd edn. Addison-Wesley: Reading, Mass.

G Koopman, B. O. (1977). Intuition in mathematical operations research. *Oper. Res.*, **25**, 189–206.

✓ Koopman, B. O. (1979). An operational critique of detection. *Oper. Res.*, **27**, 115–133.

Kronmal, R. A., and Peterson, A. V. (1979). On the alias method for generating random variables from a discrete distribution. *Amer. Stat.*, **33**, 214–218.

Kronmal, R. A., and Peterson, A. V. (1981). A variant of the acceptance/rejection method for computer generation of random variables. *JASA*, **76**, 446–451.

Landauer, J. P. (1976). Hybrid digital/analog computer systems. *Computer*, **9**, 15–24.

Landwehr, C. E. (1980). An abstract type for statistics collection in SIMULA. *ACM Trans. Prog. Lang. Syst.*, **2**, 544–563.

Larmouth, J. (1981). Fortran 77 portability. *Software-Pract. Exper.*, **11**, 1071–1117.

Lavenberg, S. S., Moeller, T. L., and Welch, P. D. (1982). Statistical results on multiple control variables with application to queueing network simulation. *Oper. Res.*, **30**, 182–202.

Lavenberg, S. S., and Sauer, C. H. (1977). Sequential stopping rules for the regenerative method of simulation. *IBM J. Res. Develop.*, **21**, 545–558.

Lavenberg, S. S., and Welch, P. D. (1981). A perspective on the use of control variables to increase the efficiency of Monte Carlo simulations. *Manage. Sci.*, **27**, 322–335.

Law, A. M. (1975). Efficient estimators for simulated queueing systems. *Manage. Sci.*, **22**, 30–41.

Law, A. M. (1977). Confidence intervals in discrete-event simulation: a comparison of replication and batch means. *Nav. Res. Logistics Quart.*, **24**, 667–678.

Law, A. M. (1979). *Statistical Analysis of Simulation Output Data With Simscript II.5.* CACI Inc.: Los Angeles.

Law, A. M., and Carson, J. S. (1979). A sequential procedure for determining the length of a steady-state simulation. *Oper. Res.*, **27**, 1011–1025.

Law, A. M., and Kelton, W. D. (1984). Confidence intervals for steady-state simulation, I: A survey of fixed sample size procedures. *Oper. Res.*, **32**, 1221–1239.

Law, A. M., and Kelton, W. D. (1982). Confidence intervals for steady-state simulation, II: A survey of sequential procedures. *Manage. Sci.*, **28**, 550–562.

✓ L'Ecuyer, P. (1986). Efficient and portable combined pseudorandom number generators. *Commun. ACM* (to appear).

Lee, Y. T., and Requicha, A. A. G. (1982a). Algorithms for computing the volume and other integral properties of solids, I: Known methods and open issues. *Commun. ACM*, **25**, 635–641.

Lee, Y. T., and Requicha, A. A. G. (1982b). Algorithms for computing the volume and other integral properties of solids, II: A family of algorithms based on representation conversion and cellular approximation. *Commun. ACM*, **25**, 642–650.

Lehmann, E. L. (1966). Some concepts of dependence. *Ann. Math. Stat.*, **37**, 1137–1153.

Leringe, Ö., and Sundblad, Y. (1975). *Simula för den som kan Algol* (English translation: "Simula for those who know Algol"). Tekniska Högskolans Studentkår Kompendieförmedlingen: Stockholm.

Lewis, P. A. W., and Shedler, G. S. (1979a). Simulation of nonhomogeneous Poisson processes by thinning. *Nav. Res. Logistics Quart.*, **26**, 403–414.

Lewis, P. A. W., and Shedler, G. S. (1979b). Simulation of nonhomogeneous Poisson processes with degree-two exponential polynomial rate function. *Oper. Res.*, **27**, 1026–1040.

Lewis, T. G., and Payne, W. H. (1973). Generalized feedback shift register pseudorandom number algorithm. *J. ACM*, **20**, 456–468.

✓ Lilliefors, H. W. (1967). On the Kolmogorov–Smirnov test for normality with mean and variance unknown. *JASA*, **62**, 399–402.

Lilliefors, H. W. (1969). On the Kolmogorov–Smirnov test for the exponential distribution with mean unknown. *JASA*, **64**, 387–389.

Ling, R. E., and Roberts, H. V. (1980). *User's Manual for IDA*. Scientific Press: Palo Alto, Calif.

✓ Lurie, D., and Hartley, H. O. (1972). Machine-generation of order statistics for Monte Carlo computations. *Amer. Stat.*, **26**, 26–27.

McCormack, W. M., and Sargent, R. G. (1981). Analysis of future event set algorithms for discrete-event simulation. *Commun. ACM*, **24**, 801–812.

McKeon, R. (1941). *The Basic Works of Aristotle*. Random House: New York.

MacLaren, M. D., Marsaglia, G., and Bray, T. A. (1964). A fast procedure for generating exponential random variables. *Commun. ACM*, **7**, 298–300.

MacWilliams, F. J., and Sloane, N. J. A. (1976). Pseudo-random sequences and arrays. *Proc. IEEE*, **64**, 1715–1729.

Mahl, R., and Boussard, J. C. (1977). *Algorithmique et structures de données*. Laboratoire d'informatique, Université de Nice: Nice.

Mann, H. B., and Wald, A. (1942). On the choice of the number of class intervals in the application of the chi-square test. *Ann. Math. Stat.*, **13**, 306–317.

Marsaglia, G. (1961). Expressing a random variable in terms of uniform random variables. *Ann. Math. Stat.*, **32**, 894–899.

Marsaglia, G. (1964). Generating a variable from the tail of the normal distribution. *Technometrics*, **6**, 101–102.

Marsaglia, G. (1968). Random numbers fall mainly in the planes. *Proc. Nat. Acad. Sci.*, **60**, 25–28.

Marsaglia, G. (1977). The squeeze method for generating gamma variates. *Comput. Math. Appl.*, **3**, 321–325.

Marsaglia, G., and Bray, T. A. (1968). On-line random number generators and their use in combinations. *Commun. ACM*, **11**, 757–759.

Marshall, A. W. (1956). In *Symposium on Monte Carlo Methods* (H. A. Meyer, editor), p. 14. Wiley: New York.

Meilijson, I., and Nádas, A. (1979). Convex majorization with an application to the length of critical paths. *J. Appl. Probab.*, 671–677.

Meketon, M. S., and Heidelberger, P. (1982). A renewal theoretic approach to bias reduction in regenerative simulations. *Manage. Sci.*, **26**, 173–181.

Meketon, M. S., and Schmeiser, B. (1984). Overlapping batch means: something for nothing? *Proceedings Winter Simulation Conference*, pp. 227–230.

Melamed, B. (1979). Characterizations of Poisson traffic streams in Jackson queueing networks. *Adv. Appl. Probab.*, **11**, 422–438.

Melamed, B. (1982). On Markov jump processes imbedded at jump epochs and their queueing-theoretic applications. *Math. Oper. Res.*, **7**, 111–128.

Miller, R. G. (1974). The jackknife—a review. *Biometrika*, **61**, 1–5.

Miller, R. G. (1981). *Simultaneous Statistical Inference*. Springer-Verlag: New York, Heidelberg, Berlin.

Mitchell, B. (1973). Variance reduction by antithetic variates in $G/G/1$ queueing simulations. *Oper. Res.*, **21**, 988–997.

Morse, P. M. (1977). ORSA twenty-five years later. *Oper. Res.*, **25**, 186–188.

Mullarney, A., and Johnson, G. D. (1984). *Simscript II.5 Programming Language*. CACI Inc.: Los Angeles.

Mulvey, J. M. (1980). Reducing the U.S. Treasury's taxpayer data base by optimization. *Interfaces*, **10**, 101–112.

Nance, R. E., and Overstreet, Jr., C., (1978). Some observations on the behavior of composite random number generators. *Oper. Res.*, **26**, 915–935.

Neuts, M. F. (editor) (1977a). *Algorithmic Methods in Probability* (TIMS Studies in the Management Sciences), Vol. 7. North Holland: Amsterdam.

Neuts, M. F. (1977b). The mythology of steady state. Working Paper, Dept. of Statistics and Computer Science, University of Delaware, Newark. Presented at the ORSA-TIMS Meeting, Atlanta, Ga., November 7–9, 1977.

Neuts, M. F. (1981). *Matrix-Geometric Solutions in Stochastic Models*. Johns Hopkins University Press: Baltimore.

Niederreiter, H. (1978). Quasi-Monte Carlo methods and pseudo-random numbers. *Bull. Amer. Math. Soc.*, **84**, 957–1042.

✓Niederreiter, H., and McCurley, K. (1979). Optimization of functions by quasi-random search methods. *Computing*, **22**, 119–123.

Niederreiter, H., and Peart, P. (1986). Localization of search in quasi-Monte Carlo methods for global optimization. *SIAM J. Sci. Statist. Comput.*, **7**, 660–664.

Nozari, A. (1986). Confidence intervals based on steady-state continuous-time statistics. *Oper. Res. Lett.* (to appear).

Odeh, R. E., and Evans, J. O. (1974). Algorithm AS 70: percentage points of the normal distribution, *Appl. Stat.*, **23**, 96–97.

Ohlin, M. (1977). Next random—a method of fast access to any number in the random generator cycle. *Simula Newslett.*, **6**, 18–20.

Parzen, E. (1962). *Stochastic Processes*. Holden-Day: San Francisco.

Paulson, A. S., Holcomb, E. W., and Leitch, R. A. (1975). The estimation of parameters of the stable laws. *Biometrika*, **62**, 163–170.

Payne, W. H. (1970). FORTRAN Tausworthe pseudorandom number generator. *Commun. ACM*, **13**, 57.

Payne, W. H., Rabung, J. R., and Bogyo, T. P. (1969). Coding the Lehmer pseudo-random number generator. *Commun. ACM*, **12**, 85–86.

Pike, M. C., and Hill, I. D. (1966). Algorithm 291—logarithm of gamma function. *Commun. ACM*, **9**, 684–685.

Press, W. H., Flannery, B. P., Teukolsky, S. A., and Vetterling, W. T. (1986). *Numerical Recipes*, Cambridge University Press: Cambridge.

Priestley, M. B. (1981). *Special Analysis and Time Series*. Academic Press: London.

Pritsker, A. A. B. (1974). *The GASP IV Simulation Language*. Wiley: New York.

Pritsker, A. A. B. (1977). *Modeling and Analysis Using Q-GERT Networks*. Halsted Press (a division of Wiley): New York.

Pritsker, A. A. B. (1986). *Introduction to Simulation and SLAM II*, Wiley: New York.

Pritsker, A. A. B., and Kiviat, P. J. (1969). *Simulation with Gasp II: A Fortran Based Simulation Language*. Prentice-Hall: Englewood Cliffs, N.J.

Proceedings of the Winter Simulation Conference (1981) (available from: Association for Computing Machinery, 1133 Avenue of the Americas, New York, N.Y. 10036).

Purdom, P. W. (1978). Tree size by partial backtracking. *SIAM J. Comput.*, **7**, 481–491.

Ramberg, J., and Schmeiser, B. (1974). An approximate method for generating asymmetric random variables. *Commun. ACM*, **17**, 78–82.

Rand Corporation (1955). *A Million Random Digits with 100,000 Normal Deviates*. Free Press: Glencoe, Ill.

Renyi, A. (1953). On the theory of order statistics. *Acta Math. Acad. Sci. Hung.*, **4**, 191–231.

Reiman, M. I., and Weiss, A. (1986). Sensitivity analysis for simulations via likelihood ratios. Technical Report. AT & T Bell Laboratories: Murray Hill, New Jersey.

Robbins, H., and Munro, S. (1951). A stochastic approximation method. *Ann. Math. Stat.*, **22**, 400–407.

Rosenblatt, M. (1975). Multiply schemes and shuffling. *Math. Comp.*, **29**, 929–934.

Ross, S. M. (1970). *Applied Probability Models with Optimization Applications*. Holden-Day: San Francisco.

Ross, S. M. (1985). *Introduction to Probability Models*, Academic Press: Orlando, FL.

Rothery, P. (1982). The use of control variates in Monte Carlo estimation of power. *Appl. Stat.*, **31**, 125–129.

Rothkopf, M. (1969). A model of rational competitive bidding. *Manage. Sci.*, **15**, 362–373.

Rothkopf, M., and Oren, S. S. (1979). A closure approximation for the nonstationary $M/M/s$ queue. *Manage. Sci.*, **25**, 522–534.

Russell, E. C. (1983). *Building Simulation Models with Simscript II.5*. CACI Inc.: Los Angeles.

Sargent, R. G. (1980). Verification and validation of simulation models. Working Paper No. 80–013, College of Engineering, Syracuse University, Syracuse, New York.

Scheffé, H. (1959). *The Analysis of Variance*. Wiley: New York.

Schmeiser, B. (1980). Generation of variates from distribution tails. *Oper. Res.*, **28**, 1012–1017.

Schmeiser, B. (1982). Batch size effects in the analysis of simulation output. *Oper. Res.*, **30**, 556–568.

Schmeiser, B., and Babu, A. J. G. (1980). Beta variate generation via exponential majorizing functions. *Oper. Res.*, **28**, 917–926.

Schmeiser, B., and Kachitvichyanukul, V. (1981). Poisson random variate generation. Research Memorandum 81–4, School of Industrial Engineering, Purdue University, West Lafayette, Ind.

Schmeiser, B., and Lal, R. (1980). Squeeze methods for generating gamma variates. *JASA*, **75**, 679–682.

Schmeiser, B., and Lal, R. (1982). Bivariate gamma random vectors. *Oper. Res.*, **30**, 355–374.

Schmeiser, B., and Kachitvichyanukul (1986). Correlation induction without the inverse transformation. *Proceedings Winter Simulation Conference*, Washington, D.C.

Schrage, L. (1979). A more portable Fortran random number generator. *ACM Trans. Math. Software*, **5**, 132–138.

Schriber, T. J. (1974). *Simulation Using GPSS*. Wiley: New York.

Schruben, L. W. (1978). Communication to the editor (reply to Fox (1978a)). *Manage. Sci.*, **24**, 862.

Schruben, L. W. (1982). Detecting initialization bias in simulation output. *Oper. Res.*, **30**, 569–590.

Schruben, L. W. (1983a). Analysis of simulation event incidence graphs. *Comm. ACM*, **26**, 957–963.

Schruben, L. W. (1983b). Confidence interval estimation using standardized time series. *Oper. Res.*, **31**, 1090–1108.

Schruben, L. W. (1986a). Simulation optimization using frequency domain methods. *Proceedings Winter Simulation Conference*, Washington, D.C.

Schruben, L. W. (1986b). Sequential simulation run control using standardized time series. *ASA-ACM Interface Meeting*, Colorado.

Schruben, L. W., and Margolin, B. H. (1978). Pseudorandom number assignment in statistically designed simulation and distribution sampling experiments. *JASA*, **73**, 504–525.

Schucany, W. R., (1972). Order statistics in simulation. *J. Stat. Comp. Simul.*, **1**, 281–286.

Shapiro, S. S., and Wilk, M. B. (1965). An analysis of variance test for normality (complete samples). *Biometrika*, **52**, 591–611.

Shapiro, S. S., Wilk, M. B., and Chen, H. J. (1968). A comparative study of various tests of normality. *JASA*, **63**, 1343–1372.

Shearn, D. C. S. (1975). Discrete event simulation in Algol 68. *Software—Pract. Exper.*, **5**, 279–293.

Sheil, B. A. (1981). The psychological study of programming. *Comput. Surv.*, **13**, 101–120.

Siegmund, D. (1976). Importance sampling in the Monte Carlo study of sequential tests. *Ann. Stat.*, **4**, 673–684.

Siklósi, K. (1975). *Simula Simulation*. Tekniska Högskolans Studentkår Kompendieförmedlingen: Stockholm.

Smith, R. L. (1984). Efficient Monte Carlo procedures for generating points uniformly distributed over bounded regions. *Oper. Res.*, **32**, 1296–1308.

Sobol', I. M. (1974). Pseudorandom numbers for constructing discrete Markov chains by the Monte Carlo method. *USSR Comput. Maths. Math. Phys.*, **14**, 36–45.

Sobol', I. M. (1982). On an estimate of the accuracy of a simple multidimensional search. *Soviet Math. Dokl.*, **26**, 398–401.

✓Solis, F. J., and Wets, R. J.-B. (1981). Minimization by random search techniques. *Math. Oper. Res.*, **6**, 19–30.

Stahnke, W. (1973). Primitive binary polynomials. *Math. Comp.*, **27**, 977–980.

Starr, N. (1966). The performance of a sequential procedure for the fixed-width interval estimation of the mean. *Ann. Math. Stat.*, **37**, 36–50.

Stidham, Jr., S., (1970). On the optimality of single-server queueing systems. *Oper. Res.*, **18**, 708–732.

Strang, G. (1986). *Introduction to Applied Mathematics*. Wellesley–Cambridge Press: Wellesley, MA.

Strauch, R. E. (1970). When a queue looks the same to an arriving customer as to an observer. *Manage. Sci.*, **17**, 140–141.

Sunder, S. (1978). Personal communication.

Tadikamalla, P. R. (1978). Computer generation of gamma random variables, II. *Commun. ACM*, **21**, 925–927.

Tadikamalla, P. R. (1980). Random sampling from the exponential power distribution. *JASA*, **75**, 683–686.

Tadikamalla, P. R., and Johnson, M. E. (1981). A complete guide to gamma variate generation. *Amer. J. Math. Manage. Sci.*, **1**, 213–236.

Tannenbaum, A. S. (1981). *Computer Networks*. Prentice-Hall: Englewood Cliffs, N.J.

Tausworthe, R. C. (1965). Random numbers generated by linear recurrence modulo two. *Math. Comput.*, **19**, 201–209.

Theil, H. (1961). *Economic Forecasts and Policy*. North-Holland: Amsterdam.

Thesen, A. (1977). The evolution of a new discrete event simulation language for inexperienced users (WIDES). *Software—Pract. Exper.*, **7**, 519–533.

Toothill, J. P. R., Robinson, W. D., and Adams, A. G. (1971). The runs up-and-down performance of Tausworthe pseudo-random number generators. *J. ACM*, **18**, 381–399.

Van Slyke, R. (1963). Monte Carlo methods and the PERT problem. *Oper. Res.*, **11**, 839–860.

Vaucher, J. G. (1973). La programmation avec Simula 67. Document de travail no. 36, Département d'informatique et de recherche opérationnelle, Université de Montréal, Montréal.

Vaucher, J. G. (1975). Prefixed procedures: a structuring concept for operations. *INFOR*, **13**, 287–295.

Vaucher, J. G. (1976). On the distribution of event times for the notices in a simulation event list. *INFOR*, **15**, 171–182.

Vaucher, J. G., and Duval, P. (1975). A comparison of simulation event list algorithms. *Commun. ACM*, **18**, 223–230.

Von Neumann, J. (1951). Various techniques used in connection with random digits. *National Bureau of Standards Applied Mathematics*, Series 12, pp. 36–38.

Wagner, H. M. (1975). *Principles of Operations Research*, 2nd edn. Prentice-Hall: Englewood Cliffs, N.J.

Wahba, G. (1980). Automatic smoothing of the log periodogram. *JASA*, **75**, 122–132.

Walker, A. J. (1977). An efficient method for generating discrete random variables with general distributions. *ACM Trans. Math. Software*, **3**, 253–256.

Weissman, I. (1978a). Estimation of parameters and large quantiles based on the k largest observations. *JASA*, **73**, 812–815.

Weissman, I. (1978b). Private communication.

Welsh, J., and McKeag, M. (1980). *Structured System Programming*. Prentice-Hall: Englewood Cliffs, N.J.

West, J. (1979). *SIMGRAPH—A Graphics Package for Simscript II.5.* CACI Inc.: Los Angeles.

Whitt, W. (1976). Bivariate distributions with given marginals. *Ann. Stat.*, **4**, 1280–1289.

Whitt, W. (1983). Untold horrors of the waiting room: what the steady-state distribution will never tell about the queue-length process. *Manage. Sci.*, **29**, 395–408.

Wichmann, B. A., and Hill, I. D. (1982). An efficient and portable pseudo-random number generator. *Appl. Stat.*, **31**, 188–190.

Wilson, J. R. (1984). Variance reduction techniques for digital simulation. *Amer. J. Math. Management Sci.*, **4**, 277–312.

Wolff, R. W. (1982). Poisson arrivals see time averages. *Oper. Res.*, **30**, 223–231.

Woolsey, R. E. D., and Swanson, H. S. (1975). *Operations Research for Immediate Applications.* Harper & Row: New York.

Wright, R. D., and Ramsay, Jr., T. E. (1979). On the effectiveness of common random numbers. *Manage. Sci.*, **25**, 649–656.

Xerox (1972). *SL-1 Reference Manual.* Publication 90 16 76B, Xerox Corporation, El Segundo, Calif.

Yakowitz, S., Krimmel, J. E., and Szidarovszky, F. (1978). Weighted Monte Carlo integration. *SIAM J. Numer. Anal.*, **15**, 1289–1300.

Yuen, C. (1982). The inversion method in random variate generation. M.Sc. Thesis, School of Computer Science, McGill University, Montreal.

Zierler, N. (1969). Primitive trinomials whose degree is a Mersenne exponent. *Inf. Cont.*, **15**, 67-69.

Author Index

Subject Index